Lecture Notes in Applied and Computational Mechanics

Volume 45

AF788100

Series Editors

Prof. Dr.-Ing. Friedrich Pfeiffer
Prof. Dr.-Ing. Peter Wriggers

Lecture Notes in Applied and Computational Mechanics

Edited by F. Pfeiffer and P. Wriggers

Further volumes of this series found on our homepage: springer.com

Vol. 45: Shevchuk I.V.
Convective Heat and Mass Transfer in Rotating Disk Systems
248 p. 2009 [978-3-642-00717-0]

Vol. 44: Ibrahim R.A.; Babitsky V.I.; Okuma M. (Eds.)
Vibro-Impact Dynamics of Ocean Systems and Related Problems
300 p. 2009 [978-3-642-00628-9]

Vol. 43: Ibrahim R.A.
Vibro-Impact Dynamics: Modeling, Mapping and Applications
320 p. 2009 [978-3-642-00274-8]

Vol.42: Hashiguchi K.
Elastoplasticity Theory
432 p. 2009 [978-3-642-00272-4]

Vol. 41: Browand F.; Ross J.; McCallen R. (Eds.)
Aerodynamics of Heavy Vehicles II: Trucks, Buses, and Trains
486 p. 2009 [978-3-540-85069-4]

Vol. 40: Pfeiffer F.
Mechanical System Dynamics
578 p. 2008 [978-3-540-79435-6]

Vol. 39: Lucchesi, M.; Padovani, C.; Pasquinelli, G.; Zani, N.
Masonry Constructions: Mechanical Models and Numerical Applications
176 p. 2008 [978-3-540-79110-2]

Vol. 38: Marynowski, K.
Dynamics of the Axially Moving Orthotropic Web
140 p. 2008 [978-3-540-78988-8]

Vol. 37: Chaudhary, H.; Saha, S.K.
Dynamics and Balancing of Multibody Systems
200 p. 2008 [978-3-540-78178-3]

Vol. 36: Leine, R.I.; van de Wouw, N.
Stability and Convergence of Mechanical Systems with Unilateral Constraints
250 p. 2008 [978-3-540-76974-3]

Vol. 35: Acary, V.; Brogliato, B.
Numerical Methods for Nonsmooth Dynamical Systems: Applications in Mechanics and Electronics
545 p. 2008 [978-3-540-75391-9]

Vol. 34: Flores, P.; Ambrósio, J.; Pimenta Claro, J.C.; Lankarani Hamid M.
Kinematics and Dynamics of Multibody Systems with Imperfect Joints: Models and Case Studies
186 p. 2008 [978-3-540-74359-0]

Vol. 33: Niesłony, A.; Macha, E.
Spectral Method in Multiaxial Random Fatigue
146 p. 2007 [978-3-540-73822-0]

Vol. 32: Bardzokas, D.I.; Filshtinsky, M.L.; Filshtinsky, L.A. (Eds.)
Mathematical Methods in Electro-Magneto-Elasticity
530 p. 2007 [978-3-540-71030-1]

Vol. 31: Lehmann, L. (Ed.)
Wave Propagation in Infinite Domains
186 p. 2007 [978-3-540-71108-7]

Vol. 30: Stupkiewicz, S. (Ed.)
Micromechanics of Contact and Interphase Layers
206 p. 2006 [978-3-540-49716-5]

Vol. 29: Schanz, M.; Steinbach, O. (Eds.)
Boundary Element Analysis
571 p. 2006 [978-3-540-47465-4]

Vol. 28: Helmig, R.; Mielke, A.; Wohlmuth, B.I. (Eds.)
Multifield Problems in Solid and Fluid Mechanics
571 p. 2006 [978-3-540-34959-4]

Vol. 27: Wriggers P., Nackenhorst U. (Eds.)
Analysis and Simulation of Contact Problems
395 p. 2006 [978-3-540-31760-9]

Vol. 26: Nowacki, J.P.
Static and Dynamic Coupled Fields in Bodies with Piezoeffects or Polarization Gradient
209 p. 2006 [978-3-540-31668-8]

Vol. 25: Chen C.-N.
Discrete Element Analysis Methods of Generic Differential Quadratures
282 p. 2006 [978-3-540-28947-0]

Vol. 24: Schenk, C.A., Schuëller. G
Uncertainty Assessment of Large Finite Element Systems
165 p. 2006 [978-3-540-25343-3]

Vol. 23: Frémond M., Maceri F. (Eds.)
Mechanical Modelling and Computational Issues in Civil Engineering
400 p. 2005 [978-3-540-25567-3]

Vol. 22: Chang C.H.
Mechanics of Elastic Structures with Inclined Members: Analysis of Vibration, Buckling and Bending of X-Braced Frames and Conical Shells
190 p. 2004 [978-3-540-24384-7]

Vol. 21: Hinkelmann R.
Efficient Numerical Methods and Information-Processing Techniques for Modeling Hydro- and Environmental Systems
305 p. 2005 [978-3-540-24146-1]

Vol. 20: Zohdi T.I., Wriggers P.
Introduction to Computational Micromechanics
196 p. 2005 [978-3-540-22820-2]

Vol. 19: McCallen R., Browand F., Ross J. (Eds.)
The Aerodynamics of Heavy Vehicles: Trucks, Buses, and Trains
567 p. 2004 [978-3-540-22088-6]

Convective Heat and Mass Transfer in Rotating Disk Systems

Igor V. Shevchuk

With 116 Figures and 43 Tables

Springer

Dr. Igor V. Shevchuk
MBtech Powertrain GmbH
Salierstr. 38
70736 Fellbach-Schmiden
Germany
ivshevch@i.com.ua

ISSN 1613-7736 e-ISSN 1860-0816
ISBN 978-3-642-26058-2 e-ISBN 978-3-642-00718-7
DOI 10.1007/978-3-642-00718-7
Springer Heidelberg Dordrecht London New York

© Springer-Verlag Berlin Heidelberg 2009
Softcover reprint of the hardcover 1st edition 2009
This work is subject to copyright. All rights are reserved, whether the whole or part of the material is concerned, specifically the rights of translation, reprinting, reuse of illustrations, recitation, broadcasting, reproduction on microfilm or in any other way, and storage in data banks. Duplication of this publication or parts thereof is permitted only under the provisions of the German Copyright Law of September 9, 1965, in its current version, and permission for use must always be obtained from Springer. Violations are liable to prosecution under the German Copyright Law.
The use of general descriptive names, registered names, trademarks, etc. in this publication does not imply, even in the absence of a specific statement, that such names are exempt from the relevant protective laws and regulations and therefore free for general use.

Cover design: WMX-Design, Heidelberg

Printed on acid-free paper

Springer is part of Springer Science+Business Media (www.springer.com)

To my wife Nataliya and sons Vladimir, Aleksandr and Nikolay

Preface

The book is devoted to investigation of a series of problems of convective heat and mass transfer in rotating-disk systems. Such systems are widespread in scientific and engineering applications. As examples from the practical area, one can mention gas turbine and computer engineering, disk brakes of automobiles, rotating-disk air cleaners, systems of microclimate, extractors, dispensers of liquids, evaporators, circular saws, medical equipment, food process engineering, etc. Among the scientific applications, it is necessary to point out rotating-disk electrodes used for experimental determination of the diffusion coefficient in electrolytes. The system consisting of a fixed disk and a rotating cone that touches the disk by its vertex is widely used for measurement of the viscosity coefficient of liquids.

For time being, large volume of experimental and computational data on parameters of fluid flow, heat and mass transfer in different types of rotating-disk systems have been accumulated, and different theoretical approaches to their simulation have been developed. This obviously causes a need of systematization and generalization of these data in a book form.

Three books are widely known currently, which are completely or partially devoted to the considered subject. The classical books of L.A. Dorfman "Hydrodynamic Resistance and the Heat Loss of Rotating Solids" (Oliver and Boyd, Edinburgh, UK, 1963) and V.G. Levich "Physicochemical Hydrodynamics" (Prentice-Hall, Inc., Englewood Cliffs, N.J.: 1962) for decades became desktop books for the specialists in the fields of convective heat transfer at air flow in rotating-disk systems and experimental determination of the diffusion coefficient in electrolytes with the help of the rotating-disk electrode technique, respectively. The fundamental monograph of J.M. Owen, R.H. Rogers "Flow and Heat Transfer in Rotating-Disc Systems" (Research Studies Press Ltd., UK, 1989 and 1995) represents an in-depth insight into the modern state-of-the-art of investigations in the field of secondary air cooling systems of gas turbines including data for a free rotating disk, rotor–stator systems, as well as rotating cavities formed by parallel co-rotating disks.

For the last two decades, considerable advance has been done in experimental and theoretical research of scientific and practical problems of convective heat and mass transfer, which the above-mentioned books are devoted to. However, degree of critical analysis and generalizations of the accumulated data, both in these books and in newly published works of different authors, are frequently insufficient even at the level of similarity equations. A series of problems were successfully solved with the help of integral methods. However, theoretical foundations of the known integral methods have appeared insufficiently developed that in a number of cases resulted in essential errors of the solutions obtained on the basis of these methods. In a number of works, modelling approaches using exact self-similar solutions of the Navier–Stokes and energy equations have been worked out. However, for many problems in rotating-disk systems, possible self-similar forms of the solutions have not been found that essentially narrows down capabilities of theoretical modelling.

A number of other important scientific and practical problems are not elucidated in the aforementioned books. Among them, the following problems of convective heat transfer of a disk rotating in air are of interest from the point of view of this book: (a) non-stationary conjugate heat transfer; (b) impingement of uniform flow or a single co-axial jet onto an orthogonal disk; (c) flow and heat transfer in a gap between a rotating disk and/or a cone touching the disk by its vertex; (d) flow in a rotating-disk air cleaner. Also actual are problems of convective heat and mass transfer at Prandtl and Schmidt numbers: (e) moderately exceeding unity as applied to the technique of experimental measurement of mass transfer rate for naphthalene sublimation in air and (f) much exceeding unity with reference to problems of electrochemistry.

The problems mentioned above became motivation to undertake investigations that laid down the basis for preparation of this book.

The present book consists of eight chapters. The main attention in the book is given to heat transfer in air flow, except for Chap. 8, where problems of heat and mass transfer at Prandtl numbers or Schmidt larger than unity are considered.

Chapter 1 includes characterization of several known types of rotating-disk systems, description of forces that act on flow and general notations of momentum, continuity, energy and convective diffusion equations in different coordinate systems.

In Chap. 2, differential equations of motion and energy are written as applied to rotating-disk systems, methods of their solution known in the literature are briefly described, an integral method developed by the author is outlined and a general solution is written for the cases of disk rotation in a fluid rotating as a solid body and simultaneous imposed accelerating radial flow.

Chapter 3 represents analysis and generalization of the data and models of different authors for a free rotating disk. With the help of the integral method developed by the author, analytical and numerical solutions are obtained possessing essentially higher accuracy, than the solutions known before.

In Chap. 4, self-similar solutions of the problem of non-stationary heat convection, as well as analytical and numerical solutions of the problem of conjugate non-stationary heat transfer of the disk are represented. Peculiarities of application of transient experimental techniques for determination of heat transfer coefficients are also discussed.

Chapter 5 is devoted to analysis of the solutions obtained with the help of the integral method developed by the author for the case of disk rotation in a fluid rotating as a solid body without imposed radial flow, and also for accelerating radial flow (due to its orthogonal impingement) without imposed external rotation.

In Chap. 6, hydrodynamics and heat transfer are modelled for outward underswirled and overswirled radial flow between parallel co-rotating disks (the integral method), and also aerodynamics and heat transfer in a rotating-disk air cleaner (with the help of CFD).

In Chap. 7, a self-similar solution of a problem of laminar heat transfer in a gap between a rotating disk and/or a cone, as well as that for outward swirling flow in a stationary conical diffuser is presented.

Chapter 8 contains analysis and generalization of the data of different authors for problems of convective heat and mass transfer at Prandtl and Schmidt numbers exceeding unity. Recommendations as applied to the technique of experimental measurement of mass transfer rate for naphthalene sublimation in air are developed. In the integral method developed by the author, effects of large Prandtl and Schmidt numbers are taken into account.

The author deeply acknowledges financial support of Alexander von Humboldt Foundation (Germany) in the form of a Research Fellowship taken by the author at Technische Universität Dresden in 2003–2005, which enabled him to prepare the present book. For the three years that passed since then, the author has refined Chap. 8 and introduced some editing to other chapters in view of the new publications, which have been published for this time. The author would like to thank all the colleagues, whom he has collaborated with during the time of performing the research that laid foundation of the book, for their contribution, useful advices and fruitful discussions.

Stuttgart, Germany Igor V. Shevchuk

Contents

1 General Characteristic of Rotating-Disk Systems 1
1.1 Industrial Applications of Rotating-Disk Systems 1
1.2 Acting Forces 2
1.3 Differential Equations of Continuity, Momentum and Heat Transfer . 4
1.4 Differential Equation of Convective Diffusion 9

2 Modelling of Fluid Flow and Heat Transfer in Rotating-Disk Systems . 11
2.1 Differential and Integral Equations 11
2.1.1 Differential Navier–Stokes and Energy Equations 11
2.1.2 Differential Boundary Layer Equations 13
2.1.3 Integral Boundary Layer Equations 14
2.2 Differential Methods of Solution 15
2.2.1 Self-Similar Solution 15
2.2.2 Approximate Analytical Methods for Laminar Flow Based on Approximations of Velocity Profiles 17
2.2.3 Numerical Methods 17
2.3 Integral Methods of Solution 18
2.3.1 Momentum Boundary Layer 18
2.3.2 Thermal Boundary Layer 22
2.4 Integral Method for Modelling Fluid Flow and Heat Transfer in Rotating-Disk Systems 23
2.4.1 Structure of the Method 23
2.4.2 Turbulent Flow: Improved Approximations of the Velocity and Temperature Profiles 24
2.4.3 Models of Surface Friction and Heat Transfer 25
2.4.4 Integral Equations with Account for the Models for the Velocity and Temperature Profiles 27
2.5 General Solution for the Cases of Disk Rotation in a Fluid Rotating as a Solid Body and Simultaneous Accelerating Imposed Radial Flow 29

3 Free Rotating Disk 33
3.1 Laminar Flow 33
3.2 Transition to Turbulent Flow and Effect of Surface Roughness...... 37

3.3 Turbulent Flow ... 41
3.3.1 Parameters of the Turbulent Boundary Layer ... 41
3.3.2 Surface Heat Transfer: Experimental and Theoretical Data of Different Authors ... 45
3.3.3 Effect of Approximation of the Radial Velocity Profile on Parameters of Momentum and Thermal Boundary Layers ... 48
3.3.4 Numerical Computation of Turbulent Flow and Heat Transfer for an Arbitrary Distribution of the Wall Temperature ... 54
3.4 Generalized Analytical Solution for Laminar and Turbulent Regimes Based on the Novel Model for the Enthalpy Thickness ... 58
3.5 Inverse Problem of Restoration of the Wall Temperature Distribution at a Specified Arbitrary Power Law for the Nusselt Number ... 61
3.5.1 Solution of the Problem ... 61
3.5.2 Limiting Case of the Solution ... 64
3.5.3 Properties of the Solution for Temperature Head ... 65
3.5.4 Analysis of the Solution ... 66
3.6 Theory of Local Modelling ... 72
3.6.1 Solution of the Problem ... 72
3.6.2 Other Interpretations ... 74

4 Unsteady Laminar Heat Transfer of a Free Rotating Disk ... 77
4.1 Transient Experimental Technique for Measuring Heat Transfer over Rotating Disks ... 77
4.2 Self-Similar Navier–Stokes and Energy Equations ... 79
4.3 Exact Solution for Surface Heat Transfer of an Isothermal Rotating Disk ... 82
4.4 Numerical Solution of an Unsteady Conjugate Problem of Hydrodynamics and Heat Transfer of an Initially Isothermal Disk 85
4.4.1 Computational Domain and Grid ... 85
4.4.2 Validation for Steady-State Fluid Flow and Heat Transfer ... 86
4.4.3 Unsteady Fluid Flow and Heat Transfer ... 88
4.5 Unsteady Conjugate Laminar Heat Transfer of a Rotating Non-uniformly Heated Disk ... 91
4.5.1 Problem Statement ... 91
4.5.2 Self-Similar Solution of the Transient Laminar Convective Heat Transfer Problem ... 92
4.5.3 Solution of the Unsteady Two-Dimensional Problem of Heat Conduction in a Disk ... 93
4.5.4 Analysis of the Solutions for Unsteady Heat Conduction in a Disk ... 94

5 External Flow Imposed over a Rotating Disk ... 101
5.1 Rotation of a Disk in a Fluid Rotating as a Solid Body Without Imposed Radial Flow ... 101
5.1.1 Turbulent Flow ... 101
5.1.2 Laminar Flow ... 106
5.2 Accelerating Radial Flow Without Imposed External Rotation ... 118
5.2.1 Flow Impingement onto an Orthogonal Rotating Disk: Experimental and Computational Data of Different Authors ... 118
5.2.2 Turbulent Flow ... 123
5.2.3 Laminar Flow ... 125
5.3 Non-symmetric Flow over a Parallel Rotating Disk ... 143

6 Outward Underswirled and Overswirled Radial Flow Between Parallel Co-rotating Disks ... 147
6.1 Flow in the Ekman Layers ... 147
6.2 Radial Outflow Between Parallel Co-rotating Disks ... 148
6.2.1 Flow Structure, Experiments and Computations of Different Authors ... 148
6.2.2 Computation of the Radial Variation of the Swirl Parameter Using the Integral Method ... 152
6.2.3 Local Nusselt Numbers ... 157
6.2.4 Effect of the Radial Distribution of the Disk Surface Temperature ... 161
6.3 Effect of the Flow Overswirl ... 164
6.4 Aerodynamics and Heat Transfer in a Rotating-Disk Air Cleaner ... 168
6.4.1 General Characteristics of the Problem ... 168
6.4.2 Geometrical and Regime Parameters of the Air Cleaner ... 169
6.4.3 Parameters of the Computational Scheme ... 171
6.4.4 Results of Simulations ... 171

7 Laminar Fluid Flow and Heat Transfer in a Gap Between a Disk and a Cone that Touches the Disk with Its Apex ... 179
7.1 General Characterization of the Problem ... 179
7.2 Navier–Stokes and Energy Equations in the Self-similar Form ... 181
7.3 Rotating Disk and/or Cone ... 184
7.3.1 Numerical Values of Parameters in the Computations ... 184
7.3.2 Cone Rotation at a Stationary Disk ... 185
7.3.3 Disk Rotation at a Stationary Cone ... 187
7.3.4 Co-rotating Disk and Cone ... 188
7.3.5 Counter-Rotating Disk and Cone ... 188
7.4 Radially Outward Swirling Flow in a Stationary Conical Diffuser ... 189

8 Heat and Mass Transfer of a Free Rotating Disk for the Prandtl and Schmidt Numbers Larger than Unity ... 193
8.1 Laminar Flow ... 193

8.2 Transitional and Turbulent Flows for the Prandtl or Schmidt Numbers Moderately Different from Unity 201
8.3 Transitional and Turbulent Flows at High Prandtl and Schmidt Numbers 208
8.4 An Integral Method for Modelling Heat and Mass Transfer in Turbulent Flow for the Prandtl and Schmidt Numbers Larger than Unity 214
8.4.1 Prandtl and Schmidt Numbers Moderately Different from Unity 214
8.4.2 High Prandtl and Schmidt Numbers 217

References 225

Index 235

Nomenclature

a	— thermal diffusivity, m^2/s; radial velocity gradient on the outer boundary of the boundary layer, Eq. (2.32), 1/s;
$A = ad_j / V_j$	— non-dimensional radial velocity gradient on the outer boundary of the boundary layer,
b	— outer radius of a disk, m;
$Bi_1 = \alpha_1 b / \lambda_w$	— Biot number in unsteady heat transfer of the cylindrical surface of a rotating disk;
$Bi_2 = 0.5\alpha_2 s / \lambda_w$, $Bi = 0.5\alpha s / \lambda_w$	— Biot number in unsteady heat transfer of the flat surface of a rotating disk;
C	— concentration, mol/m^3;
$c_f / 2 = \tau_w / (\rho V_*)^2$	— surface friction coefficient;
$C_M = 4M / (\rho\, \omega^2 b^5)$	— moment coefficient of two flat sides of a rotating disk;
c_p	— isobaric specific heat, J/(kg·K);
$C_w = \dot{m} / (\mu b)$	— non-dimensional radial mass flowrate through a cavity between two rotating disks;
$c_{0*} = (T_{w,i} - T_\infty)_{n_*=0}$	— constant temperature head on the surface at the initial moment of time $t=0$ in unsteady heat transfer problem, K;
d	— disk diameter, m;
d_j	— nozzle diameter, m;
D_m	— diffusion coefficient, m^2/s;
$\boldsymbol{F}$	— mass force per unit volume (boldface denotes vector parameter), N/m^3;
F_x, F_y, F_z	— mass force components in Cartesian coordinates (per unit volume), N/m^3;
F_r, F_φ, F_z	— mass force components in cylindrical polar coordinates (per unit volume), N/m^3;
F, G, H, P	— self-similar functions, Eqs. (2.31);
$Fo = 4a_w t / s^2$	— Fourier number;

$F_t(t)=\dfrac{T_w(t,r)-T_\infty}{T_{w,i}(r)-T_\infty}$ — non-dimensional disk surface temperature in the unsteady heat transfer problem;

g — acceleration of gravity, m/s^2;

h — height of a conical gap, m;

h_j — nozzle-to-disk distance, m;

$\boldsymbol{j}$ — acceleration of a mass force, m/s^2;

$K_H=\dfrac{\int_0^{\delta_T} v_r(T-T_\infty)\,dz}{(T_w-T_\infty)\int_0^{\delta_T} v_r dz}$ — shape-factor of the temperature profile (modified enthalpy thickness);

$K_m=\delta^{-1}\int_0^{\delta}\dfrac{v_r}{\omega r}dz$ — non-dimensional radial mass flowrate through the boundary layer;

$K_V=\dfrac{\int_0^{\infty} v_r(v_\varphi-v_{\varphi,\infty})dz}{(\omega r-v_{\varphi,\infty})\int_0^{\infty} v_r dz}$ — shape-factor of the velocity profile;

$M=-2\pi\int_0^{b} r^2\tau_{w\varphi}dr$ — moment of one side of a rotating disk, Pa·m^3;

$\dot{m}$ — total radial mass flowrate through the cavity between two rotating disks, kg/s;

$\dot{m}_d=2\pi r\rho\int_0^{\delta} v_r\,dz$ — mass flowrate through the momentum boundary layer over a rotating disk, kg/s;

$\dot{m}_{d,T}=2\pi r\rho\int_0^{\delta_T} v_r\,dz$ — mass flowrate through the thermal boundary layer over a rotating disk, kg/s;

n — exponent in the power-law approximation of the velocity profiles;

n_T — exponent in the power-law approximation of the temperature profiles;

n_* — exponent in the power-law approximation of the temperature head on a surface, Eqs. (2.34)–(2.36);

$N=v_{r,\infty}/(\omega r)$ — non-dimensional radial velocity in potential flow outside of the boundary layer;

$Nu=\dfrac{q_w r}{\lambda(T_w-T_\infty)}$ — local Nusselt number;

$Nu_b=\dfrac{q_w b}{\lambda(T_w-T_\infty)}$ — local Nusselt number at the outer radius of a rotating disk;

$Nu_d = \dfrac{q_w d}{\lambda(T_w - T_\infty)}$ — local Nusselt number based on the diameter of a rotating disk;

$Nu_{dj} = \dfrac{q_w d_j}{\lambda(T_w - T_\infty)}$ — local Nusselt number based on the nozzle diameter at flow impingement onto a rotating disk;

$Nu_{\mathrm{av}} = q_{w,\mathrm{av}} b / [\lambda (T_w - T_\infty)_{\mathrm{av}}]$ — average Nusselt number;

$Nu_m = \dfrac{\alpha_m r}{D_m}$ — mass transfer Nusselt number;

$Nu_{m,\mathrm{av}} = \dfrac{\alpha_{m,av} b}{D_m}$ — average mass transfer Nusselt number;

p — pressure, Pa;

$Pr = \mu c_p / \lambda$ — Prandtl number;

q — heat flux per unit area, $\mathrm{W/m^2}$;

$q_w = -\lambda \left(\dfrac{dT}{dz} \right)_{z=0}$ — wall value of heat flux per unit area, $\mathrm{W/m^2}$;

$q_{w,\mathrm{av}} = \int_0^b q_w r dr \Big/ \int_0^b r dr$ — surface-averaged wall value of heat flux per unit area, $\mathrm{W/m^2}$;

$Re_a = a d^2 / \nu$ — Reynolds number of radial flow at flow impingement;

$Re_{\omega d} = \omega d^2 / \nu$ — rotational Reynolds number based on the disk diameter;

$Re_\omega = \omega r^2 / \nu$ — local rotational Reynolds number;

$Re_\Omega = \Omega r^2 / \nu$ — local rotational Reynolds number for rotating cones in cone–disk systems;

$Re_\varphi = \omega b^2 / \nu$ — rotational Reynolds number at the outer radius of a disk;

$Re_j = V_j d_j / \nu$ — Reynolds number based on an impingement velocity;

$Re_{V*} = V_* \delta / \nu$ — Reynolds number based on the velocity V_*;

$Re_T^{**} = \dfrac{\omega r \delta_T^{**}}{\nu}$ — enthalpy Reynolds number;

r, φ, z — cylindrical polar coordinates, m or rad;

s — spacing (height) between rotating disks, m; thickness of a disk in the problem of unsteady conjugate heat transfer;

$Sc = \nu / D_m$ — Schmidt number;

$Sh = \dfrac{\alpha_m r}{D_m}$ — Sherwood number;

$Sh_{\mathrm{av}} = \dfrac{\alpha_{m,\mathrm{av}} b}{D_m}$ — average Sherwood number;

$St = \dfrac{q_w}{\rho c_p V_* (T_w - T_\infty)}$ — Stanton number;

t — time, s;

T — temperature, K;

T_w — wall temperature; instantaneous disk surface temperature, K;

$T_{w,i}$ — initial value of the wall temperature in unsteady heat transfer, K;

T_∞ — temperature in potential flow outside of the boundary layer, K;

$T^+=(T_w-T)\rho_\infty V_\tau/q_w$ — local temperature in wall coordinates;

$(T_w - T_\infty)_{av} = \int_0^b (T_w-T_\infty) r dr \Big/ \int_0^b r dr$ — surface-averaged temperature head, K;

$tan\varphi = \dfrac{v_r}{\omega r - v_\varphi}$ — tangent of the flow swirl angle;

$V = \left[v_r^2 + (v_\varphi - \omega r)^2\right]^{1/2}$ — total velocity, m/s;

$V^+ = V/V_\tau$ — total velocity in wall coordinates;

$V_\tau = (\tau_w/\rho)^{1/2}$ — friction velocity, m/s;

v_r, v_φ, v_z — velocity components in cylindrical coordinates, m/s;

$\bar{v}_r = \dfrac{v_r}{\omega r - v_{\varphi,\infty}}$ — non-dimensional radial velocity component;

$\bar{v}_\varphi = \dfrac{v_\varphi - \omega r}{v_{\varphi,\infty} - \omega r}$ — non-dimensional tangential velocity component;

$V_* = \omega r |\beta - 1| (1 + \alpha^2)^{1/2}$ — characteristic velocity, m/s;

V_j — axial flow velocity at infinity or at the outlet of a nozzle, m/s;

u, v, w — velocity components in Cartesian coordinates, m/s;

x, y, z — Cartesian coordinates, m/s;

$x=r/b$ — non-dimensional radial coordinate;

$y=z/(0.5s)$ — non-dimensional axial coordinate in the problem of heat conduction inside a disk;

$\tilde{z} = z/h$ — non-dimensional axial coordinate in conical gaps;

$z^+ = z V_\tau/\nu$ — wall-law coordinate;

$\alpha = -\tau_{wr}/\tau_{w\varphi}$ — wall value of the tangent of the flow swirl angle; heat transfer coefficient, W/(m^2·K);

α_m — mass transfer coefficient, m/s;

$\alpha_{m,av} = \dfrac{2}{b^2}\int_0^b h_m r dr$ — surface-averaged mass transfer coefficient, m/s;

$\beta = \mathrm{v}_{\varphi,\infty}/(\omega r)$	— parameter of flow swirl, i.e. non-dimensional tangential velocity component in potential flow outside of the boundary layer;
δ	— thickness of a momentum boundary layer, m;
$\bar{\delta} = \delta / b$	— non-dimensional thickness of a momentum boundary layer;
δ_T	— thickness of a thermal boundary layer;
$\Delta = \delta_T/\delta$	relative thickness of a thermal/diffusion boundary layer;
δ^*	— displacement thickness, m;
δ^{**}	— momentum thickness, m;
$\delta_T^{**} = \int_0^{\delta_T} \frac{\mathrm{v}_r}{\omega r} \frac{T - T_\infty}{T_w - T_\infty} dz$	— enthalpy thickness, m;
$\bar{\delta}_T^{**} = \delta_T^{**} / \delta$	— non-dimensional enthalpy thickness;
$\Delta T = T_w - T_\infty$	— temperature head on a surface, K;
$\Delta T = T_{w,i}(r) - T_\infty$	— temperature head on a surface at the initial moment of time t=0 in unsteady conditions, K;
$\Delta T_t(t,r) = T_w(t,r) - T_\infty$	— instantaneous value of the temperature head on a surface in the unsteady heat transfer problem, K;
$\Delta T_{x=1}$	— temperature head on a surface at x=1, K;
$\overline{\Delta T} = \Delta T / \Delta T_{x=1}$	— relative non-dimensional temperature head on a surface;
ζ	— self-similar variable, Eq. (2.31);
$\Theta = \frac{T - T_w}{T_\infty - T_w}$	— non-dimensional temperature;
$\theta = \frac{T - T_\infty}{T_w - T_\infty}$	— non-dimensional temperature;
$\vartheta(t,r,z) = (T - T_\infty) / c_{0*}$	— non-dimensional temperature inside a disk in the unsteady heat transfer problem;
$\kappa = tan\varphi_\infty = \mathrm{v}_{r,\infty}/(\omega r - \mathrm{v}_{\varphi,\infty})$	— tangent of the flow swirl angle in potential flow;
λ	— thermal conductivity, W/(m·K);
μ	— dynamic viscosity, Pa·s;
ν	— kinematic viscosity, m^2/s;
$\xi = z/\delta$	— non-dimensional coordinate;
$\xi_T = z / \delta_T$	— non-dimensional coordinate;
ρ	— density, kg/m^3;
τ	— shear stress, Pa;
$\tau_w = (\tau_{wr}^2 + \tau_{w\varphi}^2)^{1/2}$	— total shear stress on the wall, Pa;

$\tau_{wr} = \mu(dv_r / dz)_{z=0}$	—	radial shear stress on the wall, Pa;
$\tau_{w\varphi} = \mu(dv_\varphi / dz)_{z=0}$	—	tangential shear stress on the wall, Pa;
$\varphi_w = arctan[v_r/(\omega r - v_\varphi)]$	—	swirl angle;
χ	—	Reynolds analogy parameter, Eq. (2.57);
ω	—	angular velocity of rotation of a disk (or co-rotating disks), 1/s;
Ω	—	angular velocity of rotation of a fluid in rotating-disk systems; angular velocity of rotation of a cone in cone-disk systems, 1/s.

Subscripts

av	—	average value;
c	—	centrifugal forces (accelerations);
Cor	—	Coriolis forces (accelerations);
E	—	Ekman layers;
i	—	initial moment of time; inlet to a channel (cavity);
j	—	impinging jet;
lam	—	laminar flow;
max	—	value at a point of maximum;
t	—	turbulent parameters; instantaneous value of a parameter in the problem of unsteady heat transfer of a rotating disk;
turb	—	turbulent flow;
T	—	parameters of a thermal boundary layer;
tr	—	parameters at the point of abrupt transition from laminar to turbulent flow;
tr1	—	parameters at the point of the beginning of transition from laminar to turbulent flow;
tr2	—	parameters at the point of the end of transition from laminar to turbulent flow;
w	—	wall value (at $z=0$); thermophysical properties of material of a wall;
0	—	standard conditions: free rotating disk at $\mathrm{v}_{r,\infty}=0$ and $\mathrm{v}_{\varphi,\infty}=0$;
1	—	boundary of a viscous or heat conduction sub-layer; outer cylindrical surface of a disk in the unsteady heat transfer problem;
2	—	flat surface of a disk in the unsteady heat transfer problem;
∞	—	potential flow outside of a boundary layer;
$\overline{u'\mathrm{v}'}$	—	time-averaged pulsation turbulent values.

Chapter 1
General Characteristic of Rotating-Disk Systems

1.1 Industrial Applications of Rotating-Disk Systems

Rotating-disk systems are widely used in gas turbine engineering, aircraft engines, computer disk drives, car breaks systems, rotational air cleaners, extractors, atomisers, evaporators, microclimate systems, chemical engineering, electrochemistry, medical equipment, food processing technologies, etc. Widely spread are the cases of disk rotation in an infinite resting fluid or fluid rotating with another angular velocity (Fig. 1.1a), impinging jet cooling of a disk (Fig. 1.1b), co-rotating or contra-rotating parallel disks with and without forced radial through-flow in a gap between them (Fig. 1.1c), flow between a rotor and a stator (Fig. 1.1d), closed non-ventilated cavities of gas turbines formed by two disks and two cylindrical surfaces (Fig. 1.1e), air cooling systems with inlet flow pre-swirl (Fig. 1.1f), shrouded rotating disks, flow of a liquid thin film over a disk surface, etc.

Flows in gaps between rotating surfaces of different geometries are widely used in power engineering, chemical, oil and food processing industry, medical equipment, aircraft engines, viscosimetry, etc. In particular, one can mention flows in the gaps between a disk and a cone whose apex touches the disk, with both of them rotating independently (Fig. 1.2a), as well as swirl flows in a stationary conical diffuser (Fig. 1.2b).

In practice, one can find flows with constant and also varying angular velocity of rotation, as well as flows complicated with additional influencing factors.

In this monograph, a series of problems are considered that encompass fluid flow, heat and mass transfer over disks rotating in a resting, rotating or radially accelerating fluid; unsteady conjugate heat transfer of a rotating disk (Fig. 1.1a); disk cooling by means of an impinging jet (Fig. 1.1b); forced radial flow in the cavities between co-rotating parallel disks (Fig. 1.1c); air cooling systems with inlet flow pre-swirl (Fig. 1.1f, left); gaps between a rotating disk and/or a rotating cone (Fig. 1.2a) and swirl flows in a stationary conical diffuser (Fig. 1.2b).

I.V. Shevchuk, *Convective Heat and Mass Transfer in Rotating Disk Systems*, Lecture Notes in Applied and Computational Mechanics 45, DOI 10.1007/978-3-642-00718-7_1,
© Springer-Verlag Berlin Heidelberg 2009

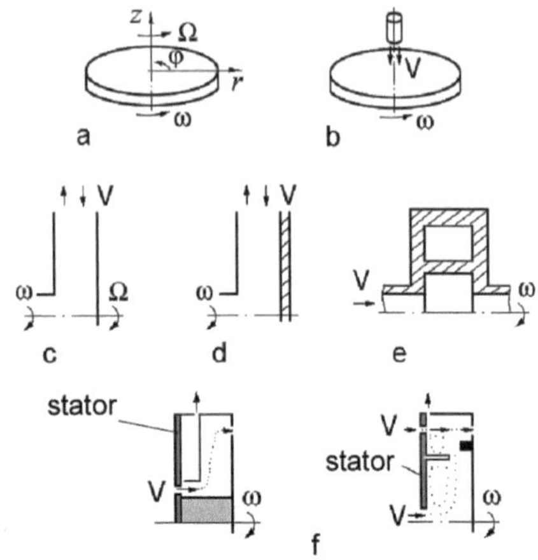

Fig. 1.1 Rotating-disk systems: (**a**) a disk in a resting or rotating fluid, (**b**) impinging jet cooling of a disk, (**c**) parallel co-rotating disks with and without forced radial throughflow, (**d**) flow between a rotor and a stator, (**e**) non-ventilated cavities of gas turbines, (**f**) air cooling systems with inlet flow pre-swirl

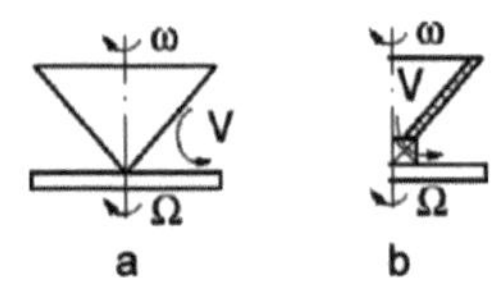

Fig. 1.2 Swirl flow (**a**) in a gap between a rotating disk and/or a cone and (**b**) in a stationary conical diffuser

1.2 Acting Forces

It is known that two types of forces act on fluid particles: mass forces (or, in other words, body forces) and surface forces [113]. Mass forces act on each fluid particle and include in the most general case gravity, inertia, electrostatic forces, magnetic or electrical fields, etc. Surface forces act on elementary parts of a surface; they include pressure, internal friction (viscosity), forces acting on a surface from the side of flow and forces of reaction from the body onto the flow.

Mass forces are caused by force fields, such as gravitational, inertial or electromagnetic [113]. Gravitational forces are a result of the global gravity of the Earth. Inertial forces emerge at accelerating or decelerating translation motion of a system, in which fluid flow takes place. Inertial forces can also be a result of rotation of a system as a whole or of a fluid only. Electromagnetic fields emerging in flow of an electrically conducting fluid in a magnetic field are not considered in this monograph.

In rotating systems, inertial forces are external with respect to fluid flow, and their strength is determined by conditions of motion both of the system and of fluid flow itself. When inertial forces emerge as a result of streamline curvature in fluid flow in a stationary geometry (curvilinear or swirl flows), their value and direction depend on the velocity distribution in the flow and are ultimately determined by pressure and viscous forces.

Gravitational and inertial mass forces can be expressed by a relation

$$\boldsymbol{F} = \boldsymbol{j}\rho, \tag{1.1}$$

where $\boldsymbol{j}$ is acceleration determining a mass force (for gravitational force, $\boldsymbol{j} = \boldsymbol{g}$ on the Earth surface). Here and throughout Chap. 1, mass forces are considered per unit volume, while boldface is used for vectors.

Centrifugal forces are directed outwards from a rotation axis and orthogonal to it. They are caused either by streamline curvature or by system rotation and can be determined by the following relation:

$$\boldsymbol{F}_c = \rho\boldsymbol{\omega} \times (\boldsymbol{\omega} \times \boldsymbol{R}) = -\rho\boldsymbol{\omega} \cdot (\boldsymbol{R} \cdot \boldsymbol{\omega}) + \rho\boldsymbol{R}\,\omega^2. \tag{1.2}$$

The parameter $\boldsymbol{R}$ included in $\boldsymbol{F}_c$ is a local radius-vector of a fluid particle relative to the rotation axis, and the symbol ´denotes a vector product of vectors. Scalar product of vectors $\mathbf{R}\cdot\boldsymbol{\omega}$ is equal to zero, since vectors $\boldsymbol{R}$ and $\boldsymbol{\omega}$ are orthogonal to each other. In curvilinear flow, where system rotation is absent, a conventional local velocity of rotation at each specific point can be defined as $\boldsymbol{\omega} = \boldsymbol{V}/\boldsymbol{R}$, which results in the relation

$$\boldsymbol{F}_c = \rho\boldsymbol{R}\,(\boldsymbol{V}/\boldsymbol{R})^2 = \rho\,\boldsymbol{V}^2/\boldsymbol{R}, \tag{1.3}$$

where $\boldsymbol{V}$ is fluid flow velocity relative to the system (i.e. relative velocity).

Coriolis forces emerge in systems rotating as a whole, if the vectors of angular velocity of rotation $\boldsymbol{\omega}$ and the relative velocity $\boldsymbol{V}$ do not coincide. Coriolis force in a rotating coordinate system is determined by the following relation [51, 162]:

$$\boldsymbol{F}_{Cor} = 2\rho\boldsymbol{\omega} \times \boldsymbol{V}. \tag{1.4}$$

Coriolis force is directed perpendicular to the conventional surface, formed by vectors $\boldsymbol{\omega}$ and $\boldsymbol{V}$, in that direction from which, after matching the origins of the vectors $\boldsymbol{F}_{Cor}$, $\boldsymbol{\omega}$ and $\boldsymbol{V}$, the shortest turn from $\boldsymbol{\omega}$ to $\boldsymbol{V}$ would appear to be going counter-clockwise.

Mass forces (a) can serve as a main source of fluid flow, (b) result in secondary flows (e.g. recirculation) or (c) cause a stabilizing effect onto a fluid. However, these effects can take place only in a non-uniform field of mass forces, i.e. under condition of their spatial variation in the system. Difference of mass forces between particular locations in the system is called an excessive mass force:

$$\Delta\boldsymbol{F} = \boldsymbol{F}_2 - \boldsymbol{F}_1 = \rho_2\boldsymbol{j}_2 - \rho_1\boldsymbol{j}_1. \tag{1.5}$$

The excessive mass force emerges due to non-uniformity of density distribution in a fluid and/or non-uniformity in the acceleration causing the mass force. Simultaneous influence of the above factors is also possible.

Analysis performed by Shchukin [162] showed that "the character of fluid flow can be affected only by the mass forces, whose value is different from the pressure gradient caused by these mass forces and counteracting with them". This means that the difference between the mass force and the counteracting pressure gradient is equal to the difference between the mass forces in two different locations of the system and is in fact the aforementioned excessive mass force.

The field of mass forces can be simple or complex; in the latter case, mass forces of different nature act simultaneously in the system. On the Earth, all phenomena take place in the gravitational field, which is therefore considered to be a simple field, while any other field of mass forces will be always complemented with gravitational forces. However, gravitational force is very often insignificant in comparison with inertial mass forces, and therefore gravitational forces are quite often neglected in physical models.

The vector of a mass force can make different angles with the surface counteracting with fluid flow. If the mass force vector makes an angle of 90 degree with a surface, the mass force field is called transverse. When the mass force vector is directed along the surface, the field is called longitudinal. One should also discern steady and unsteady fields of mass forces.

Excessive mass forces can cause active influence on fluid flow (disturbing the flow, causing secondary flows and increasing turbulence level in turbulent flow regime) or conservative influence (stabilizing the flow, suppressing different occasional perturbations and turbulent pulsations). If mass forces comply with the inequality $\mathrm{grad}|\boldsymbol{F}|>0$, this is an evidence of conservative effect of the mass forces on fluid flow. For $\mathrm{grad}|\boldsymbol{F}|<0$, mass forces cause active effect on fluid flow [85, 88, 162].

1.3 Differential Equations of Continuity, Momentum and Heat Transfer

Mathematical modelling of substance transfer processes starts from a statement of a boundary problem, which includes differential or integral–differential equations, which in the general case describe laws of momentum, heat and mass transfer, continuity equation, equation of state, plus boundary and initial conditions.

Equations of momentum transfer and continuity were obtained based on linear dependence of stresses on the strain rate (the Newton's law). In a vector form in a rotating coordinate system, these equations for incompressible sub-sonic flow of a fluid with constant physical properties neglecting viscous dissipation effects have the following form [51, 85, 158]:

$$\rho\frac{D\boldsymbol{V}}{Dt} = \rho\left[\underbrace{\frac{\partial \boldsymbol{V}}{\partial t}}_{\mathrm{I}} + \underbrace{(\boldsymbol{V}\mathrm{grad})\,\boldsymbol{V}}_{\mathrm{II}}\right] \tag{1.6}$$

$$= \underbrace{\rho\boldsymbol{F}}_{\mathrm{III}} - \underbrace{\mathrm{grad}\,\boldsymbol{p}}_{\mathrm{IV}} + \underbrace{\mathrm{div}\,\boldsymbol{\Pi}}_{\mathrm{V}} - \underbrace{2\,\rho\boldsymbol{\omega}\times\boldsymbol{V}}_{\mathrm{VI}} - \underbrace{\rho\boldsymbol{\omega}\times(\boldsymbol{\omega}\times\boldsymbol{R})}_{\mathrm{VII}}\,,$$

$$\mathrm{div}(\rho\boldsymbol{V}) = 0. \tag{1.7}$$

In Eq. (1.6), the value *D/Dt* is substantial derivative (or total derivative) with respect to time, which includes a local (term I) and a convective (term II) derivative. For steady-state processes, the local derivative is equal to zero. Term III represents *mass forces, caused by gravitational, electrostatic, magnetic field, etc., but does not include inertial forces (i.e. centrifugal and Coriolis forces).* Term IV reflects pressure effect on the flow, and term V represents friction forces. The stress tensor (viscous and turbulent stresses) is denoted as $\boldsymbol{\Pi}$. Term VI represents the Coriolis force $\boldsymbol{F}_{Cor}$, while term VII denotes the centrifugal force caused by system rotation $\boldsymbol{F}_C$. The vector $\boldsymbol{V}$ is the relative velocity (the fluid flow velocity relative to the rotating coordinate system).

If Eqs. (1.6) and (1.7) are rewritten in a stationary coordinate system, then the vector $\boldsymbol{V}$ represents absolute velocity, while terms VI and VII are discarded. In this case, the centrifugal force can be taken into account by Lamé coefficients in a concrete curvilinear coordinate system, as well as by turbulent viscosity in case of turbulent flow.

In order to simplify notation of Eq. (1.6) in the case where the mass forces $\boldsymbol{F}$ possess potential A, i.e. $\boldsymbol{F}=gradA$, they can be presented, together with the centrifugal force and pressure, in a form of a modified (reduced) pressure:

$$\boldsymbol{p}^* = \boldsymbol{p} + \rho A - \frac{1}{2}\rho\,(\boldsymbol{\omega}\times\boldsymbol{R})(\boldsymbol{\omega}\times\boldsymbol{R})^2. \tag{1.8}$$

In this case, the equation of momentum transfer takes the following form:

$$\frac{\partial \boldsymbol{V}}{\partial t} + (\boldsymbol{V}\mathrm{grad})\,\boldsymbol{V} + \mathbf{2}\,\rho\boldsymbol{\omega}\times\boldsymbol{V} = \rho\,\boldsymbol{F} - \frac{1}{\rho}\mathrm{grad}\,\boldsymbol{p}^* - \frac{1}{\rho}\mathrm{div}\,\boldsymbol{\Pi}. \tag{1.9}$$

The stress tensor $\boldsymbol{\Pi}$ in Eqs. (1.1) and (1.6) looks as follows:

$$\boldsymbol{\Pi} = \begin{pmatrix} \boldsymbol{\tau}_{11} & \boldsymbol{\tau}_{21} & \boldsymbol{\tau}_{31} \\ \boldsymbol{\tau}_{21} & \boldsymbol{\tau}_{22} & \boldsymbol{\tau}_{23} \\ \boldsymbol{\tau}_{31} & \boldsymbol{\tau}_{23} & \boldsymbol{\tau}_{33} \end{pmatrix}^2, \tag{1.10}$$

where components τ_{ik} represent a sum of respective viscous and turbulent stresses.

For laminar flow, Navier–Stokes, continuity and energy equations accounting for mass forces in a *Cartesian coordinate system* have the following form [158, 162]:

$$\rho\left(\frac{\partial u}{\partial t} + u\frac{\partial u}{\partial x} + \mathrm{v}\frac{\partial u}{\partial y} + w\frac{\partial u}{\partial z}\right) = F_x - \frac{\partial p}{\partial x} + \mu\left(\frac{\partial^2 u}{\partial x^2} + \frac{\partial^2 u}{\partial y^2} + \frac{\partial^2 u}{\partial z^2}\right), \tag{1.11}$$

$$\rho\left(\frac{\partial \mathrm{v}}{\partial t} + u\frac{\partial \mathrm{v}}{\partial x} + \mathrm{v}\frac{\partial \mathrm{v}}{\partial y} + w\frac{\partial \mathrm{v}}{\partial z}\right) = F_y - \frac{\partial p}{\partial y} + \mu\left(\frac{\partial^2 \mathrm{v}}{\partial x^2} + \frac{\partial^2 \mathrm{v}}{\partial y^2} + \frac{\partial^2 \mathrm{v}}{\partial z^2}\right), \tag{1.12}$$

$$\rho\left(\frac{\partial w}{\partial t} + u\frac{\partial w}{\partial x} + \mathrm{v}\frac{\partial w}{\partial y} + w\frac{\partial w}{\partial z}\right) = F_z - \frac{\partial p}{\partial z} + \mu\left(\frac{\partial^2 w}{\partial x^2} + \frac{\partial^2 w}{\partial y^2} + \frac{\partial^2 w}{\partial z^2}\right), \tag{1.13}$$

$$\frac{\partial u}{\partial x} + \frac{\partial \mathrm{v}}{\partial y} + \frac{\partial w}{\partial z} = 0, \tag{1.14}$$

$$\frac{\partial T}{\partial t} + u\frac{\partial T}{\partial x} + \mathrm{v}\frac{\partial T}{\partial y} + w\frac{\partial T}{\partial z} = a\left(\frac{\partial^2 T}{\partial x^2} + \frac{\partial^2 T}{\partial y^2} + \frac{\partial^2 T}{\partial z^2}\right). \tag{1.15}$$

For turbulent flow, Navier–Stokes and energy equations in a *Cartesian coordinate system* have the following form [86, 87, 158]:

$$\rho\left(\frac{\partial u}{\partial t}+u\frac{\partial u}{\partial x}+\mathrm{v}\frac{\partial u}{\partial y}+w\frac{\partial u}{\partial z}\right)=F_x-\frac{\partial p}{\partial x}+\mu\left(\frac{\partial^2 u}{\partial x^2}+\frac{\partial^2 u}{\partial y^2}+\frac{\partial^2 u}{\partial z^2}\right)$$
$$+\rho\left(\frac{\partial\overline{u'^2}}{\partial x}+\frac{\partial\overline{u'\mathrm{v}'}}{\partial y}+\frac{\partial\overline{u'w'}}{\partial z}\right) \tag{1.16}$$

$$\rho\left(\frac{\partial \mathrm{v}}{\partial t}+u\frac{\partial \mathrm{v}}{\partial x}+\mathrm{v}\frac{\partial \mathrm{v}}{\partial y}+w\frac{\partial \mathrm{v}}{\partial z}\right)=F_y-\frac{\partial p}{\partial y}+\mu\left(\frac{\partial^2 \mathrm{v}}{\partial x^2}+\frac{\partial^2 \mathrm{v}}{\partial y^2}+\frac{\partial^2 \mathrm{v}}{\partial z^2}\right)$$
$$+\rho\left(\frac{\partial\overline{\mathrm{v}'^2}}{\partial y}+\frac{\partial\overline{u'\mathrm{v}'}}{\partial x}+\frac{\partial\overline{\mathrm{v}'w'}}{\partial z}\right) \tag{1.17}$$

$$\rho\left(\frac{\partial w}{\partial t}+u\frac{\partial w}{\partial x}+\mathrm{v}\frac{\partial w}{\partial y}+w\frac{\partial w}{\partial z}\right)=F_z-\frac{\partial p}{\partial z}+\mu\left(\frac{\partial^2 w}{\partial x^2}+\frac{\partial^2 w}{\partial y^2}+\frac{\partial^2 w}{\partial z^2}\right)$$
$$+\rho\left(\frac{\partial\overline{w'^2}}{\partial z}+\frac{\partial\overline{u'w'}}{\partial x}+\frac{\partial\overline{\mathrm{v}'w'}}{\partial y}\right) \tag{1.18}$$

$$\frac{\partial u}{\partial x}+\frac{\partial \mathrm{v}}{\partial y}+\frac{\partial w}{\partial z}=0, \tag{1.19}$$

$$\frac{\partial T}{\partial t}+u\frac{\partial T}{\partial x}+\mathrm{v}\frac{\partial T}{\partial y}+w\frac{\partial T}{\partial z}=a\left(\frac{\partial^2 T}{\partial x^2}+\frac{\partial^2 T}{\partial y^2}+\frac{\partial^2 T}{\partial z^2}\right)$$
$$-\left(\frac{\partial\overline{u'T'}}{\partial x}+\frac{\partial\overline{\mathrm{v}'T'}}{\partial y}+\frac{\partial\overline{w'T'}}{\partial z}\right). \tag{1.20}$$

For laminar flow, Navier–Stokes, continuity and energy equations accounting for mass forces in a *cylindrical polar coordinate system* look as follows [86, 87, 158]:

$$\rho\left(\frac{\partial \mathrm{v}_r}{\partial t}+\mathrm{v}_r\frac{\partial \mathrm{v}_r}{\partial r}+\frac{\mathrm{v}_\varphi}{r}\frac{\partial \mathrm{v}_r}{\partial \varphi}+\mathrm{v}_z\frac{\partial \mathrm{v}_r}{\partial z}-\frac{\mathrm{v}_\varphi^2}{r}\right)=F_r-\frac{\partial p}{\partial r}$$
$$+\mu\left(\nabla^2\mathrm{v}_r-\frac{\mathrm{v}_r}{r^2}-\frac{2}{r^2}\frac{\partial \mathrm{v}_\varphi}{\partial \varphi}\right), \tag{1.21}$$

$$\rho\left(\frac{\partial \mathrm{v}_\varphi}{\partial t}+\mathrm{v}_r\frac{\partial \mathrm{v}_\varphi}{\partial r}+\frac{\mathrm{v}_\varphi}{r}\frac{\partial \mathrm{v}_\varphi}{\partial \varphi}+\mathrm{v}_z\frac{\partial \mathrm{v}_\varphi}{\partial z}+\frac{\mathrm{v}_r\mathrm{v}_\varphi}{r}\right)=F_\varphi-\frac{\partial p}{\partial \varphi}$$
$$+\mu\left(\nabla^2\mathrm{v}_\varphi+\frac{2}{r^2}\frac{\partial \mathrm{v}_r}{\partial \varphi}-\frac{\mathrm{v}_\varphi}{r^2}\right), \tag{1.22}$$

$$\rho\left(\frac{\partial \mathrm{v}_z}{\partial t}+\mathrm{v}_r\frac{\partial \mathrm{v}_z}{\partial r}+\frac{\mathrm{v}_\varphi}{r}\frac{\partial \mathrm{v}_z}{\partial \varphi}+\mathrm{v}_z\frac{\partial \mathrm{v}_z}{\partial z}\right)=F_z-\frac{1}{\rho}\frac{\partial p}{\partial z}+\mu\left(\nabla^2\mathrm{v}_z\right), \tag{1.23}$$

$$\frac{\partial(r\,\mathrm{v}_r)}{\partial r}+\frac{1}{r}\frac{\partial(r\,\mathrm{v}_\varphi)}{\partial \varphi}+\frac{\partial(r\,\mathrm{v}_z)}{\partial z}=0, \tag{1.24}$$

$$\frac{\partial T}{\partial t}+\mathrm{v}_r\frac{\partial T}{\partial r}+\frac{\mathrm{v}_\varphi}{r}\frac{\partial T}{\partial \varphi}+\mathrm{v}_z\frac{\partial T}{\partial z}=a\left[\frac{1}{r}\frac{\partial}{\partial r}\left(r\frac{\partial T}{\partial r}\right)+\frac{\partial}{\partial z}\left(\frac{\partial T}{\partial z}\right)+\frac{1}{r^2}\frac{\partial}{\partial \varphi}\left(\frac{\partial T}{\partial \varphi}\right)\right], \tag{1.25}$$

where

$$\nabla^2=\left(\frac{\partial^2}{\partial r^2}+\frac{1}{r}\frac{\partial}{\partial r}+\frac{1}{r^2}\frac{\partial^2}{\partial \varphi^2}+\frac{\partial^2}{\partial z^2}\right).$$

For turbulent flow, Navier–Stokes and energy equations accounting for mass forces in a *cylindrical polar coordinate system* look as follows [86, 87, 158]:

$$\begin{aligned}&\rho\left(\frac{\partial \mathrm{v}_r}{\partial t}+\mathrm{v}_r\frac{\partial \mathrm{v}_r}{\partial r}+\frac{\mathrm{v}_\varphi}{r}\frac{\partial \mathrm{v}_r}{\partial \varphi}+\mathrm{v}_z\frac{\partial \mathrm{v}_r}{\partial z}-\frac{\mathrm{v}_\varphi^2}{r}\right)=F_r-\frac{\partial p}{\partial r}\\&+\mu\left(\nabla^2\mathrm{v}_r-\frac{\mathrm{v}_r}{r^2}-\frac{2}{r^2}\frac{\partial \mathrm{v}_\varphi}{\partial \varphi}\right)+\frac{1}{r}\frac{\partial}{\partial r}\left(-\rho\overline{\mathrm{v}'^2_r}\right)\\&+\frac{1}{r}\frac{\partial}{\partial \varphi}\left(-\rho\overline{\mathrm{v}'_r\mathrm{v}'_\varphi}\right)+\frac{\partial}{\partial z}\left(-\rho\overline{\mathrm{v}'_r\mathrm{v}'_z}\right)-\frac{1}{r}\left(-\rho\overline{\mathrm{v}'^2_\varphi}\right),\end{aligned} \tag{1.26}$$

$$\begin{aligned}&\rho\left(\frac{\partial \mathrm{v}_\varphi}{\partial t}+\mathrm{v}_r\frac{\partial \mathrm{v}_\varphi}{\partial r}+\frac{\mathrm{v}_\varphi}{r}\frac{\partial \mathrm{v}_\varphi}{\partial \varphi}+\mathrm{v}_z\frac{\partial \mathrm{v}_\varphi}{\partial z}+\frac{\mathrm{v}_r\mathrm{v}_\varphi}{r}\right)=F_\varphi-\frac{\partial p}{\partial \varphi}\\&+\mu\left(\nabla^2\mathrm{v}_\varphi+\frac{2}{r^2}\frac{\partial \mathrm{v}_r}{\partial \varphi}-\frac{\mathrm{v}_\varphi}{r^2}\right)+\frac{1}{r}\frac{\partial}{\partial \varphi}\left(-\rho\overline{\mathrm{v}'^2_\varphi}\right)+\frac{\partial}{\partial r}\left(-\rho\overline{\mathrm{v}'_r\mathrm{v}'_\varphi}\right)\\&+\frac{\partial}{\partial z}\left(-\rho\overline{\mathrm{v}'_\varphi\mathrm{v}'_z}\right)+\frac{2}{r}\left(-\rho\overline{\mathrm{v}'_r\mathrm{v}'_\varphi}\right),\end{aligned} \tag{1.27}$$

$$\begin{aligned}&\rho\left(\frac{\partial \mathrm{v}_z}{\partial t}+\mathrm{v}_r\frac{\partial \mathrm{v}_z}{\partial r}+\frac{\mathrm{v}_\varphi}{r}\frac{\partial \mathrm{v}_z}{\partial \varphi}+\mathrm{v}_z\frac{\partial \mathrm{v}_z}{\partial z}\right)=F_z-\frac{\partial p}{\partial z}+\mu\left(\nabla^2\mathrm{v}_z\right)\\&+\frac{1}{r}\frac{\partial}{\partial r}\left(-\rho\overline{\mathrm{v}'_r\mathrm{v}'_z}\right)+\frac{1}{r}\frac{\partial}{\partial \varphi}\left(-\rho\overline{\mathrm{v}'_\varphi\mathrm{v}'_z}\right)+\frac{\partial}{\partial z}\left(-\rho\overline{\mathrm{v}'^2_z}\right),\end{aligned} \tag{1.28}$$

$$\frac{\partial(r\,\mathrm{v}_r)}{\partial r}+\frac{1}{r}\frac{\partial(r\,\mathrm{v}_\varphi)}{\partial \varphi}+\frac{\partial(r\,\mathrm{v}_z)}{\partial z}=0, \tag{1.29}$$

$$\begin{aligned}&\frac{\partial T}{\partial t}+\mathrm{v}_r\frac{\partial T}{\partial r}+\frac{\mathrm{v}_\varphi}{r}\frac{\partial T}{\partial \varphi}+\mathrm{v}_z\frac{\partial T}{\partial z}=\frac{1}{r}\frac{\partial}{\partial r}\left[r\left(a\frac{\partial T}{\partial r}-\overline{\mathrm{v}'_rT'}\right)\right]\\&+\frac{\partial}{\partial z}\left(a\frac{\partial T}{\partial z}-\overline{\mathrm{v}'_zT'}\right)+\frac{1}{r^2}\frac{\partial}{\partial \varphi}\left(a\frac{\partial T}{\partial \varphi}-\overline{\mathrm{v}'_\varphi T'}\right).\end{aligned} \tag{1.30}$$

For an axisymmetric steady-state fluid flow and unsteady heat transfer, derivatives of all functions with respect to the φ-coordinate, as well as derivatives with respect to time in the equations of momentum transfer, are equal to zero: $\partial/\partial\varphi \equiv \partial/\partial t \equiv 0$. In these conditions, Eqs. (1.21), (1.22), (1.23), (1.24) and (1.25) take the following form:

$$\mathrm{v}_r\frac{\partial \mathrm{v}_r}{\partial r}+\mathrm{v}_z\frac{\partial \mathrm{v}_r}{\partial z}-\frac{\mathrm{v}_\varphi^2}{r}=\frac{1}{\rho}F_r-\frac{1}{\rho}\frac{\partial p}{\partial r}+\nu\left(\frac{\partial^2\mathrm{v}_r}{\partial r^2}+\frac{1}{r}\frac{\partial \mathrm{v}_r}{\partial r}-\frac{\mathrm{v}_r}{r^2}+\frac{\partial^2\mathrm{v}_r}{\partial z^2}\right), \quad (1.31)$$

$$\mathrm{v}_r\frac{\partial \mathrm{v}_\varphi}{\partial r}+\mathrm{v}_z\frac{\partial \mathrm{v}_\varphi}{\partial z}+\frac{\mathrm{v}_r\mathrm{v}_\varphi}{r}=\frac{1}{\rho}F_\varphi+\nu\left(\frac{\partial^2\mathrm{v}_\varphi}{\partial r^2}+\frac{1}{r}\frac{\partial \mathrm{v}_\varphi}{\partial r}-\frac{\mathrm{v}_\varphi}{r^2}+\frac{\partial^2\mathrm{v}_\varphi}{\partial z^2}\right), \quad (1.32)$$

$$\mathrm{v}_r\frac{\partial \mathrm{v}_z}{\partial r}+\mathrm{v}_z\frac{\partial \mathrm{v}_z}{\partial z}=\frac{1}{\rho}F_z-\frac{1}{\rho}\frac{\partial p}{\partial z}+\nu\left(\frac{\partial^2\mathrm{v}_z}{\partial r^2}+\frac{1}{r}\frac{\partial \mathrm{v}_z}{\partial r}+\frac{\partial^2\mathrm{v}_z}{\partial z^2}\right), \quad (1.33)$$

$$\frac{\partial(r\,\mathrm{v}_r)}{\partial r}+\frac{\partial(r\,\mathrm{v}_z)}{\partial z}=0, \quad (1.34)$$

$$\frac{\partial T}{\partial t}+\mathrm{v}_r\frac{\partial T}{\partial r}+\mathrm{v}_z\frac{\partial T}{\partial z}=a\frac{1}{r}\frac{\partial}{\partial r}\left(r\frac{\partial T}{\partial r}\right)+a\frac{\partial^2 T}{\partial z^2}, \quad (1.35)$$

while Eqs. (1.26), (1.27), (1.28), (1.29) and (1.30) can be transformed in the following way:

$$\begin{aligned}\mathrm{v}_r\frac{\partial \mathrm{v}_r}{\partial r}+\mathrm{v}_z\frac{\partial \mathrm{v}_r}{\partial z}-\frac{\mathrm{v}_\varphi^2}{r}=&\frac{1}{\rho}F_r-\frac{1}{\rho}\frac{\partial p}{\partial r}+\nu\left(\frac{\partial^2\mathrm{v}_r}{\partial r^2}+\frac{1}{r}\frac{\partial \mathrm{v}_r}{\partial r}-\frac{\mathrm{v}_r}{r^2}+\frac{\partial^2\mathrm{v}_r}{\partial z^2}\right)\\&+\frac{1}{r}\frac{\partial}{\partial r}\left(-\overline{\mathrm{v}'^2_r}\right)+\frac{\partial}{\partial z}\left(-\overline{\mathrm{v}'_r\mathrm{v}'_z}\right)-\frac{1}{r}\left(-\overline{\mathrm{v}'^2_\varphi}\right),\end{aligned} \quad (1.36)$$

$$\begin{aligned}\mathrm{v}_r\frac{\partial \mathrm{v}_\varphi}{\partial r}+\mathrm{v}_z\frac{\partial \mathrm{v}_\varphi}{\partial z}+\frac{\mathrm{v}_r\mathrm{v}_\varphi}{r}=&\frac{1}{\rho}F_\varphi+\nu\left(\nabla^2\mathrm{v}_\varphi-\frac{\mathrm{v}_\varphi}{r^2}\right)+\frac{\partial}{\partial r}\left(-\overline{\mathrm{v}'_r\mathrm{v}'_\varphi}\right)\\&+\frac{\partial}{\partial z}\left(-\overline{\mathrm{v}'_\varphi\mathrm{v}'_z}\right)+\frac{2}{r}\left(-\overline{\mathrm{v}'_r\mathrm{v}'_\varphi}\right),\end{aligned} \quad (1.37)$$

$$\mathrm{v}_r\frac{\partial \mathrm{v}_z}{\partial r}+\mathrm{v}_z\frac{\partial \mathrm{v}_z}{\partial z}=\frac{1}{\rho}F_z-\frac{1}{\rho}\frac{\partial p}{\partial z}+\nu\left(\nabla^2\mathrm{v}_z\right)+\frac{1}{r}\frac{\partial}{\partial r}\left(-\overline{\mathrm{v}'_r\mathrm{v}'_z}\right)+\frac{\partial}{\partial z}\left(-\overline{\mathrm{v}'^2_z}\right), \quad (1.38)$$

$$\frac{\partial T}{\partial t}+\mathrm{v}_r\frac{\partial T}{\partial r}+\mathrm{v}_z\frac{\partial T}{\partial z}=\frac{1}{r}\frac{\partial}{\partial r}\left[r\left(a\frac{\partial T}{\partial r}-\overline{\mathrm{v}'_rT'}\right)\right]+\frac{\partial}{\partial z}\left(a\frac{\partial T}{\partial z}-\overline{\mathrm{v}'_zT'}\right). \quad (1.39)$$

Transformation of Eqs. (1.11)–(1.30) to a rotating coordinate system is performed below for each rotating geometry separately.

1.4 Differential Equation of Convective Diffusion

The equation of convective diffusion of a substance in a gas (or fluid) is analogous to the energy equation with the only difference that the role of temperature in the convective diffusion equation is played by concentration C, while the thermal diffusivity coefficient is substituted with the diffusion coefficient D_m. Given below are different forms of the convective diffusion equation for constant physical properties of a substance.

For laminar and turbulent flow, the convective diffusion equations in a *Cartesian coordinate system* have the following forms, respectively [82, 105]:

$$\frac{\partial C}{\partial t} + u\frac{\partial C}{\partial x} + \mathrm{v}\frac{\partial C}{\partial y} + w\frac{\partial C}{\partial z} = D_m\left(\frac{\partial^2 C}{\partial x^2} + \frac{\partial^2 C}{\partial y^2} + \frac{\partial^2 C}{\partial z^2}\right), \tag{1.40}$$

$$\begin{aligned}\frac{\partial C}{\partial t} + u\frac{\partial C}{\partial x} + \mathrm{v}\frac{\partial C}{\partial y} + w\frac{\partial C}{\partial z} &= D_m\left(\frac{\partial^2 C}{\partial x^2} + \frac{\partial^2 C}{\partial y^2} + \frac{\partial^2 C}{\partial z^2}\right)\\ &\quad - \left(\frac{\partial \overline{u'C'}}{\partial x} + \frac{\partial \overline{\mathrm{v}'C'}}{\partial y} + \frac{\partial \overline{w'C'}}{\partial z}\right).\end{aligned} \tag{1.41}$$

For laminar and turbulent flow, the convective diffusion equations in a *cylindrical polar coordinate system* look as follows, respectively [82, 105]:

$$\frac{\partial C}{\partial t} + \mathrm{v}_r\frac{\partial C}{\partial r} + \frac{\mathrm{v}_\varphi}{r}\frac{\partial C}{\partial \varphi} + \mathrm{v}_z\frac{\partial C}{\partial z} = D_m\left(\frac{1}{r}\frac{\partial}{\partial r}\left(r\frac{\partial C}{\partial r}\right) + \frac{\partial^2 C}{\partial z^2} + \frac{1}{r^2}\frac{\partial^2 C}{\partial \varphi^2}\right), \tag{1.42}$$

$$\begin{aligned}\frac{\partial C}{\partial t} + \mathrm{v}_r\frac{\partial C}{\partial r} + \frac{\mathrm{v}_\varphi}{r}\frac{\partial C}{\partial \varphi} + \mathrm{v}_z\frac{\partial C}{\partial z} &= D_m\left(\frac{1}{r}\frac{\partial}{\partial r}\left(r\frac{\partial C}{\partial r}\right) + \frac{\partial^2 C}{\partial z^2} + \frac{1}{r^2}\frac{\partial^2 C}{\partial \varphi^2}\right)\\ &\quad - \left(\frac{1}{r}\frac{\partial r\overline{\mathrm{v}'_r C'}}{\partial r} + \frac{\partial \overline{\mathrm{v}'_z C'}}{\partial z} + \frac{1}{r^2}\frac{\partial \overline{\mathrm{v}'_\varphi C'}}{\partial \varphi}\right).\end{aligned} \tag{1.43}$$

For an axisymmetric steady-state fluid flow and unsteady mass transfer, derivatives of all functions with respect to the φ-coordinate in the convective diffusion equation are equal to zero: $\partial/\partial\varphi \equiv 0$. In these conditions, Eqs. (1.42) and (1.43) take the following forms, respectively:

$$\frac{\partial C}{\partial t} + \mathrm{v}_r\frac{\partial C}{\partial r} + \mathrm{v}_z\frac{\partial C}{\partial z} = D_m\left(\frac{1}{r}\frac{\partial}{\partial r}\left(r\frac{\partial C}{\partial r}\right) + \frac{\partial^2 C}{\partial z^2}\right), \tag{1.44}$$

$$\frac{\partial C}{\partial t} + \mathrm{v}_r\frac{\partial C}{\partial r} + \mathrm{v}_z\frac{\partial C}{\partial z} = D_m\left(\frac{1}{r}\frac{\partial}{\partial r}\left(r\frac{\partial C}{\partial r}\right) + \frac{\partial^2 C}{\partial z^2} + \right) - \left(\frac{1}{r}\frac{\partial r\overline{\mathrm{v}'_r C'}}{\partial r} + \frac{\partial \overline{\mathrm{v}'_z C'}}{\partial z}\right). \tag{1.45}$$

The equation of convective diffusion of a substance in a gas (or fluid) is used in Chap. 8 to consider heat and mass transfer over rotating disks for the Prandtl and Schmidt numbers larger than unity.

Chapter 2
Modelling of Fluid Flow and Heat Transfer in Rotating-Disk Systems

2.1 Differential and Integral Equations

2.1.1 Differential Navier–Stokes and Energy Equations

We will consider here stationary axisymmetric fluid flow over disks rotating with a sufficiently high angular velocity so that effects of gravitational forces on momentum transfer are rather low. In a stationary cylindrical coordinate system arranged in such a way that a disk or a system of disks rotate around its axis of symmetry coinciding with the axis z, while the point $z = 0$ is located on a surface of the disk (Fig. 2.1), laminar fluid flow and heat transfer are described by Eqs. (1.31), (1.32), (1.33), (1.34) and (1.35) simplified accounting for the conditions $F_r = F\varphi = F_z = 0$ [41, 138, 139]:

$$\mathrm{v}_r \frac{\partial \mathrm{v}_r}{\partial r} + \mathrm{v}_z \frac{\partial \mathrm{v}_r}{\partial z} - \frac{\mathrm{v}_\varphi^2}{r} = -\frac{1}{\rho}\frac{\partial p}{\partial r} + \nu\left(\frac{\partial^2 \mathrm{v}_r}{\partial r^2} + \frac{1}{r}\frac{\partial \mathrm{v}_r}{\partial r} - \frac{\mathrm{v}_r}{r^2} + \frac{\partial^2 \mathrm{v}_r}{\partial z^2}\right), \tag{2.1}$$

$$\mathrm{v}_r \frac{\partial \mathrm{v}_\varphi}{\partial r} + \mathrm{v}_z \frac{\partial \mathrm{v}_\varphi}{\partial z} + \frac{\mathrm{v}_r \mathrm{v}_\varphi}{r} = \nu\left(\frac{\partial^2 \mathrm{v}_\varphi}{\partial r^2} + \frac{1}{r}\frac{\partial \mathrm{v}_\varphi}{\partial r} - \frac{\mathrm{v}_\varphi}{r^2} + \frac{\partial^2 \mathrm{v}_\varphi}{\partial z^2}\right), \tag{2.2}$$

$$\mathrm{v}_r \frac{\partial \mathrm{v}_z}{\partial r} + \mathrm{v}_z \frac{\partial \mathrm{v}_z}{\partial z} = -\frac{1}{\rho}\frac{\partial p}{\partial z} + \nu\left(\frac{\partial^2 \mathrm{v}_z}{\partial r^2} + \frac{1}{r}\frac{\partial \mathrm{v}_z}{\partial r} + \frac{\partial^2 \mathrm{v}_z}{\partial z^2}\right), \tag{2.3}$$

$$\frac{\partial \mathrm{v}_r}{\partial r} + \frac{\mathrm{v}_r}{r} + \frac{\partial \mathrm{v}_z}{\partial z} = 0, \tag{2.4}$$

$$\frac{\partial T}{\partial t} + \mathrm{v}_r \frac{\partial T}{\partial r} + \mathrm{v}_z \frac{\partial T}{\partial z} = a\frac{1}{r}\frac{\partial}{\partial r}\left(r\frac{\partial T}{\partial r}\right) + a\frac{\partial^2 T}{\partial z^2}. \tag{2.5}$$

For turbulent flow with account for the conditions $F_r = F\varphi = F_z = 0$, Eqs. (1.36), (1.37), (1.38) and (1.39) take the following form [41, 138, 139]:

I.V. Shevchuk, *Convective Heat and Mass Transfer in Rotating Disk Systems*, Lecture Notes in Applied and Computational Mechanics 45, DOI 10.1007/978-3-642-00718-7_2,

© Springer-Verlag Berlin Heidelberg 2009

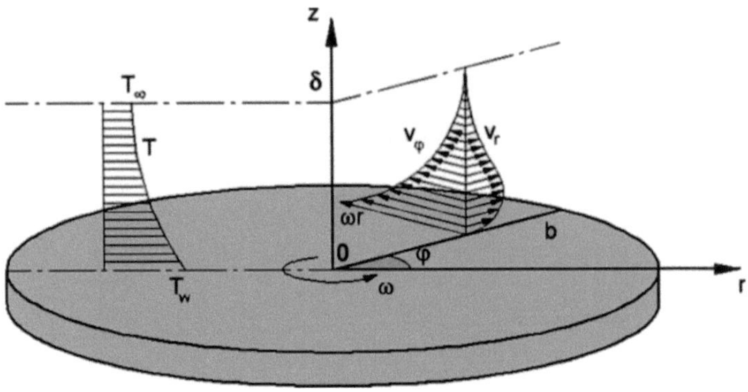

Fig. 2.1 Geometrical arrangement and main parameters of the problem of fluid flow and heat transfer over a rotating disk in still air.

$$\begin{aligned} \mathrm{v}_r\frac{\partial \mathrm{v}_r}{\partial r}+\mathrm{v}_z\frac{\partial \mathrm{v}_r}{\partial z}-\frac{\mathrm{v}_\varphi^2}{r} &= -\frac{1}{\rho}\frac{\partial p}{\partial r}+\nu\left(\nabla^2\mathrm{v}_r-\frac{\mathrm{v}_r}{r^2}\right)\\ &+\frac{1}{r}\frac{\partial}{\partial r}\left(-\overline{\mathrm{v}'^2_r}\right)+\frac{\partial}{\partial z}\left(-\overline{\mathrm{v}'_r\mathrm{v}'_z}\right)-\frac{1}{r}\left(-\overline{\mathrm{v}'^2_\varphi}\right), \end{aligned} \quad (2.6)$$

$$\begin{aligned} \mathrm{v}_r\frac{\partial \mathrm{v}_\varphi}{\partial r}+\mathrm{v}_z\frac{\partial \mathrm{v}_\varphi}{\partial z}+\frac{\mathrm{v}_r\mathrm{v}_\varphi}{r} &= \nu\left(\nabla^2\mathrm{v}_\varphi-\frac{\mathrm{v}_\varphi}{r^2}\right)+\frac{\partial}{\partial r}\left(-\overline{\mathrm{v}'_r\mathrm{v}'_\varphi}\right)\\ &+\frac{\partial}{\partial z}\left(-\overline{\mathrm{v}'_\varphi\mathrm{v}'_z}\right)+\frac{2}{r}\left(-\overline{\mathrm{v}'_r\mathrm{v}'_\varphi}\right), \end{aligned} \quad (2.7)$$

$$\mathrm{v}_r\frac{\partial \mathrm{v}_z}{\partial r}+\mathrm{v}_z\frac{\partial \mathrm{v}_z}{\partial z}=-\frac{1}{\rho}\frac{\partial p}{\partial z}+\nu\left(\nabla^2\mathrm{v}_z\right)+\frac{1}{r}\frac{\partial}{\partial r}\left(-\overline{\mathrm{v}'_r\mathrm{v}'_z}\right)+\frac{\partial}{\partial z}\left(-\overline{\mathrm{v}'^2_z}\right), \quad (2.8)$$

$$\frac{\partial T}{\partial t}+\mathrm{v}_r\frac{\partial T}{\partial r}+\mathrm{v}_z\frac{\partial T}{\partial z}=\frac{1}{r}\frac{\partial}{\partial r}\left[r\left(a\frac{\partial T}{\partial r}-\overline{\mathrm{v}'_rT'}\right)\right]+\frac{\partial}{\partial z}\left(a\frac{\partial T}{\partial z}-\overline{\mathrm{v}'_zT'}\right). \quad (2.9)$$

In a rotating coordinate system connected with a disk, Eqs. (2.1), (2.2) and (2.3) for laminar flow can be transformed as follows [138, 139]:

$$\mathrm{v}_r\frac{\partial \mathrm{v}_r}{\partial r}+\mathrm{v}_z\frac{\partial \mathrm{v}_r}{\partial z}-\frac{\mathrm{v}_\varphi^2}{r}-2\omega\mathrm{v}_\varphi-\omega^2r=-\frac{1}{\rho}\frac{\partial p}{\partial r}+\nu\left(\frac{\partial^2\mathrm{v}_r}{\partial r^2}+\frac{1}{r}\frac{\partial \mathrm{v}_r}{\partial r}-\frac{\mathrm{v}_r}{r^2}+\frac{\partial^2\mathrm{v}_r}{\partial z^2}\right), \quad (2.10)$$

$$\mathrm{v}_r\frac{\partial \mathrm{v}_\varphi}{\partial r}+\mathrm{v}_z\frac{\partial \mathrm{v}_\varphi}{\partial z}+\frac{\mathrm{v}_r\mathrm{v}_\varphi}{r}+2\omega\mathrm{v}_r=\nu\left(\frac{\partial^2\mathrm{v}_\varphi}{\partial r^2}+\frac{1}{r}\frac{\partial \mathrm{v}_\varphi}{\partial r}-\frac{\mathrm{v}_\varphi}{r^2}+\frac{\partial^2\mathrm{v}_\varphi}{\partial z^2}\right), \quad (2.11)$$

$$\mathrm{v}_r\frac{\partial \mathrm{v}_z}{\partial r}+\mathrm{v}_z\frac{\partial \mathrm{v}_z}{\partial z}=-\frac{1}{\rho}\frac{\partial p}{\partial z}+\nu\left(\frac{\partial^2\mathrm{v}_z}{\partial r^2}+\frac{1}{r}\frac{\partial \mathrm{v}_z}{\partial r}+\frac{\partial^2\mathrm{v}_z}{\partial z^2}\right). \quad (2.12)$$

The terms $2\omega\mathrm{v}_\varphi$ and $2\omega\mathrm{v}_r$ are the projections of the Coriolis forces onto the axes r and φ, respectively, while the term $\omega^2 r$ is the projection of centrifugal force onto the axis r (all divided by ρ). For turbulent flow, Eqs. (2.6), (2.7) and (2.8) can be rewritten in the same way [138, 139].

2.1.2 Differential Boundary Layer Equations

In the boundary layer approximation, it is assumed that [41, 138, 139]

(a) the velocity component v_z is by order of magnitude lower than the components v_r and v_φ;
(b) the rate of variation of the velocity, pressure and temperature in the direction of the axis z is much larger than the rate of their variation in the direction of the axis r;
(c) static pressure is constant in the z-direction.

The continuity equation (2.4) remains invariable, while the other equations of the system (2.6), (2.7), (2.8) and (2.9) take the following form [41, 138, 139]:

$$v_r \frac{\partial v_r}{\partial r} + v_z \frac{\partial v_r}{\partial z} - \frac{v_\varphi^2}{r} = -\frac{1}{\rho}\frac{\partial p}{\partial r} + \frac{1}{\rho}\frac{\partial \tau_r}{\partial z}, \quad (2.13)$$

$$v_r \frac{\partial v_\varphi}{\partial r} + v_z \frac{\partial v_\varphi}{\partial z} + \frac{v_r v_\varphi}{r} = \frac{1}{\rho}\frac{\partial \tau_\varphi}{\partial z}, \quad (2.14)$$

$$\frac{1}{\rho}\frac{\partial p}{\partial z} = 0, \quad (2.15)$$

$$\frac{\partial T}{\partial t} + v_r \frac{\partial T}{\partial r} + v_z \frac{\partial T}{\partial z} = -\frac{1}{\rho c_p}\frac{\partial q}{\partial z}, \quad (2.16)$$

$$\tau_r = \mu \frac{\partial v_r}{\partial z} - \rho \overline{v'_r v'_\varphi}, \quad (2.17)$$

$$\tau_\varphi = \mu \frac{\partial v_\varphi}{\partial z} - \rho \overline{v'_\varphi v'_z}, \quad (2.18)$$

$$q = -(\lambda \frac{\partial T}{\partial z} - \rho c_p \overline{T' v'_z}). \quad (2.19)$$

In Eqs. (2.17), (2.18) and (2.19), only those components of the turbulent shear stresses and heat fluxes are taken into account, whose input into momentum and heat transfer is the most important. It is natural that the pressure in the boundary layer is assumed to be equal to its value outside of the boundary layer in the region of potential flow $p=p_\infty$.

The equation of the stationary thermal boundary layer looks as follows:

$$v_r \frac{\partial T}{\partial r} + v_z \frac{\partial T}{\partial z} = -\frac{1}{\rho c_p}\frac{\partial q}{\partial z}. \quad (2.20)$$

The system of Eqs. (2.13), (2.14), (2.15), (2.16), (2.17), (2.18) and (2.19) is completed by an equation in the region of potential flow outside of the boundary layer, where velocity components and pressure p_∞ are considered invariable in the z-direction:

$$\frac{1}{2}\frac{d\mathrm{v}_{r,\infty}^2}{dr} - \frac{\mathrm{v}_{\varphi,\infty}^2}{r} = -\frac{1}{\rho}\frac{dp_\infty}{dr}. \tag{2.21}$$

2.1.3 Integral Boundary Layer Equations

These equations (which are in fact integral–differential equations) for stationary fluid flow and heat transfer can be derived from Eqs. (2.13), (2.14), (2.15), (2.17), (2.18), (2.19) and (2.20) with allowance for Eqs. (2.4) and (2.21) in the following form [41, 138, 139]:

$$\frac{d}{dr}\left[r\int_0^\delta \mathrm{v}_r\left(\mathrm{v}_{r,\infty}-\mathrm{v}_r\right)dz\right] + r\frac{d\mathrm{v}_{r,\infty}}{dr}\int_0^\delta\left(\mathrm{v}_{r,\infty}-\mathrm{v}_r\right)dz - \int_0^\delta\left(\mathrm{v}_{\varphi,\infty}^2-\mathrm{v}_\varphi^2\right)dz = r\tau_{wr}/\rho, \tag{2.22}$$

$$\frac{d}{dr}\left[r^2\int_0^\delta \mathrm{v}_r\left(\mathrm{v}_\varphi-\mathrm{v}_{\varphi,\infty}\right)dz\right] + \frac{\dot m_d}{2\pi\rho}\frac{d}{dr}\left(r\mathrm{v}_{\varphi,\infty}\right) = -r^2\tau_{w\varphi}/\rho, \tag{2.23}$$

or

$$\frac{d}{dr}\left[r^2\int_0^\delta \mathrm{v}_r\mathrm{v}_\varphi dz\right] + r\mathrm{v}_{\varphi,\infty}\frac{d}{dr}\left(\frac{\dot m_d}{2\pi\rho}\right) = -r^2\tau_{w\varphi}/\rho, \tag{2.24}$$

$$\frac{d}{dr}\left[r\int_0^{\delta_T} \mathrm{v}_r\left(T-T_\infty\right)dz\right] + \frac{dT_\infty}{dr}\cdot\frac{\dot m_{d,T}}{2\pi\rho} = rq_w/(\rho c_p). \tag{2.25}$$

Equations (2.22), (2.23) and (2.25) can be rewritten as follows [41, 138, 139, 180]:

$$\frac{d}{dr}\left(\mathrm{v}_{r,\infty}^2 r\delta\overline{\delta}_r^{**}\right) + \mathrm{v}_{r,\infty}r\delta\frac{d\mathrm{v}_{r,\infty}}{dr}\overline{\delta}_r^{*} - \mathrm{v}_{\varphi,\infty}^2\delta\overline{\delta}_\varphi^{**} = r\tau_{wr}/\rho, \tag{2.26}$$

$$\frac{d}{dr}[\delta r^2(\omega r)^2\overline{\delta}_{\varphi r}^{**}] + \frac{\dot m_d}{2\pi\rho}\frac{d}{dr}(r\mathrm{v}_{\varphi,\infty}) = -r^2\tau_{w\varphi}/\rho, \tag{2.27}$$

$$\frac{d}{dr}[\omega r^2\delta\overline{\delta}_T^{**}(T_w-T_\infty)] + \frac{dT_\infty}{dr}\cdot\frac{\dot m_{d,T}}{2\pi\rho} = rq_w/(\rho c_p), \tag{2.28}$$

where

$$\overline{\delta}_r^* = \int_0^1 (1-\tilde{v}_r)\, d\xi, \quad \overline{\delta}_r^{**} = \int_0^1 \tilde{v}_r (1-\tilde{v}_r)\, d\xi, \quad \overline{\delta}_\varphi^{**} = \int_0^1 \left(1 - \frac{v_\varphi^2}{v_{\varphi,\infty}^2}\right) d\xi, \tag{2.29}$$

$$\overline{\delta}_{\varphi r}^{**} = \int_0^1 \frac{v_r(v_\varphi - v_{\varphi,\infty})}{(\omega\, r)^2}\, d\xi, \quad \tilde{v}_r = v_r/v_{r,\infty}. \tag{2.30}$$

2.2 Differential Methods of Solution

2.2.1 Self-Similar Solution

For laminar flows over a single rotating disk, exact solutions of the Navier–Stokes and energy equations were obtained in works [33, 41, 55, 58, 80, 106, 138, 139, 158, 199] using the following self-similar variables:

$$\begin{aligned} &v_r = (a+\omega) r F(\zeta), \quad v_z = \sqrt{(a+\omega)\nu}\, H(\zeta), \quad v_\varphi = (a+\omega) r G(\zeta), \\ &p = -\rho\nu\omega P(\zeta), \quad \theta = (T - T_\infty)/(T_w - T_\infty), \quad \zeta = z\sqrt{(a+\omega)/\nu}, \end{aligned} \tag{2.31}$$

under the boundary conditions

$$\zeta \to \infty: \quad v_{r,\infty} = ar, \quad v_{z,\infty} = -2az, \quad v_{\varphi,\infty} = \Omega r, \quad \beta = \Omega/\omega = \text{const}, \quad \theta = 0, \tag{2.32}$$

$$\zeta = 0: \quad F = H = 0, \quad G = 1, \quad \theta = 1, \tag{2.33}$$

$$\zeta = 0: \quad T_w = T_{ref} + c_{0w} r^{n_*}, \quad T_\infty = T_{ref} + c_{0\infty} r^{n_*} \text{ or } T_\infty = T_{ref} + \beta c_{0w} r^{n_*}, \tag{2.34}$$

where c_0, c_{0w}, $c_{0\infty}$ and n_* are constants. Boundary conditions (2.34) can be transformed as follows:

$$\Delta T = T_w - T_\infty = c_0 r^{n_*} \quad (\text{for } c_0 = c_{0w} - c_{0\infty}), \tag{2.35}$$

or

$$\Delta T = c_{0w}(1-\beta)r^{n_*}. \tag{2.36}$$

Taking into account Eq. (2.21) for the radial pressure gradient in the region of potential flow, Eqs. (2.1), (2.2), (2.3) and (2.4) and (2.20) take the following self-similar form:

$$F^2 - G^2 + F'H = \frac{N^2-\beta^2}{(1+N)^2} + F'', \tag{2.37}$$

$$2FG + G'H = G'', \tag{2.38}$$

$$HH' = P' + H'', \tag{2.39}$$

$$2F + H' = 0, \tag{2.40}$$

$$\theta'' - \Pr\left(n_* F\theta + H\theta'\right) = 0, \tag{2.41}$$

where $N=a/\omega=const$ (and, naturally, turbulence components were neglected). It is impossible to assign simultaneously non-zero values of β and N in Eq. (2.37), because in this case the derivative of the component $F(\zeta)$ does not tend to zero on the outer boundary of the boundary layer. However, Eq. (2.37) still holds either for $N\neq0$ and $\beta=0$ or for $\beta\neq0$ and $N=0$.

In the past, solutions of the Eqs. (2.37), (2.38), (2.39), (2.40) and (2.41) have been obtained with the help of individually developed computer codes based on expansions in power or exponential series [80], use of the shooting method [58, 106, 138, 199], etc. Currently, standard computer mathematics software like MathCAD, etc. allows programming solutions of the systems of equations like Eqs. (2.37), (2.38), (2.39), (2.40) and (2.41) with the help of the user interface options.

As shown in works [41, 138, 139], a self-similar form of the energy equation with account for dissipation effects imposes restriction onto the boundary conditions (2.34), (2.35) and (2.36): in this case, one can use only the value of the exponent $n_*=2$. Since effects of radial heat conduction and energy dissipation in air cooling systems at sub-sonic speeds are negligible, the advantage to use arbitrary n_* values in the thermal boundary layer equation (2.41) by far compensates very minor losses involved because of neglecting the aforementioned terms in the energy equation.

Exact solutions of Eqs. (2.37), (2.38), (2.39), (2.40) and (2.41) provide a reliable database useful, among other applications, in validations of CFD codes and models aimed at solving much more complicated physical problems. Use of the self-similar solutions also enables obtaining approximate analytical solutions for problems with boundary conditions different from Eqs. (2.32), (2.33), (2.34), (2.35) and (2.36).

2.2.2 Approximate Analytical Methods for Laminar Flow Based on Approximations of Velocity Profiles

The method of Slezkin-Targ was used in the work [41] to model laminar fluid flow over a free rotating disk at $\beta=0$ and $N=0$, as well as for the case of $N=const$ and $\beta=0$. Velocity profiles derived for the case of $\beta=0$ and $N=0$ were described by the sixth-order power polynomials, which in view of the necessity to further develop the method would inevitably result in obtaining inconvenient and cumbersome relations for the Nusselt number. Decreasing the order of the approximating polynomial to the third order resulted, in the case of $N=const$ and $\beta=0$, in noticeable errors in calculations of surface friction, which were equal to approximately 25% at $N=5$ and increased sharply with the further increasing values of N.

The author of the work [4] obtained an approximate solution for the velocity components in laminar fluid flow over a free rotating disk in a form of a rather complex combination of exponential and logarithmic functions. The method was not extended to include the heat transfer problem as well as to take into consideration the boundary condition (2.32). It should be expected that the development of the method [4] in this direction would result in obtaining even more inconvenient and cumbersome relations, in particular, for the Nusselt number, than those resulting from the approach of Slezkin-Targ [41].

The approximate solution [83] for porous injection through a rotating disk has a form of a combined expansion in power and exponential series. The authors did not generalize their method for more complex cases; however, it is obvious that their approach has the same deficiencies as the methods of [4, 41].

It can be thus concluded that velocity, pressure and temperature profiles in laminar boundary layers over a rotating disk are so much complicated from the mathematical point of view that a search of their rather accurate analytical approximation is inexpedient. As shown below, a combination of an integral method with the data of self-similar solutions can result in obtaining rather simple and accurate approximate analytical solutions for surface friction coefficients and Nusselt numbers.

2.2.3 Numerical Methods

Authors of [136] solved boundary layer equations (2.13), (2.14), (2.15), (2.16), (2.17), (2.18), (2.19) and (2.20) with the help of a finite-difference method employing a modified algebraic model of turbulent viscosity by Cebeci-Smith [22]. For the case of laminar steady-state heat transfer with tangential non-uniformity of heating of the disk surface, Eq. (1.30) was reduced to a two-dimensional equation in works [205, 206] using modified variables (2.31). A steady-state axisymmetric problem with a localized heat source was modelled in the work [137] with the help of Eq. (2.9). In both aforementioned cases, a finite-difference technique was used. Equations (2.6), (2.7), (2.8) and (2.9) were used to model both laminar and turbulent flow and heat transfer in cavities between parallel rotating disks [78, 79, 145, 148,

151]. Authors of [148] applied a finite-difference method for a two-dimensional laminar steady-state problem. In works [25, 78, 79, 90, 145, 225], an in-house computer code based on the finite-volume method was used to solve Eqs. (2.6), (2.7), (2.8) and (2.9) closed with low Reynolds number k–ε models of turbulence [101, 128, 129]. Large eddy simulation (LES) approach was used in the work [224] to model a three-dimensional stationary turbulent flow over a rotating disk, and turbulent flow and heat transfer over a single disk in air flow parallel to the disk surface (non-symmetrical flow) in works [220, 221]. Author of [38] used a finite-difference method to solve the Navier–Stokes equations closed with a k–ε turbulence model by [100] as applied to three-dimensional flow of a carrying phase (air) in a rotating-disk grinder of solid particles.

Authors of [151, 152] used a commercial CFD code Fluent to solve Eqs. (2.6), (2.7), (2.8) and (2.9) closed with an RNG k–ε turbulence model. Commercial CFD code Phoenics was used by the authors of [188] to simulate stationary turbulent one-phase flow in a rotating-disk air cleaner using a standard k–ε turbulence model. In the work [61], modelling of laminar conjugate transient heat transfer of a free rotating disk was performed using a commercial CFD code CFX.

Numerical methods are the most universal tool of mathematical modelling inevitably used in simulations of problems with complex geometry, arbitrary boundary conditions, etc. Relative complications in use of such methods are significant time consumption for one run (often tens of hours), generation of a computational mesh (often weeks), sometimes lack of convergence of the numerical solution, etc. Therefore, use of such methods is not justified for relatively simple problems that can be solved by means of simpler approaches.

A general disadvantage of the differential methods (which one can nevertheless comply with) is obtaining solution in a numerical form that is usually a certain inconvenience in comparison with analytical solutions.

2.3 Integral Methods of Solution

2.3.1 Momentum Boundary Layer

The essence of integral methods consists in solving Eqs. (2.22), (2.23), (2.24), (2.25), (2.26), (2.27) and (2.28) closed with models for the velocity profiles and shear stress components on the wall for the momentum boundary layer, as well as for the temperature profiles (or enthalpy thickness) and wall heat flux for the thermal boundary layer.

The most perfect integral method among those known in the literature is the method of [138, 139], which developed ideas of the authors of [41, 80]. A key point of the method [138, 139] consists in use of a generalized form of the models that takes into account existence of laminar or turbulent flow via assigning particular numerical values to certain parameters of the model. This is a confirmation of the idea expressed by Loytsyanskiy still in the year of 1945 [113], who said that there exists “an analogy between basic characteristics of laminar and turbulent boundary layers”.

In the boundary layer, the radial v_r and tangential v_φ velocity components correlate with each other. This correlation is described by the formula [163]

$$\overline{v}_r = \overline{v}_\varphi tan\varphi. \tag{2.42}$$

For flows, where radial velocity of the flow outside of the boundary layer can be neglected, i.e. for $v_{r,\infty}=0$, authors of [138, 139] specified the velocity profiles as

$$\overline{v}_\varphi = 1 - g(\xi), \quad \overline{v}_r = \alpha f(\xi), \tag{2.43}$$

where functions $g(\xi)$ and $f(\xi)$ expressed in terms of the independent variable $\xi=z/\delta$ were assumed to be self-similar (independent of the coordinate r).

For a laminar boundary layer, functions $g(\xi)$ and $f(\xi)$ can be expressed as

$$g(\xi) = G_0(\xi), \quad f(\xi) = F_0(\xi)/\alpha_0, \tag{2.44}$$

where functions $G_0(\xi)$ and $F_0(\xi)$ are tabulated [41, 138, 139] being a solution of the problem of fluid flow over a free rotating disk in the self-similar form of Eqs. (2.37), (2.38), (2.39) and (2.40) at $N=0$ and $\beta=0$.

For a turbulent boundary layer, the following power-law approximations were used in the works [41, 138, 139]:

$$g(\xi) = 1 - \xi^n, \tag{2.45}$$

$$f(\xi) = \xi^n(1-\xi), \quad tan\varphi = \alpha(1-\xi), \tag{2.46}$$

where n=1/5–1/10. The models (2.45) and (2.46) were formulated by von Karman [80] in 1921 for the first time by analogy with the model for turbulent flow in a round pipe and over a flat plate [158]. The exponent n is assumed to be known and selected depending on the characteristic Reynolds number (Figs. 2.2, 2.3 and 2.4).

Model (2.45) for the turbulent boundary layer was also used by the authors of [7, 122, 130]. A more accurate approximation of the function $f(\xi)$ in turbulent flow is the expression used by the authors of [32, 70, 118, 196]:

$$f(\xi) = \xi^n(1-\xi)^2, \quad tan\varphi = \alpha(1-\xi)^2, \tag{2.47}$$

which, however, has not been used so widely because of somewhat increased complexity in integrating the terms in the left-hand side of Eqs. (2.22), (2.23) and (2.24).

For flows at $N=const$, the following relation was used in [212]:

$$tan\varphi = \alpha + (N-\alpha)\,\xi. \tag{2.48}$$

Authors of [41, 69] used the following expressions:

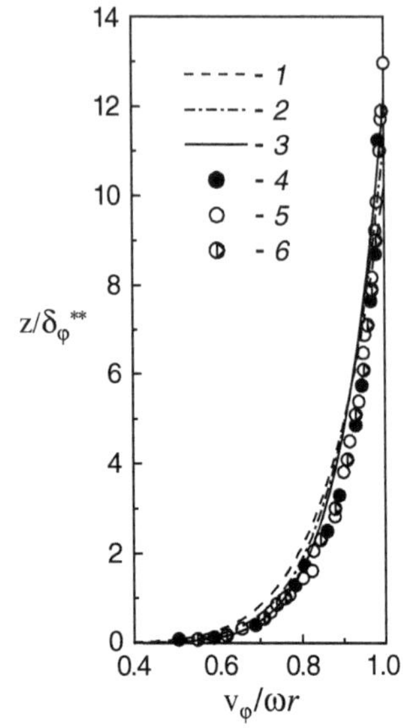

Fig. 2.2 Profiles of the non-dimensional tangential velocity in the turbulent boundary layer over a free rotating disk. Calculation by Eq. (2.45) [163], *1* – n=1/7, *2* – 1/8, *3* – 1/9. Experiments: *4* – Re_ω=(0.4–1.6)·10^6 [111], *5* – (0.6–1.0)·10^6 [63], *6* – (0.65–1.0)·10^6 [46]. Here $\delta_\varphi^{**} = \int_0^b \frac{v_\varphi}{\omega r}\left(1 - \frac{v_\varphi}{\omega r}\right) dz$ (definition of [26, 46, 47, 63, 111]).

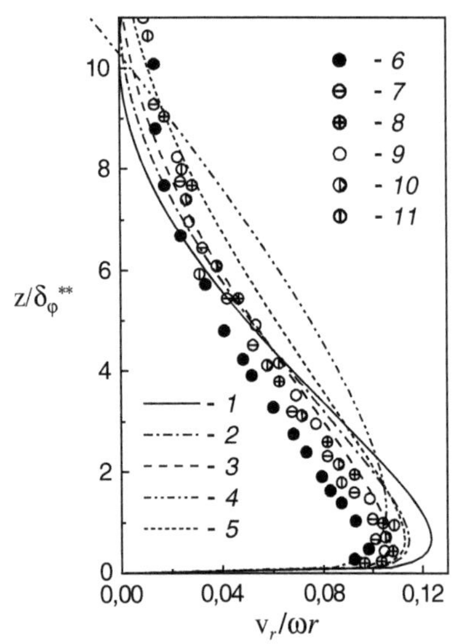

Fig. 2.3 Profiles of the non-dimensional radial velocity in the turbulent boundary layer over a free rotating disk. Calculation by Eq. (2.47) (or Eq. (2.63)) [163], *1* – n=1/7, *2* – 1/8, *3* – 1/9, *4* – 1/7, von Karman's Eq. (2.46) [80], *5* – 1/7, Eq. (2.50) for b=0.7, c=1.2, α=0.2003. Experiments: *6* – Re_ω=0.4·10^6, *7* – 0.65·10^6, *8* – 0.94·10^6, *9* – 1.6·10^6 [111], *10* – 0.6·10^6, *11* – 1.0·10^6 [63]

$$f(\xi) = \xi^n(1 - \xi^{n/m}), \quad tan\varphi = \alpha(1 - \xi^{n/m}), \tag{2.49}$$

where constants n and m could be varied independently. However, model (2.49) has not been further developed because of its excessive complexity.

Authors of [1, 2] used a trigonometric function to approximate the tangent of the flow swirl angle

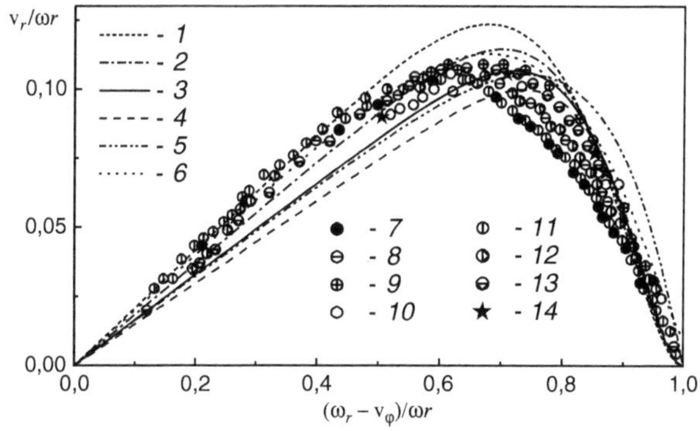

Fig. 2.4 Correlation between the radial and tangential components in the boundary layer. Calculation by Eq. (2.64) at $L=2$ (curves *1–4*) or $L=1$ (curve *5*) [163]: *1* – $n=1/7$, *2* – 1/8, *3* – 1/9, *4* – 1/10, *5* – 1/7, von Karman's method [80], *6* – 1/7, Eq. (2.50) for $b=0.7$, $c=1.2$, $\alpha=0.2003$. Experiments: *7* – $Re_\omega=0.4\cdot10^6$, *8* – $0.65\cdot10^6$, *9* – $0.94\cdot10^6$, *10* – $1.6\cdot10^6$ [111], *11* – $0.4\cdot10^6$, *12* – $0.6\cdot10^6$, *13* – $1.0\cdot10^6$ [63], *14* – $2.0\cdot10^6$ [21].

$$tan\varphi = \alpha[(1 - sin^b(c\xi)], \tag{2.50}$$

where constants b and c take the values $b=0.7$, $c=0.12$ at $n=1/7$, and $b=0.697$, $c=0.117$ at $n=1/8$. Authors [1, 2] asserted that the model (2.50) allows to attain better agreement with experiments for a free rotating disk than with von Karman's approach (2.45), (2.46).

Analysis shows, however, that the values of the constants b and c given in [1] are erroneous. For example, keeping the value $b=0.7$ at $n=1/7$ invariable, it is necessary to use the value of $c=1.2$ (Figs. 2.3 and 2.4). However, the model [1, 2] has not been further developed to include heat transfer, apparently, because resulting expressions for the Nusselt number obtained on the base of Eq. (2.50) would have been too cumbersome.

In works [73–75], it was assumed for $N=const$ that

$$tan\varphi = \alpha\,(1 - \xi) + \kappa, \tag{2.51}$$

where $\kappa = \dot{m}/\left[2\pi\rho sr(1 - \beta)\omega r\right]$. This approximation is worse than Eq. (2.48), because it does not agree with the apparent condition $tan\varphi_w = \alpha$ and complicates the solution of Eqs. (2.22) and (2.23).

The author of [72] used a relation $tan\,\varphi = cN$, where $N = \dot{m}/(2\pi\rho sr)$, with the constant c varying from the value 1.0 at the inlet to a cavity between parallel co-rotating disks to 1.22 in the region of stabilized flow. This relation was used for large values of $v_{r,\infty}$ and $v_{\varphi,\infty}\approx0$. The disadvantages of this approach are lack of agreement with the boundary condition $tan\varphi_w = \alpha$ and involvement of a new empirical constant c.

Thus, models (2.49), (2.50) and (2.51) have shown worse performance than Eqs. (2.42), (2.43), (2.44), (2.45), (2.46), (2.47) and (2.48).

After integration of the terms in the left-hand side of Eqs. (2.22) and (2.23) with account for Eqs. (2.42), (2.43), (2.44), (2.45), (2.46), (2.47) and (2.48), the unknowns to be found are parameters $\alpha(r)$ and $\delta(r)$ for a prescribed distribution of $\beta(r)$ or $\alpha(r)$ and $\beta(r)$ for a prescribed $\delta(r)$. Under conditions N=*const* or β=*const*, the value of α is also constant, while the value of δ is either also constant in laminar flow or has a form of a power-law function of r in turbulent flow [41, 138, 139].

Apart from the works [138, 139], no other author has developed an integral method for laminar flow for an arbitrary distribution of $\beta(r)$.

In frames of the approach with power-law approximation of the velocity profiles, shear stresses τ_{wr} and $\tau_{w\varphi}$ in the right-hand side of Eqs. (2.22), (2.23) and (2.24) are determined by the following equations [41, 138, 139]:

$$\tau_{wr} = -\alpha\tau_{w\varphi}, \quad \tau_{w\varphi} = -\mathrm{sgn}(1-\beta)\tau_w(1+\alpha^2)^{1/2}, \tag{2.52}$$

$$c_f = C_n^{-2/(n+1)} \cdot Re_{V_*}^{-2n/(n+1)}, \tag{2.53}$$

$$C_n = 2.28 + 0.924/n. \tag{2.54}$$

The constant C_n takes values 8.74, 9.71, 10.6 and 11.5 for n=1/7, 1/8, 1/9 and 1/10, respectively [41, 80, 130, 138, 139]. Approximation (2.54) was obtained in the work [69].

In works [73–75], a modified velocity $V_* = \omega r\,|\beta - 1|\,[1 + (\alpha+\kappa)^2\,]^{1/2}$ was used that, with no real justification of such a choice, only complicated all the mathematical derivations.

One should also mention the model [52], which employed logarithmic approximations for the velocity profiles. In frames of this approach, one has near the wall

$$\mathrm{v}_r = \alpha\omega r + \frac{2.5\alpha V_\tau}{(1+\alpha^2)^{1/2}}\ln(\xi), \quad \mathrm{v}_\varphi = -\frac{2.5V_\tau}{(1+\alpha^2)^{1/2}}\ln(\xi). \tag{2.55}$$

Approach (2.55) was validated only for the free disk case; the heat transfer problem has not been solved. An algebraic equation for the moment coefficient C_M obtained in the work [52] (see Sect. 3.3) is transcendental. Namely because of its excessive complexity and inconvenience, the logarithmic approach [52] has not been widely used further.

2.3.2 Thermal Boundary Layer

In the majority of the known works [41, 68, 72–77, 80, 114, 130, 133, 135, 138, 139, 196], heat transfer modelling in frames of the integral method has been performed with the help of a theory of local modelling. For the first time, this theory was applied to rotating-disk systems by Dorfman [41], who used for this purpose the method of Loytsyanskiy [113]. According to the theory of local modelling, a heat transfer law determining variation of the Stanton number has the following form [41, 82]:

$$St = M_s Re_T^{**-\sigma} Pr^{-n_s}, \tag{2.56}$$

where M_s, σ and n_s are universal constants independent of the Prandtl number and temperature distribution T_w on the disk surface. For turbulent flow, the constants in Eq. (2.56) have the values σ=0.25, n_s=0.5 and $M_s=7.246\cdot10^{-3}$; for laminar flow, these constants are equal to σ=1.0, n_s=1.0 and M_s=0.07303 [41, 89, 138, 139]. Relation (2.56) allows closing the thermal boundary layer equation (2.28), where the value δ_T^{**} becomes the only unknown.

Instead of the enthalpy thickness δ_T^{**}, the authors of the work [138, 139] employed the Reynolds analogy parameter χ defined as

$$\frac{q_w}{\tau_{w\varphi}} = \chi \frac{c_p(T_\infty - T_w)}{\omega r(1-\beta)}. \tag{2.57}$$

Using model (2.56) and definition (2.57), the authors of [138, 139] solved Eq. (2.28) to find the unknown value of χ.

For a long time, the work [122] had been the only one that involved a power-law approximation of the temperature profile in turbulent flow

$$\Theta = \frac{T - T_w}{T_\infty - T_w} = \xi_T^{n_T}, \quad \theta = \frac{T - T_\infty}{T_w - T_\infty} = 1 - \Theta = 1 - \xi_T^{n_T} \tag{2.58}$$

at n_T=1/5. Authors of [122] used an unjustified additional correlation $\Delta = \delta_T/\delta = 6$ at T_w=*const* and did not offer a model to derive dependence of the parameter Δ on the factors affecting heat transfer, which could become an alternative to the approach based on Eq. (2.56).

2.4 Integral Method for Modelling Fluid Flow and Heat Transfer in Rotating-Disk Systems

2.4.1 Structure of the Method

A series of original results of investigations into fluid flow and heat transfer in rotating-disk systems presented in this monograph were obtained with the help of an improved integral method developed by the author of the monograph [163–181, 184, 189, 190]. This method will be referred to as ***the present integral method*** throughout the monograph.

The present integral method is based on using the following:

(a) integral equations (2.22), (2.23), (2.24), (2.25), (2.26), (2.27) and (2.28);
(b) improved models of the turbulent velocity and temperature profiles;
(c) a novel model for the enthalpy thickness in laminar flow;

(d) closing relations for the shear stresses and heat fluxes on the disk surface;
(e) boundary conditions for the velocity and temperature outside of the boundary layer, as well as for the disk surface temperature.

The key point of the present integral method consists in using the same approach to model both laminar and turbulent flow regimes. The only difference consists in the particular values of the certain numerical coefficients involved in the model. Such an approach is in fact a development of the idea of Loytsyanskiy [113], who noticed an analogy between the main parameters of the laminar and turbulent boundary layers subjected to the same boundary conditions. This idea has already proved its fruitfulness in the problems of convective heat transfer in rotating-disk systems. It has already been used by the authors of the works [41, 80, 138, 139]; however, the approach employed in these works in view of its intrinsic imperfectness lead to significant inaccuracy in modelling of some heat transfer regimes (see Sect. 3.2).

An important feature of the present integral method is understanding the fact that any power-law approximation of the velocity and temperature profiles in laminar flow requires using polynomials of not less than the seventh order. In its turn, this leads to deriving cumbersome and inconvenient relations for the Nusselt number and the friction coefficient. On the other hand, simple power-law approximations of the velocity and temperature profiles for turbulent flow result in quite simple and lucid relations for the rest of the boundary layer characteristics including the Nusselt number and the enthalpy thickness. Having obtained, on this basis, the mathematical form of the necessary parameters for turbulent flow, it is rather easy to extend these formulas onto laminar flow under an assumption that certain coefficients are deemed to be unknowns to be found empirically from comparisons with the self-similar exact solution. As shown below, this method proved to be the most accurate among all the known integral methods for the rotating-disk systems.

Therefore, the logical sequence of developing the integral method below is as follows. First, the integral method for turbulent flow will be developed and thoroughly validated. Second, the integral method will be extended and validated for laminar flow conditions.

2.4.2 Turbulent Flow: Improved Approximations of the Velocity and Temperature Profiles

For modelling of the velocity profiles, we will use power-law approximations, namely, Eq. (2.42) for v_r, Eqs. (2.43) (the first one) and (2.45) for v_φ. The function $tan\varphi$ is specified in the form of a quadratic parabola, whose coefficients a, b and c can be found using the boundary conditions on a disk and on the outer boundary of the boundary layer:

$$tan\varphi = a + b\xi + c\xi^2, \tag{2.59}$$

$$\xi = 0, \quad tan\varphi = tan\varphi_w = \alpha, \tag{2.60}$$

$$\xi = 1, \quad tan\varphi = tan\varphi_\infty = \mathrm{v}_{r,\infty}/(\omega r - \mathrm{v}_{\varphi,\infty}) = N/(1-\beta) = \kappa, \qquad (2.61)$$

$$\xi = 1, \quad d(tan\varphi)/d\xi = 0. \qquad (2.62)$$

Based on this, one can obtain that

$$a = \alpha, \quad b = -2(\alpha - \kappa), \quad c = \alpha - \kappa, \quad (tan\varphi - \kappa)/(\alpha - \kappa) = (1-\xi)^2. \quad (2.63)$$

Profiles of the velocity components $\overline{\mathrm{v}}_r$ and $\overline{\mathrm{v}}_\varphi$ for a free rotating disk (κ=0) computed by Eqs. (2.42) and (2.63) are depicted in Figs. 2.2 and 2.3.

Wall values of the tangent of the flow swirl angle α depending on the exponent n were determined with the help of the present integral method and documented below in Table 3.4 devoted to a free rotating disk. Also presented in Table 3.4 are numerical data for α obtained using von Karman's method [80]. Advantages of the power-law model for the $\overline{\mathrm{v}}_r$ and $\overline{\mathrm{v}}_\varphi$ profiles jointly with the quadratic approximation of $tan\varphi$ in the form of Eq. (2.63) are obvious: it provides computational results that agree well with the experimental data in the outer region of the boundary layer, with the profiles at n=1/9 being in best agreement with the experimental data [63, 111] (Figs. 2.2 and 2.3). The same conclusion follows from Fig. 2.4, where the radial and tangential velocities are interrelated using an equation obtained from Eqs. (2.42), (2.43), (2.45) and (2.63) [163]

$$\overline{\mathrm{v}}_r = \alpha\, \overline{\mathrm{v}}_\varphi\, (1 - \overline{\mathrm{v}}_\varphi^{1/n})^L, \qquad (2.64)$$

where L=2 for the present integral method and L=1 for the von Karman's method [80] (although, in the near-wall region, the closest agreement between the profiles is observed for n=1/7–1/8). From Eq. (2.64), a maximum in the dependence of $\overline{\mathrm{v}}_r$ on $\overline{\mathrm{v}}_\varphi$ can also be obtained in the following form [163]:

$$\overline{\mathrm{v}}_{\varphi,\max} = \xi_{\max}^n, \quad \xi_{\max} = n/(n+L). \qquad (2.65)$$

Equation (2.63) is a generalization of the quadratic relation (2.47) proposed for the case κ=0. Somewhat worse agreement of Eq. (2.47) with experimental data mentioned in the works [118, 196] may have been caused by a less accurate choice of the constants n and α.

Temperature profiles in the boundary layer have been approximated with the power law, Eq. (2.58), which agrees well with known experiments (see Fig. 2.5).

2.4.3 Models of Surface Friction and Heat Transfer

Relations for the shear stresses $\tau_{w\varphi}$, τ_{wr} and wall heat flux q_w will be found with the help of a two-layer model of the velocity and temperature profiles in the wall

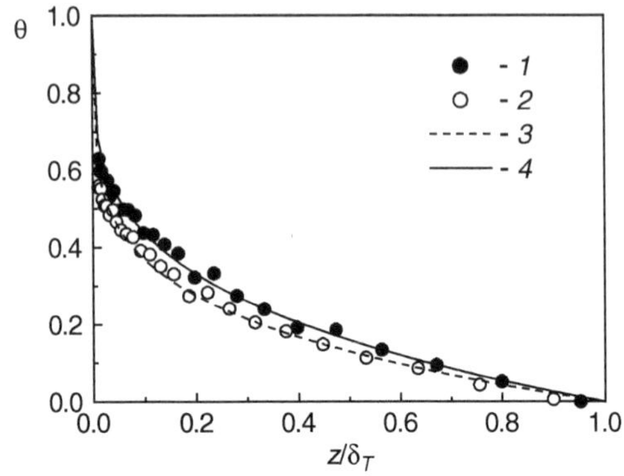

Fig. 2.5 Profiles of the non-dimensional temperature θ in the turbulent boundary layer over a free rotating disk. Experiments [46], q_w=*const*, $Re_\omega=1.0\cdot10^6$: *1* – inner heater on, *2* – inner heater off. Calculations by Eq. (2.58): *3* – n_T=1/5, *4* – 1/4.

coordinates. It is obvious that in the region where the power-law profiles (2.45) and (2.58) are in force, one can rewrite these profiles as follows:

$$V^+ = \xi^n/\sqrt{c_f/2}, \quad T^+ = \xi_T^{n_T}\sqrt{c_f/2}/St. \tag{2.66}$$

Since these power-law equations do not hold in the viscous sub-layer, they are replaced here with the following equations:

$$V^+ = z^+, \quad T^+ = \Pr z^+. \tag{2.67}$$

Splicing relations (2.66) and (2.67) at the boundary of the viscous sub-layer (at the coordinate z_1^+) and heat conduction sub-layer (at the coordinate z_{1T}^+) and transforming, one can finally obtain formulas for the surface friction coefficient and the Stanton number

$$c_f/2 = (z_1^+)^{2(n-1)/(n+1)} \cdot Re_{V_*}^{-2n/(n+1)}, \tag{2.68}$$

$$St = (z_1^+)^{n_T-1} Re_{V*}^{-n_T} (c_f/2)^{(1-n_T)/2} \Delta^{-n_T} (z_{1T}^+/z_1^+)^{n_T-1} Pr^{-n_T}. \tag{2.69}$$

The value of z_1^+ most often used in a modified form of the coefficient $C_n = (z_1^+)^{1-n}$ is a constant depending only on the exponent n (see Eq. (2.54)) [41, 130, 138, 139]. The interrelation among shear stresses τ_w, $\tau_{w\varphi}$ and τ_{wr} is defined by Eqs. (2.52). Taking this into account

$$\begin{aligned}\tau_{wr}/\rho &= C_n^{-2/(n+1)} \mathrm{sgn}(1-\beta)(\nu/\delta)^{2n/(n-1)}(\omega r|1-\beta|)^{2/(n-1)}\alpha(1+\alpha^2)^{0.5(1-n)/(1+n)},\\ \tau_{w\varphi}/\rho &= -C_n^{-2/(n+1)} \mathrm{sgn}(1-\beta)(\nu/\delta)^{2n/(n-1)}(\omega r|1-\beta|)^{2/(n-1)}(1+\alpha^2)^{0.5(1-n)/(1+n)}.\end{aligned} \tag{2.70}$$

The ratio (z_{1T}^+/z_1^+) is a function of the Prandtl number only; the quantity Δ (the unknown to be found) depends on the boundary condition for $T_w(r)$ and the Prandtl number Pr. Let us denote $(z_{1T}^+/z_1^+)^{n_T-1} Pr^{-n_T} = Pr^{-n_p}$, where the constant n_p remains unknown as yet.

In the majority of the solutions below, an assumption $n_T = n$ will be used. In this case, formulas for the Stanton and the Nusselt number take the following form:

$$St = (c_f/2)\Delta^{-n} Pr^{-n_p}, \tag{2.71}$$

$$Nu = St\frac{V_* r}{\nu}Pr = St Re_\omega Pr\,|\beta - 1|\,(1+\alpha^2)^{1/2}. \tag{2.72}$$

2.4.4 Integral Equations with Account for the Models for the Velocity and Temperature Profiles

Integration of Eqs. (2.22) and (2.23) with respect to the coordinate z with allowance for relations (2.42), (2.43), (2.44) and (2.45) and (2.63) yields the following result [163]:

$$\frac{d}{dr}\left\{\delta r(\omega r)^2(1-\beta)^2\left[\kappa\,(A_1\alpha + A_2\kappa) - \left(B_1\alpha^2 + B_2\alpha\kappa + B_3\kappa^2\right)\right]\right\} +$$
$$\delta\omega\, r^2(1-\beta)\frac{d\,(N\omega r)}{dr}\left[\kappa - (A_1\alpha + A_2\kappa)\right] + \rho\delta\,(\omega r)^2\left(C_1 + C_2\beta + C_3\beta^2\right)$$
$$= r\tau_{wr}/\rho, \tag{2.73}$$

$$\begin{aligned}&\tfrac{d}{dr}\left\{\delta\omega^2 r^4(1-\beta)\left[\alpha(D_1+\beta D_2) + \kappa\,(D_3+\beta D_4)\right]\right\}\\ &-(\omega r)^2\beta\tfrac{d}{dr}\left[\delta\omega r^2(1-\beta)(A_1\alpha + A_2\kappa)\right] = -r^2\tau_{w\varphi}/\rho\end{aligned}, \tag{2.74}$$

where $A_1 = 1/(n+1) - A_2$; $A_2 = 2/(n+2) - 1/(n+3)$;
$B_1 = 1/(2n+1) - 2/(n+1) + 6/(2n+3) - 2/(n+2) + 1/(2n+5)$;
$B_2 = 2/(n+1) - 10/(2n+3) + 4/(n+2) - 2/(2n+5)$; $D_1 = A_1 - D_2$;
$C_1 = -2/(n+1) - C_3$, $C_2 = -2\,(1/(2n+1) - 1/(n+1))$; $C_3 = -1 + 1/(2n+1)$;
$D_1 = A_1 - D_2$; $D_2 = 1/(2n+1) - D_4$; $D_3 = A_2 - D_4$; $D_4 = 1/(n+1) - 1/(2n+3)$.

The thermal boundary layer equation (2.25), being integrated with respect to the coordinate z with allowance for Eqs. (2.42), (2.43), (2.45), (2.58) and (2.63), can be reduced to the following form [168]:

$$\begin{aligned}&\tfrac{d}{dr}\left[\delta\omega r^2(1-\beta)\ F_1\,(T_\infty - T_w)\right] + \tfrac{dT_\infty}{dr}\delta\omega r^2(1-\beta)\ F_2 = \\ &= -StV_*\, r\,\Delta^{-n_T} Pr^{-n_P}\,(T_\infty - T_w)\end{aligned}, \tag{2.75}$$

where $F_1 = E_1$, $F_2 = E_2$ at $\Delta \le 1$; $F_1 = E_3$, $F_2 = E_4$ at $\Delta \ge 1$;
$E_1 = \Delta^{n+1}(aa_{*T} + bb_{*T}\Delta + cc_{*T}\Delta^2)$,
$a_{*T} = 1/(1+n+n_T) - 1/(1+n)$,
$b_{*T} = 1/(2+n+n_T) - 1/(2+n)$,
$c_{*T} = 1/(3+n+n_T) - 1/(3+n)$,
$E_2 = \Delta^{n+1}[a/(n+1) + b\Delta/(n+2) + c\Delta^2/(n+3)]$,
$E_3 = E_5 + \kappa E_6$,

$E_4 = \alpha A_1 + \kappa(\Delta - 1) + \kappa A_2,$
$E_5 = \alpha(-A_1 + \Delta^{-n_T} D_{2T}),$
$D_{2T} = 1/(1 + n + n_T) - D_{4T},$
$E_6 = (\Delta - \Delta^{-n_T})/(n_T + 1) - \Delta + 1 - A_2 + \Delta^{-n_T} D_{4T},$
$D_{4T} = 2/(2 + n + n_T) - 1/(3 + n + n_T).$

Mass flowrate in the boundary layer can be calculated from the following relation:

$$\dot{m}_d / (\rho\omega r^3) = 2\pi(1 - \beta)(A_1\alpha + A_2\kappa)\delta/r. \tag{2.76}$$

As applied to the problems under consideration, Eqs. (2.73), (2.74) and (2.75) contain three unknowns:

(a) *in the entraining boundary layers*: α, δ and Δ for given β and T_∞;
(b) *in the Ekman-type boundary layers*: α, β for a given distribution of δ (i.e. for a constant mass flowrate $\dot{m}_d$) and T_∞ for a given constant value Δ.

In the former case for particular boundary conditions (2.32), (2.33), (2.34), (2.35) and (2.36) (and *N=const*), Eqs. (2.73), (2.74) and (2.75) can be solved analytically at constant values α and Δ and power law of radial variation of the boundary layer thickness δ. For arbitrary boundary conditions, Eqs. (2.73), (2.74) and (2.75) are solved numerically; for the sake of this, they are reduced to a form convenient for integration using the Runge-Kutte method [164, 169]:

$$\begin{cases} \alpha' = (\Phi_1\Phi_4 + \Phi_2)/(1 - \Phi_1\Phi_3), \\ \bar{\delta}' = (\Phi_2\Phi_3 + \Phi_4)/(1 - \Phi_1\Phi_3), \end{cases} \tag{2.77}$$

$$\Delta' = (S_1 - S_2 - S_3)/S_4. \tag{2.78}$$

Here

$\Phi_2 = \{[\mathrm{sgn}(1 - \beta)\left|c_{fr}/2\right| \bar{r}^3 Re_{V_*}^2/\bar{\delta}^2 - Z_1\bar{\delta} - G_1\bar{\delta} - G_2]/(\bar{\delta}\bar{r}) - Q_2\bar{r}^2\} / Q_1;$
$\Phi_4 = \{-\mathrm{sgn}(1 - \beta)\left|c_{f\varphi}/2\right| \bar{r}^2 Re_{V_*}^2/\bar{\delta}^2 - \bar{\delta}[\alpha Q'_3 + \bar{Q}'_4 + (\beta Re_\omega)'(\alpha Q_5 + Q_6)]/Q_7\};$
$\Phi_1 = -Z_1/(\bar{\delta}Q_1);\ \Phi_3 = -\bar{\delta}Q_3/Q_7;\ G_1 = Re_\omega^2(C_1 + C_2\beta + C_3\beta^2);$
$Z_1 = Re_\omega^2(1 - \beta)^2[-B_1\alpha^2 + \alpha\kappa(A_1 - B_2) + \kappa^2(A_2 - B_3)];\ G_2$
$= Re_\omega^2(1 - \beta)\bar{\delta}[-A_1\alpha + \kappa(1 - A_2)]\bar{v}'_{r,\infty};$
$Q_1 = Re_\omega^2(1 - \beta)^2[-2\alpha B_1 + \kappa(A_1 - B_2)];$
$Q_2 = Re_{\omega i}^2\{-\alpha^2 B_1[\bar{r}^2(1 - \beta)^2]' + \alpha(A_1 - B_2)[\bar{r}(1 - \beta)\bar{v}_{r,\infty}]' + (A_2 - B_3)(\bar{v}_{r,\infty}^2)'\};$
$Q_3 = -Re_\omega^2(1 - \beta)^2 D_1;\ Q_4 = -Re_\omega^2(1 - \beta)\bar{v}_{r,\infty}D_3/\bar{r};\ Q_5 = -Re_\omega(1 - \beta)A_1;$
$Q_6 = -Re_{\omega i}\bar{r}\bar{v}_{r,\infty}A_2;$
$Q_7 = \alpha Q_3 + Q_4;\ \bar{v}_{r,\infty} = \bar{v}_{r,\infty}/(\omega a);\ Re_{\omega i} = \omega r_i^2/\nu;\ \left|c_{fr}/2\right|$
$= (c_f/2)\alpha/(1 + \alpha^2)^{1/2};$
$\left|c_{f\varphi}/2\right| = (c_f/2)/(1 + \alpha^2)^{1/2}\ \bar{\delta} = \delta/r_i;\ \bar{r} = r/r_i.$

In Eq. (2.78), one has at $\Delta \leq 1$:
$S_1 = -Re_\omega |1-\beta| (1+\alpha^2)^{1/2} St(\bar{T}_\infty - \bar{T}_w)$;
$S_2 = \bar{T}'_\infty \bar{\delta} Re_\omega \Delta^{n+1}[\alpha(1-\beta)/(n+1) - 2\Delta(\alpha(1-\beta)-N)/(n+2)$
$+ \Delta^2(\alpha(1-\beta)-N)/(n+3)]$;
$S_3 = \Delta^{n+1}L'_1 + \Delta^{n+2}L'_2 + \Delta^{n+3}L'_3$;
$S_4 = L_1(n+1)\Delta^n + L_2(n+2)\Delta^{n+1} + L_3(n+3)\Delta^{n+2}$;
$L_1 = L_0 a_{*T}\alpha(1-\beta)$;
$L_2 = L_0 b_{*T}(-2)[\alpha(1-\beta)-N]$;
$L_3 = L_0 c_{*T}[\alpha(1-\beta)-N]$;
$L_0 = \bar{\delta} Re_\omega(\bar{T}_\infty - \bar{T}_w)$.
The value of S_1 is the same for both $\Delta \geq 1$ and $\Delta \leq 1$.
At $\Delta \geq 1$:
$S_2 = \bar{T}'_\infty \bar{\delta} Re_\omega [\alpha(1-\beta)A_1 + NA_2 + N(\Delta-1)]$;
$S_3 = L'_{1*}C^*_{6T} + L'_{2*}C^*_{7T}$;
$S_4 = -n_T\Delta^{-n_T-1}D_{2T}L_{1*} + [(1+n_T\Delta^{-n_T-1})/(n_T+1) - D_{4T}n_T\Delta^{-n_T-1} - 1]L_{2*}$;
$L_{1*} = L_0(1-\beta)\alpha$;
$L_{2*} = L_0 N$;
$C^*_{6T} = -A_1 + \Delta^{-n_T}D_{2T}$;
$C^*_{7T} = (\Delta - \Delta^{-n_T})/(n_T+1) - \Delta + 1 - A_2 + D_{4T}\Delta^{-n_T}$;
$\bar{T} = T/T_{ref}$.
The primes denote here derivatives with respect to the radial coordinate $d/d\bar{r}$; r_i is a characteristic radius (very often, the inlet radius).
In the Ekman-type layers

$$\begin{cases} \alpha' = \frac{c_f}{2}\alpha(\beta-1)Re_\omega(1+\alpha^2)^{1/2}\frac{4\pi A_1 r_i}{B_1 C_w b} + \frac{d\beta}{d\bar{r}}\frac{\alpha}{\beta-1} - \frac{C_3[\beta+n/(n+1)]}{\bar{r}(\beta-1)\alpha B_1} - \frac{\alpha}{\bar{r}}, \\ \beta' = \left\{-\frac{c_f}{2}(1-\beta)^2 Re_\omega(1+\alpha^2)^{1/2}\frac{4\pi A_1 r_i}{D_1 C_w b} - \frac{2}{\bar{r}}\left[\beta\left(1-\frac{A_1}{D_1}\right)-1\right]\right\} \Big/ \left(1-\frac{A_1}{D_1}\right), \end{cases} \tag{2.79}$$

$$\frac{d\bar{T}_\infty}{d\bar{r}} = \left[St\frac{V_* r}{\nu}\frac{2\pi}{0.5C_w}\frac{r_i}{b}\frac{1}{K_H}\left(\bar{T}_\infty - \bar{T}_w\right) + \frac{d\bar{T}_w}{d\bar{r}}\right]\frac{K_H}{K_H-1}. \tag{2.80}$$

According to the recommendations of [138, 139], parameter K_H is considered to be constant in the Ekman-type layers [164, 167, 170]

$$K_H = 1 - (D_{2T}/A_1)\Delta^{-n_T} = \text{const} \quad \text{or} \quad \Delta = \text{const}. \tag{2.81}$$

2.5 General Solution for the Cases of Disk Rotation in a Fluid Rotating as a Solid Body and Simultaneous Accelerating Imposed Radial Flow

Let us consider the case where $\beta = const$, $N = const$ and $\kappa > 0$. The condition $\beta = const$ means solid-body rotation, which takes place between rotors and stators; the condition $N = const$ means accelerating radial flow that exists in the stagnation region of fluid flow impinging on an orthogonal surface; the condition $\kappa>0$ means that recirculation flow does not emerge on a rotating disk [138, 139, 196]. Under these conditions, the system of Eqs. (2.73) and (2.74) can be solved analytically in the most general form [174, 180, 189]:

$$\delta = C_\delta r^m, \quad C_\delta = \gamma(\omega/\nu)^{-2n/(3n+1)}, \quad \delta/r = \gamma Re_\omega^{-2n/(3n+1)}, \tag{2.82}$$

$$\alpha = \text{const}, \quad m = (1-n)/(3n+1), \tag{2.83}$$

$$\gamma = \gamma_* \left|1-\beta\right|^{(1-n)/(3n+1)}, \tag{2.84}$$

$$C_M = \varepsilon_M Re_\varphi^{-2n/(3n+1)}, \tag{2.85}$$

$$\dot{m}_d/(\mu r) = \varepsilon_m Re_\omega^{(n+1)/(3n+1)}, \tag{2.86}$$

$$c_f/2 = A_c Re_\omega^{-2n/(3n+1)}, \tag{2.87}$$

$$\alpha = -H_2/2H_3 + [(H_2/2H_3)^2 - H_1/H_3]^{1/2}, \tag{2.88}$$

$$\gamma_* = C_n^{-2/(3n+1)}(1+\alpha^2)^{0.5(1-n)/(3n+1)} H_9^{-(n+1)/(3n+1)}, \tag{2.89}$$

$$\varepsilon_m = \varepsilon_m^* \left|1-\beta\right|^{2(n+1)/(3n+1)}, \quad \varepsilon_m^* = 2\pi\gamma(A_1\alpha + A_2\kappa)sgn(1-\beta), \tag{2.90}$$

$$\varepsilon_M = \frac{8\pi}{5-4n/(3n+1)} C_n^{-\frac{2}{n+1}} \gamma_*^{\frac{2n}{n+1}} \left|1-\beta\right|^{\frac{2(n-1)}{3n+1}} (1+\alpha^2)^{\frac{1-n}{2(n+1)}} sgn(1-\beta), \tag{2.91}$$

$$A_c = C_n^{-2/(n+1)} \gamma^{-2n/(n+1)} (1+\alpha^2)^{-n/(n+1)} \left|\beta-1\right|^{-2n/(n+1)}, \tag{2.92}$$

where $H_1 = C_3(\beta - C_5) + (\beta-1)\kappa^2 H_4$; $H_2 = \kappa(\beta H_5 + H_6)$; $H_3 = \beta H_7 + H_8$; $H_4 = 1+(2+m)A_2-(3+m)B_3$; $H_5 = A_1(2+m)-B_2(3+m)+D_4(m+4)-A_2(2+m)$; $H_6 = -A_1(2+m)+B_2(3+m)+D_3(4+m)$; $H_7 = -(3+m)B_1+(4+m)D_2-(2+m)A_1$; $H_8 = (3+m)B_1 + (4+m)D_1$; $C_5 = C_1/C_3$;
$H_9 = \alpha[(D_1+\beta D_2)(4+m) - \beta(2+m)A_1] + \kappa[(D_3+\beta D_4)(4+m) - \beta A_2(2+m)]$.

An analytical solution of Eq. (2.75) is possible only under assumptions of $\Delta=const$, $Pr=const$, $n=n_T$ and boundary conditions (2.34), (2.35) and (2.36). It is evident that in this case $D_{2T}=D_2$ and $D_{4T}=D_4$.

Substituting Eqs. (2.36), (2.72), (2.82) and (2.83) into Eq. (2.75) and solving it jointly with Eq. (2.74), one can obtain [168] that

$$\left[F_1\,(2+m+n_*) + \frac{\beta n_*}{\beta-1}F_2\right] \Delta^n Pr^{n_p} = (4+m)C_4 + \frac{2\beta}{\beta-1}C_5. \tag{2.93}$$

The relations for the functions F_1 and F_2 are given in explanations to Eq. (2.75), while $C_4 = -(\alpha D_1 + \kappa D_2)$, $C_5 = 1/(n+1)+1/(n+2)+1/(n+3)$. Equation (2.93) has different solutions for the cases $\Delta\geq 1$ and $\Delta\leq 1$ (which is reflected by the different forms of the functions F_1 and F_2 for $\Delta\geq 1$ and $\Delta\leq 1$). Case $\Delta\geq 1$ corresponds to heat transfer in gases at $Pr \leq 1$. Case $\Delta \leq 1$ corresponds, in general, to heat transfer in liquids at $Pr \geq 1$; more details relevant to the case where $\Delta \leq 1$ can be found in Chap. 8.

Solution (2.93) in its general form for simultaneously non-zero values of β and N is a transcendental algebraic equation. A solution for the parameter Δ in an explicit form can be obtained at $N=0$ and $\Delta \geq 1$. The exponent n_{P}, which is considered universally the same for all types of fluid flow, will be found below on analysis of the free rotating disk case.

Expressions for the Nusselt and Stanton numbers have the following form:

$$St = A_c Re_{\omega}^{-2n/(3n+1)} \Delta^{-n} Pr^{-n_p}, \tag{2.94}$$

$$Nu = A_c (1 + \alpha^2)^{1/2} \, |1 - \beta| \, Re_{\omega}^{(n+1)/(3n+1)} \Delta^{-n} Pr^{1-n_p}. \tag{2.95}$$

Validations of the *present integral method* are performed in Chaps. 3, 5 and 6 for air flows. Extension of the present integral method onto the laminar flow case is performed in Chaps. 3 and 5. Chapter 3 represents results for a free rotating disk ($\beta=0$, $N=0$). Analysed in Chap. 5 are the cases of a disk rotating in a fluid that itself rotates as a solid body ($\beta=const$, $N=0$) and a disk rotating in a uniformly accelerating non-rotating fluid ($\beta=0$, $N=const$). Chapter 6 represents results for turbulent throughflow between parallel co-rotating disks. In Chap. 8, the method is validated for cases of Prandtl or Schmidt numbers larger than unity.

Chapter 3
Free Rotating Disk

3.1 Laminar Flow

A schematic of fluid flow and convective heat transfer over a free rotating disk in infinite ambience is depicted in Fig. 3.1. Velocity and temperature profiles computed numerically based on a solution of the self-similar system of Eqs. (2.37), (2.38), (2.39), (2.40) and (2.41) for $\beta=0$, $N=0$ with the help of the MathCAD software are presented in Fig. 3.2. Here $F_0(\zeta)=\mathrm{v}_r/(\omega r)$, $H_0(\zeta)=\mathrm{v}_z/(\omega\nu)^{1/2}$, $G_0(\zeta)=\mathrm{v}_\varphi/(\omega r)$ and $\zeta_0=z(\omega/\nu)^{1/2}$, while subscript "0" denotes a free disk. Self-similar velocity and temperature profiles are in excellent agreement with earlier computations [41, 138, 139, 158] and experiments [46, 63, 109]. Important characteristics of the profiles of F_0 and G_0 are their derivatives on the disk surface, as well as mass flowrate through the boundary layer [138, 139]:

$$G'_{0w} = (dG_0/d\zeta_0)_{\zeta=0} = -0.6159, \quad F'_{0w} = (dF_0/d\zeta_0)_{\zeta=0} = 0.5102, \tag{3.1}$$

$$\alpha_0 = -F'_{0w}/G'_{0w} = 0.8284, \quad \dot{m}_d/(\mu r) = 0.8845 Re_\omega^{1/2}. \tag{3.2}$$

Boundary layer thickness δ_0 is constant in laminar flow. Particular choice of the value of δ_0 depends on deviation of the value of G_0 from unity at $\zeta_0=\delta_0(\omega/\nu)^{1/2}$ specified as a criterion. For example, it was accepted in [138] that $\delta_0=5.5(\omega/\nu)^{1/2}$ for $G_0=0.01$. A choice of the absolute value of δ_0 does not play any role in the development of an integral method [138].

Moment coefficient of a free rotating disk C_M can be calculated by the relation [41]

$$C_M = 3.87 Re_\varphi^{-1/2}. \tag{3.3}$$

Agreement of Eq. (3.3) with experiments [207] is very good (see Fig. 3.3), while moderate deviation of Eq. (3.3) from experiments [84, 159] should be probably attributed to insufficient accuracy of experimental techniques used in the works [84, 159].

I.V. Shevchuk, *Convective Heat and Mass Transfer in Rotating Disk Systems*, Lecture Notes in Applied and Computational Mechanics 45, DOI 10.1007/978-3-642-00718-7_3,
© Springer-Verlag Berlin Heidelberg 2009

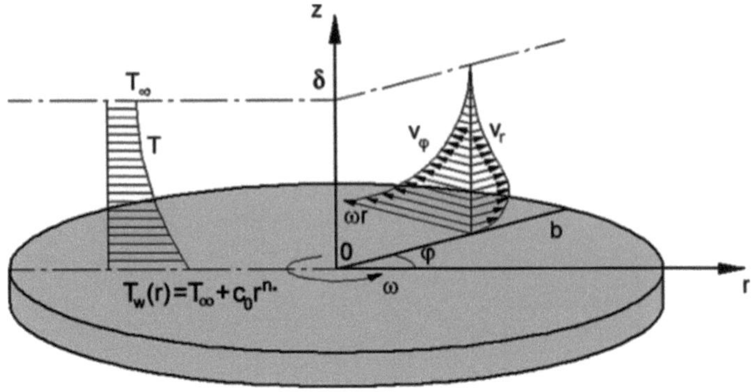

Fig. 3.1 Schematic of fluid flow and convective heat transfer over a heated free rotating disk in infinite ambience

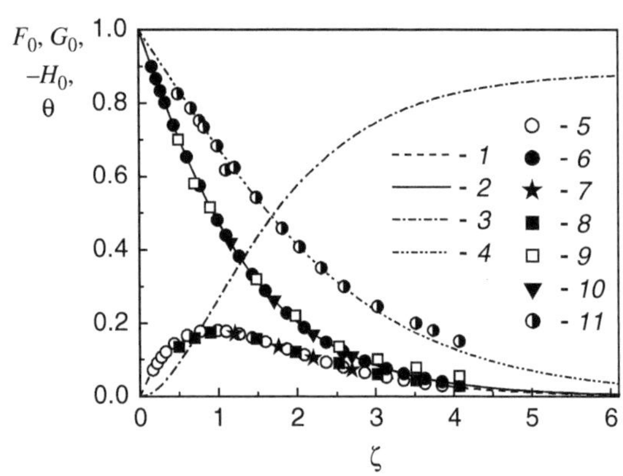

Fig. 3.2 Velocity and temperature profiles in laminar flow over a free rotating disk. Computations: *1* – F_0, *2* – G_0, *3* – $-H_0$, *4* – θ for $n_* = 0$, $Pr = 0.71$. Experiments: *5* – F_0 [63], *6* – F_0 [46], *7* – F_0 [109], *8* – G_0 [63], *9* – G_0 [46], *10* – G_0 [109], *11* – θ for $n_* = 0$, $Pr = 0.71$ [46]

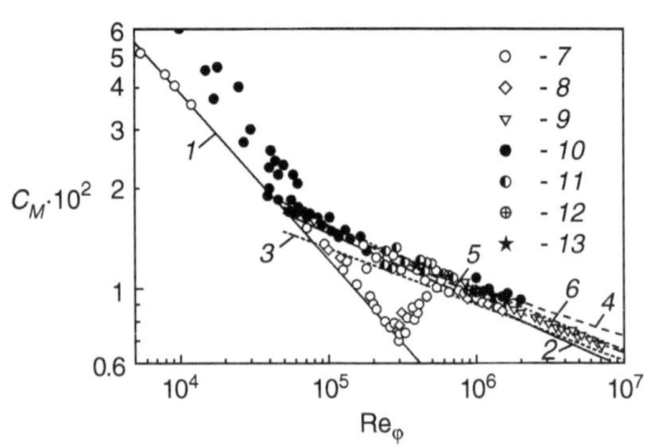

Fig. 3.3 Moment coefficient of a free rotating disk. Laminar flow: *1* – Eq. (3.3). Turbulent flow: *2* – Eq. (2.85) for $n = 1/7$, *3* – Eq. (2.85) for $n = 1/9$, *4* – Eq. (3.23), *5* – Eq. (3.24), *6* – Eq. (3.25). Experiments: *7, 8, 9* – [207], *10* – [84], *11* – [159], *12* – [111], *13* – [47]. Data *7–11* are cited according to [41, 158]; data [47, 111] were processed in [140]

Heat transfer of rotating disks can be calculated by the following equation:

$$Nu = K_1 Re_{\omega}^{n_R}, \quad Nu_{\text{av}} = K_2 Re_{\varphi}^{n_R}. \tag{3.4}$$

Coefficients K_1 and K_2 in Eq. (3.4) depend on the flow regime, Prandtl number and temperature distribution on the wall, while exponent n_R depends in general only on the flow regime. For laminar flow, $K_1 = K_2$, $n_R = 1/2$ [41, 46, 138, 139]. In accordance with the most reliable experimental data [11, 24, 32, 46, 64, 65, 97, 138, 139, 142, 143, 146, 195, 201], for an isothermal disk rotating in air ($Pr = 0.71$–0.72) the value of the constant K_1 is equal to 0.32–0.34 (see Fig. 3.4). Experimental data for the constant K_1 for non-isothermal surfaces and at other values of the Prandtl number over the range $Pr \leq 1$ to our knowledge are absent in the published literature.

Theoretical values of the constant K_1 can be obtained based on a numerical solution of the self-similar system (2.37), (2.38), (2.39), (2.40) and (2.41) for $\beta = 0$ and $N = 0$ and boundary condition (2.35)

$$K_1 = -\theta'_{\zeta=0}. \tag{3.5}$$

Numerical data for the constant K_1 obtained in such a way by the author of this monograph are represented in Table 3.1. Analogous data obtained by other authors [58, 138, 199] are less complete and contain inaccuracies or typographical errors.

For laminar flow over a rotating disk, the value of the exponent in the boundary condition (2.35) is the same $n_* = 0$ for the conditions $T_w = const$ and $q_w = const$.

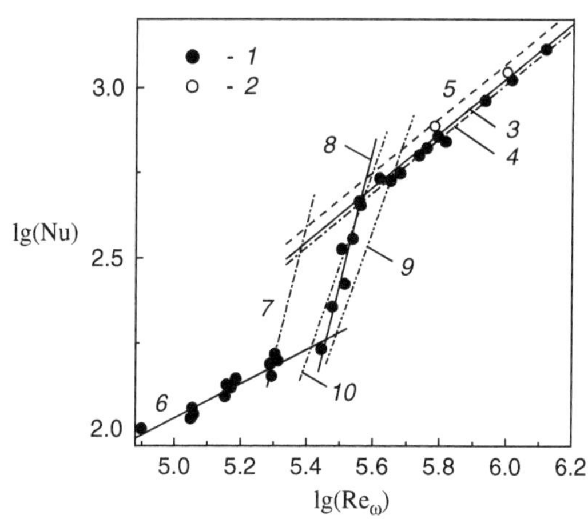

Fig. 3.4 Local Nusselt numbers over a rotating disk. Experiments [46]: *1* – $q_w = const$; *2* – $T_w \approx const$. *3–10* – calculations, Eq. (3.4). Turbulent flow, $n_R = 0.8$: *3* – $K_1 = 0.0169$ [173]; *4* – $K_1 = 0.0163$ [46]; *5* – $K_1 = 0.0187$ [171]. Laminar flow, $n_R = 0.5$: *6* – $K_1 = 0.34$ [46]. Transitional flow, $n_R = 4$: *7* – $K_1 = 10.0 \times 10^{-20}$ [146]; *8* – $K_1 = 2.65 \times 10^{-20}$ [46]. Transitional flow, $n_R = 2.8$: *9* – $K_1 = 8.01 \times 10^{-14}$ [19, 20]; *10* – $K_1 = 1.2 \times 10^{-13}$ (our suggestion)

Table 3.1 Values of the constant K_1 according to the exact solution, Eqs. (2.37), (2.38), (2.39), (2.40) and (2.41) [174, 175, 177]

Pr	$n_* = -2$	$n_* = -1.5$	$n_* = -1$	$n_* = -0.5$	$n_* = 0$	$n_* = 1$	$n_* = 2$	$n_* = 3$	$n_* = 4$
1.0	0.0	0.1305	0.2352	0.3221	0.3963	0.5180	0.6159	0.6982	0.7693
0.9	0.0	0.1217	0.2204	0.3029	0.3737	0.4905	0.5849	0.6643	0.7331
0.8	0.0	0.1124	0.2046	0.2824	0.3495	0.4608	0.5513	0.6276	0.6939
0.72	0.0	0.1045	0.1911	0.2647	0.3286	0.4352	0.5223	0.5959	0.6599
0.71	0.0	0.1035	0.1893	0.2624	0.3259	0.4319	0.5185	0.5918	0.6555
0.6	0.0	0.0917	0.1691	0.2358	0.2943	0.3929	0.4742	0.5433	0.6036
0.5	0.0	0.0802	0.1490	0.2091	0.2623	0.3531	0.4287	0.4935	0.5502
0.4	0.0	0.0675	0.1267	0.1792	0.2263	0.3078	0.3767	0.4362	0.4887
0.3	0.0	0.0536	0.1017	0.1452	0.1849	0.2550	0.3153	0.3682	0.4153
0.2	0.0	0.0381	0.0732	0.1058	0.1362	0.1912	0.2400	0.2838	0.3234
0.1	0.0	0.0204	0.0399	0.0586	0.0766	0.1104	0.1417	0.1709	0.1981

Following an increase in the value of n_*, the constant K_1 also increases significantly. For air at $Pr = 0.71$, the constant K_1 increases by 3.13 times, if the exponent n_* varies from –1 to 3 (see Table 3.1).

The exact analytical solution for the constant K_1 does not exist. The most widely known approximate analytical solution for gases was obtained by Dorfman [41]:

$$K_1 = K_2 = 0.308\,(n_* + 2)^{1/2}\,Pr^{1/2}. \tag{3.6}$$

Computations by Eq. (3.6) exceed the exact solution by 34–238% over the range of $n_* = -1.5$–0 for $Pr = 1$–0.1; the differences increase as the Prandtl number decreases.

For an isothermal disk ($T_w = \text{const}$) and the Prandtl numbers varying over the range $Pr = 0$–∞, authors of the works [26, 106] obtained the following approximations for the constant K_1, respectively:

$$K_1 = 0.6109Pr/(0.5301 + 0.3996Pr^{1/2} + Pr)^{2/3}, \tag{3.7}$$

$$K_1 = 0.6Pr/(0.56 + 0.26Pr^{1/2} + Pr)^{2/3}. \tag{3.8}$$

Maximal deviations of Eqs. (3.7) and (3.8) from the self-similar solution do not exceed 4 and 5%, respectively. In accordance with these equations, $K_1 \sim Pr$ at $Pr \to 0$, which agrees with the asymptotic solution $K_1/Pr = 0.885$ obtained in [199]. At $Pr \to \infty$, one can obtain $K_1 \approx 0.62Pr^{1/3}$, which also agrees with the data presented in the work [199]. However, Eqs. (3.7) and (3.8) are valid only for $n_* = 0$, which devalues them to a certain extent from the point of view of heat transfer problems. A more detailed analysis of Eqs. (3.7) and (3.8) is given in Chap. 8 devoted to mass transfer of rotating disks for the range of the Prandtl and Schmidt numbers $Pr \geq 1$ and $Sc \geq 1$.

3.2 Transition to Turbulent Flow and Effect of Surface Roughness

Laminar–turbulent transition over a rotating disk is accompanied by instability of laminar flow, emergence of spiral vortices and gradual development of turbulence. In one of the first experimental investigations into the conditions of laminar–turbulent transition [197], its authors observed 32 standing sinusoidal waves of perturbations that propagated spirally starting from the axis of rotation. Angle ε between the disk radius and the direction normal to the spiral line was about 14 degrees. Later, the authors of [53] made a visualization of the flow by means of a thin layer of kaolin applied on the surface of a transparent disk, whose reverse side was painted black. The number of vortices in this research varied from 28 to 31. Photos of the kaolin layer made after the experiments showed that, following an increase in the radial coordinate from the rotation axis, there existed ring-like regions of pure laminar flow, laminar flow with spiral vortices and turbulent flow.

It was shown in works [35, 49, 66, 67, 120] that the number of spiral vortices emerging, as the flow instability develops, depends on the local Reynolds number Re_ω. For example, authors of the work [49] observed 14–16 vortices with the angle ε equal to 20 degrees. A relation between the number of vortices n_v and the Reynolds number Re_ω was obtained in the work [120]:

$$n_v = 0.0698\mathrm{Re}_\omega^{1/2}. \tag{3.9}$$

This equation agrees well with experimental data of the authors of the works [49, 120, 222].

Table 3.2 summarizes data characterizing Reynolds numbers at the onset of instability, beginning and end of transition to turbulent flow obtained in experiments of different authors. Average results for each group of techniques (thermo-anemometry; visualization and acoustic measurements; methods of heat and mass transfer) are presented in Table 3.3.

Inconsistency of experimental data of different authors regarding Reynolds numbers at the onset of instability, beginning and end of transition to turbulent flow is obviously caused by the use of different criteria for determining the critical values of Re_ω. Apparently, despite the emergence of spiral vortices (registered by the experimental techniques of groups 2 and 3, Table 3.3) already for $Re_\omega = (1.14-1.4) \times 10^5$, coefficients of surface heat and mass transfer begin diverting from relations for laminar flow at higher Reynolds numbers $Re_\omega = 1.85 \times 10^5$. Besides, increase in local mass transfer in the beginning of transitional flow was not immediately reflected by an increase of average mass transfer of an *entire* disk measured in a number of investigations. Moreover, the disagreements between different sets of data may have been caused by different roughnesses, possible vibrations, etc. that could have taken place in experimental investigations. Data for the end of transition to turbulent flow averaged within each group of experimental techniques agree with each other rather well. This serves as an evidence of weak effect of different factors at the end of transition to turbulent regime.

Table 3.2 Values of the Reynolds number at the onset of instability, beginning and end of transition to turbulent flow

		Values of the Reynolds number Re_ω		
No	Experimental technique	Onset of instability	Beginning of transition	End of transition
1	Thermoanemometer [197]	2.1×10^5		3.1×10^5
2	Thermoanemometer [92]	8.8×10^4	2.5×10^5	3.2×10^5
3	Thermoanemometer [120]	8.6×10^4	$(2.6–2.8) \times 10^5$	
4	Thermoanemometer [222]	9×10^4		$(2.95–3.1) \times 10^5$
5	Thermoanemometer, visualization [226]		$(2.4–2.63) \times 10^5$	
6	Thermoanemometer [66, 67]		$(2.4–2.6) \times 10^5$	
7	Thermoanemometer [109]		$(2.5–2.64) \times 10^5$	
8	Thermoanemometer [63]		2.5×10^5	
9	Thermoanemometer [217, 218]	9.5×10^4		3.4×10^5
10	Acoustic measurements [54]	1.35×10^5	2.55×10^5	2.75×10^5
11	Visualization (kaolin) [53]	1.8×10^5		3.0×10^5
12	Visualization (naphthalene) and acoustic measurements [49]	1.5×10^5 $(0.33–1.9) \times 10^5$		2.65×10^5
13	Visualization in water [31]		$(2.8–2.86) \times 10^5$	$(3.2–4.6) \times 10^5$
14	Heat transfer coefficient [32]	2.0×10^5	2.4×10^5	
15	Heat transfer coefficient [46]		2.9×10^5	3.6×10^5
16	Heat transfer coefficient [146]		1.95×10^5	2.5×10^5
17	Heat transfer coefficient [19, 20]		2.5×10^5	3.2×10^5
18	Heat transfer coefficient [36]		2.4×10^5	
19	Mass transfer coefficient (naphthalene sublimation) [64]		2.7×10^5	
20	Mass transfer coefficient (naphthalene sublimation) [24]		1.9×10^5	2.75×10^5
21	Mass transfer coefficient (naphthalene sublimation) [208]		1.8×10^5	
22	Mass transfer coefficient (naphthalene sublimation) [97]		2.0×10^5	
23	Mass transfer coefficient (electrochemistry) [27]	1.7×10^5	2.6×10^5	3.5×10^5
24	Mass transfer coefficient (electro-chemistry) [37, 42, 126]		2.3×10^5	2.9×10^5
25	Mass transfer coefficient (electrochemistry) [39]		2.2×10^5	3.0×10^5

Analysis of the data for the moment coefficient C_M in Fig. 3.3 shows that in some experiments transition to turbulent flow started for $Re_\varphi \approx 3.0 \times 10^5$, while in other cases the so-called bypass transition to turbulent flow set on for much lower values of Re_φ. Ratio between the Reynolds numbers at the beginning and end of the transition to turbulent flow is equal to 1.21, which is close to the respective value 1.3 for flow over a flat plate [62].

Table 3.3 Average values of the Reynolds number for the boundaries of flow regimes over a rotating disk

		Values of the Reynolds number Re_ω		
No	Experimental technique	Onset of instability	Beginning of transition	End of transition
1	Thermoanemometer	1.14×10^5	2.54×10^5	3.15×10^5
2	Visualization and acoustic measurements	1.4×10^5	3.08×10^5	3.3×10^5
3	Methods of heat and mass transfer	1.85×10^5	2.3×10^5	3.1×10^5
4	Average value	1.46×10^5	2.64×10^5	3.18×10^5

Theoretical values of the Reynolds number $Re_{,\omega,}$ for which stationary instability sets on, resulting in emergence of spiral vortices, can be found based on linear instability theory. Taking into account the Coriolis forces, viscosity and streamline curvature [120] allowed for the first time obtaining a theoretical value of the Reynolds number $Re_\omega^{1/2} = 286$, that is close to the experimental data, at the onset of instability.

Detailed theoretical investigations of different types of instabilities emerging in flow over a rotating disk were performed later in works [27, 33, 34, 56, 95, 108–110, 119, 136, 144, 153, 204, 217]. The average theoretical value of the Reynolds number at the beginning of transition to turbulent flow is $Re_\omega^{1/2} = 513$ [27].

As seen from Table 3.2, in different experimental investigations the region of transition to turbulent flow corresponds to different ranges of the Reynolds number Re_ω. Respectively different was the heat and mass transfer rate measured at the same local Reynolds number Re_ω in transitional flow regime:

$$Nu = 10.0 \times 10^{-20} Re_\omega^4 \quad \text{for} \quad Re_\omega = (1.95 - 2.5) \times 10^5 \quad [146], \tag{3.10}$$

$$Nu = 2.65 \times 10^{-20} Re_\omega^4 \quad \text{for} \quad Re_\omega = (2.9 - 3.6) \times 10^5 \quad [46], \tag{3.11}$$

$$Nu = 8.01 \times 10^{-14} Re_\omega^{2.8} \quad \text{for} \quad Re_\omega = (2.6 - 3.2) \times 10^5 \quad [19, 20], \tag{3.12}$$

$$Sh = 20.0 \times 10^{-20} Re_\omega^4 \quad \text{for} \quad Re_\omega = (2.0 - 2.5) \times 10^5 \quad [24], \tag{3.13}$$

$$Sh = 3.4 \times 10^{-14} Re_\omega^3 Sc^{1/3} \quad \text{for} \quad Re_\omega = (2.0 - 3.0) \times 10^5 \quad [126]. \tag{3.14}$$

For example, Eq. (3.10) predicts heat transfer rate that formally exceeds data of Eq. (3.11) by 3.77 times. However, as follows from Fig. 3.4, earlier end of transition to developed turbulent flow predicted by Eq. (3.10) makes possible estimated differences in the heat transfer rate somewhat smaller. In reality, Nusselt numbers over the range of the Reynolds numbers $Re_\omega = (1.95{-}3.6) \times 10^5$ can be estimated with a maximal error of 2.3–2.4 times, in particular, over the range of $Re_\omega = (2.5{-}2.9) \times 10^5$.

Relation (3.12) is apparently erroneous. If its coefficients are true, the range of validity of Eq. (3.12) is $Re_\omega = (3.1-4.5) \times 10^5$ (see Fig. 3.4), and this relation contradicts the experimental data of the authors of [19, 20]. Assuming that the Reynolds number Re_ω at the beginning of transition to turbulent flow is equal to 2.6×10^5 (that agrees with experiments [19, 20]), and leaving the power exponent at the Reynolds number unchanged, one can obtain the coefficient K_1 in Eq. (3.12) to be equal to 1.2×10^{-13}. Corrected in such a way, Eq. (3.12) agrees well with experimental data [19, 20], as well as with Eq. (3.11) and experiments [46] (Fig. 3.4). A minor drawback of the corrected Eq. (3.12) is that it predicts the end of transition to turbulent flow at $Re_\omega = 3.7 \times 10^5$, which disagrees with the estimate of the authors [19, 20] themselves and agrees rather with data of the work [46] (Fig. 3.4).

Relation (3.13) is in good consistency with the experiments of the authors of [24]. However, comparing Eq. (3.13) with the empirical relations of the authors [24] for laminar flow, Eq. (8.4) for $K_1 = 0.59$, and turbulent flow, Eq. (3.29), it can be deduced that the range of validity of Eq. (3.13) should be corrected to the limits of $Re_\omega = (1.9-2.75) \times 10^5$.

Equation (3.14) was obtained for high Schmidt numbers $Sc = 1192$–2465. However, Eq. (3.14) agrees well with Eq. (3.13) also for $Sc = 2.28$.

It is obvious that additional investigations are needed in order to obtain more information regarding boundaries of transitional flow over a rotating disk, as well as heat and mass transfer rate in the transitional regime of flow.

One of the factors that cause earlier transition to the turbulent flow regime is surface roughness of the disk.

In laminar flow over a rotating disk, in analogy to flow in a pipe and over a flat plate, surface roughness practically does not affect surface friction. Heat and mass transfer for $Re_\omega > 5.6 \times 10^4$ somewhat increases, most probably, as a result of enlargement of the surface area of the disk [65, 103, 104].

Surface roughness of the disk causes significant effect on the onset of instability and boundaries of flow regimes. Experiments made in the work [218] showed that on a disk with the height of roughness elements of $h = 15\ \mu$m, the number of spiral vortices decreases from 32 to 25.

The Reynolds number characterizing onset of spiral vortices decreases from the value $Re_\omega = 0.95 \times 10^5$ on a smooth disk to $Re_\omega = 0.5 \times 10^5$ on a rough disk [218].

It was obtained in experiments [226] that on a rough disk transition to turbulent flow begins at $Re_\omega = (0.23-1.23) \times 10^5$, while, to remind, on a smooth disk the transition started at $Re_\omega = (2.4-2.63) \times 10^5$. Under these conditions, the critical Reynolds number Re_ω decreased with an increase in the angular velocity of rotation, which meant thinning of the boundary layer with simultaneous increase in the "roughness" Reynolds number $Re_r = h_r(\omega/\nu)^{1/2}$ (where h_r is the height of the roughness elements), which enforced roughness effects. A minimal value of the Reynolds number Re_r, for which effects of roughness were noticeable in experiments [226], was equal to 0.6.

The Reynolds number at the end of transition to turbulent flow also decreased from the value of $Re_\omega = 3.4 \times 10^5$ on a flat disk to $Re_\omega = 2.55 \times 10^5$ on a rough disk [218]. Similar to this, in work [64], the end of transition to turbulent flow was observed at the Reynolds number $Re_{\omega} = 2.0 \times 10^5$ on a rough disk, while on a smooth disk it was equal to $Re_{\omega} = 2.7 \times 10^5$ (type and parameters of roughness not specified).

Thus, in known experiments, the Reynolds number at the onset of instability and beginning of transition to turbulent flow decreased because of surface roughness by 2–11.5 times, while the Reynolds number at the end of transition to turbulent flow decreased by 25%.

Experiments [64] in the region of transitional flow exhibit increase in heat and mass transfer on a rough disk in comparison with a smooth one by 34% over the particular range of the Reynolds numbers Re_{ω} investigated in this work, namely, in the beginning of transition to turbulent flow.

3.3 Turbulent Flow

3.3.1 Parameters of the Turbulent Boundary Layer

The power-law profile (2.45) for the tangential velocity component v_{ϕ}, as well as quadratic approximation (2.63) (or Eq. (2.47)) for the v_r component, agrees well with reliable experimental data [63, 111] (see Figs. 2.2, 2.3 and 2.4). Agreement of Eq. (2.46) with experimental data in the outer part of the boundary layer is worse than that of Eq. (2.63).

The power-law approximation of the temperature profiles θ in the form of Eq. (2.58) agrees well with experiments [46] performed for $q_w = const$ (Fig. 2.5). The value $n_T = 1/4$ corresponds to a heating regime with the inner and outer heaters on, while the value $n_T = 1/5$ relates to a regime with the outer heater on and the inner heater off.

The present integral method (see Chap. 2) is validated in this chapter against the von Karman's method based on the approach (2.46) that has been very often used also in other integral methods. Based on the power-law approximations (2.45) and (2.46) proposed in the work [80], the other parameters of the boundary layer can be expressed by relations coinciding in their form with Eqs. (2.82), (2.83), (2.84), (2.85) and (2.86). However, relations for the constants in these equations differ from Eqs. (2.88), (2.89), (2.90) and (2.91) and can be written in the following form [138, 139]:

$$\alpha^2 = \frac{4(2+3/n)(1+2/n)(3+1/n)}{(16n^{-3}+85n^{-2}+145n^{-1}+66)/n^2}, \tag{3.15}$$

$$\gamma^{(3n+1)/(n+1)} = C_n^{-2/(n+1)} \frac{2(n+1)(2n+2)(3n+1)(n+2)(1+\alpha^2)^{(1-n)/[2(n+1)]}}{3(11n+5)\alpha n}, \tag{3.16}$$

$$\varepsilon_m = \frac{2\pi\alpha\gamma}{(1+n)(2+n)}, \quad \varepsilon_M = \frac{6\pi\alpha\gamma/n^2}{(1+1/n)(2+1/n)(1+2/n)}. \tag{3.17}$$

Numerical values of the constants in Eqs. (2.82), (2.83), (2.84), (2.85) and (2.86) computed by Eqs. (2.88), (2.89), (2.90) and (2.91) and (3.15), (3.16) and (3.17) are given in Table 3.4 for $n = 1/7$–$1/10$. Both methods are based on the same model for the tangential velocities and tangential stresses, Eqs. (2.43), (2.45) and (2.52), (2.53) and (2.54). Therefore, the numerical data obtained for $\tau_{w\varphi}$, C_M, γ and ε_M are identical for both methods at $n = idem$. The constants α and ε_m dependent on the approximation of the radial velocity profile are noticeably different in the two methods compared.

Calculated values of the wall value of the flow swirl angle $\alpha = tan\varphi_w$ are compared in Fig. 3.5 with experimental data [46, 63] and models of other authors [1, 23]. The experimental values of α decrease with increasing Reynolds number Re_ω within the interval between the values calculated by Eq. (2.88) at $n=1/7$ (the upper limit) and $n=1/9$ (the lower limit). It is well known [41, 138] that, with increasing Re_ω, the exponent n in the power-law approximation (2.45) also decreases, which inevitably

Table 3.4 Constants of the solution for a free disk

Coefficient	Equation, source	$n=1/7$	$n=1/8$	$n=1/9$	$n=1/10$
C_n	[80] and (2.54)	8.74	9.71	10.6	11.5
α	(2.88) [196]	0.2087	0.1842	0.1649	0.1493
α	(3.15) [138]	0.162	0.143	0.128	0.116
γ	(2.89) [196]	0.5299	0.4977	0.4773	0.4597
γ	(3.16) [138]	0.526	0.497	0.479	0.463
ε_m	(2.90) [196]	0.1806	0.1542	0.1355	0.1204
ε_m	(3.17) [138]	0.219	0.187	0.164	0.146
ε_M	(2.91) [196]	0.1466	0.1127	0.0901	0.0734
ε_M	(3.17) [138]	0.1458	0.1122	0.0896	0.073

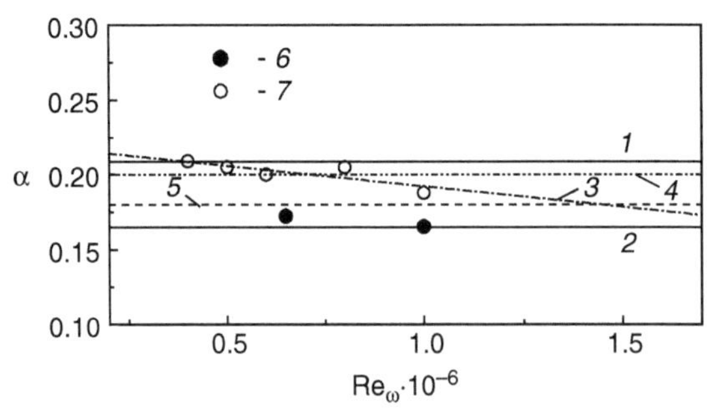

Fig. 3.5 Tangent α of the flow swirl angle at the wall. Present integral method [163, 171, 172], Eq. (3.15): *1* – $n = 1/7$; *2* – 1/9 (or 1/7 for the von Karman's method [80]); *3* – model based on original experiments of the authors [23]; *4* – α=0.2003 for n=1/7, Eq. (2.50) [1] for b=0.7, c=1.2; *5* – model [72]. Experiments: *6* – [63]; *7* – [46]

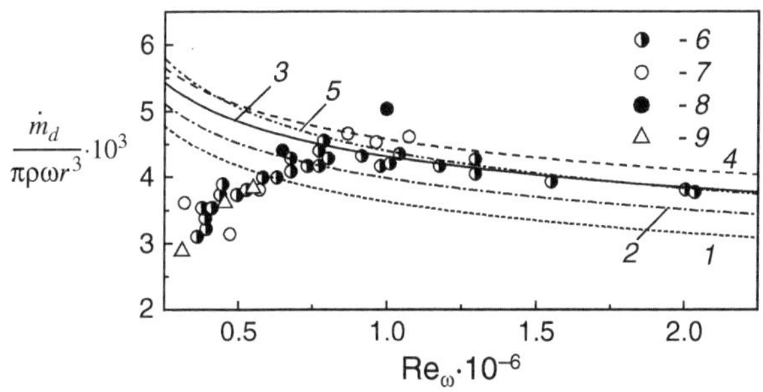

Fig. 3.6 Dimensionless mass flowrate through the boundary layer. Calculation, Eq. (2.90): *1* – n=1/7; *2* – 1/8; *3* – 1/9; *4* – 1/10; *5* – 1/7, von Karman's method (3.17) [80]. Experiments: *6* – [23]; *7* – [21]; *8* – I.V. Shevchuk based on experimental data of [46]; *9* – [69]

results in decreasing α values that is in line with the tendency observed experimentally [23, 46, 63]. The maximal value α=0.162 (at n=1/7) obtained by the von Karman's method corresponds to the lower boundary of the experimental data [23, 46, 63].

Relation (2.90) obtained for the dimensionless mass flowrate at $n = 1/9$ (the present integral method) agrees well with the experimental data [21, 23, 46] as well as with the curve obtained by the von Karman's method at $n = 1/7$ and $Re_\omega \geq 0.75 \times 10^6$ (Fig. 3.6). Equation (3.20) allows a fairly accurate estimation of the mass rate $\dot{m}_d$ over the range $Re_\omega = (0.5-0.75) \times 10^6$ at $n = 1/7$ and $n = 1/8$, although it does not correctly describe the variation of $\dot{m}_d / (\rho_\infty \omega r^3)$ as a function of Re_ω. The von Karman's method is inapplicable over this range of Re_ω at all.

Authors of the work [72] used the model (2.47) and accepted that $\alpha = 0.18$ for $n = 1/7$. As a result, they obtained that $\gamma = 0.518$, $\varepsilon_M = 0.142$ (the value of ε_m was not computed). Deviation of these values from the data of [196], Table 3.4, is rather insignificant, however, the data [196] should be considered more accurate.

In the work [69] in turbulent flow at the Reynolds numbers $Re_\omega \approx (3.1-6.64) \times 10^5$, velocity profiles were not self-similar, with the wall value of the tangent of the flow swirl angle α increasing with an increase in the Reynolds number Re_ω and local radial coordinate r/b. An account for this phenomenon was the main incentive to develop the complicated model (2.49). As a result, the following relations were obtained in [69]:

$$\alpha^2 = \frac{2(n^{-1}+3)(m^{-1}+n^{-1}+2)(2m^{-1}+n^{-1}+2)(m^{-1}+n^{-1}+1)mn}{\left[8m^{-1}(n^{-1}+1)(n^{-1}+2)(m^{-1}+n^{-1}+1)+(5n^{-1}+11)(m^{-1}+2n^{-1}+3)(2m^{-1}+n^{-1}+2)\right]}, \tag{3.18}$$

$$\gamma = (1+\alpha^2)^p \left[\frac{m^{-1}n^{-1}(m^{-1}+2n^{-1}+3)(5n^{-1}+11)\alpha}{(n^{-1}+1)(n^{-1}+2)(n^{-1}+3)(m^{-1}+n^{-1}+1)(m^{-1}+n^{-1}+2)}\right]^{-q} C_n^{-i}, \tag{3.19}$$

$$\dot{m}_d = 6.28\rho\omega r^2\alpha\delta m^{-1}n^{-1} \Big/ [(n^{-1}+1)(m^{-1}+n^{-1}+1)], \tag{3.20}$$

$$i = 2n^{-1}/(n^{-1}+3), \quad q = (1+n^{-1})/(3+n^{-1}), \quad p = (n^{-1}-1)/[2(n^{-1}+3)], \tag{3.21}$$

$$n^{-1} = 0.0079Re_{\omega}^{0.5}, \quad m^{-1} = 1.24 \times 10^{-9}(r/b)Re_{\omega}^{1.595}. \tag{3.22}$$

Solutions (3.18), (3.19), (3.20), (3.21) and (3.22) agree well with the original experiments of [69] for the velocity profiles, mass flowrate through the boundary layer and values of α. However, these achievements did not compensate the excessive complexity of the method [69], which has not been further developed to include a model for heat transfer.

The model (2.47) was also used in the work [23] jointly with an additional empirical relation for the radial mass flowrate through the boundary layer, which agreed with experiments [21, 23] for transitional and turbulent flow. In doing so, accuracy of the calculation of the turbulent flow parameters does not exceed the accuracy of the von Karman's method [80], and an improvement of agreement with the data [23] for transitional flow did not compensate the complication of the method, which has not been further developed to include effects of $v_{r,\infty}\neq0$ and $v_{\varphi,\infty}\neq0$.

Thus, the present integral method (see Chap. 2) in comparison with the von Karman's method provides much more accurate modelling not only for the radial velocity profiles, but also for the values of α and mass flowrate through the boundary layer for $Re_{\omega}=(0.5-0.8)\times10^6$. Choice of the values of n depending on the local Reynolds number Re_{ω} can be based on the data of Figs. 2.2, 2.3, 2.4, 3.5 and 3.6. The rest of the integral methods of other authors are inferior in comparison with both the present integral method and that of von Karman.

Values of the moment coefficient C_M of Eq. (2.85) for $n=1/7$ agree well with experimental data in Fig. 3.3 for $Re_{\varphi} < 2.0\times10^6$. For higher values of Re_{φ}, the curve computed by Eq. (2.85) lies somewhat lower than the experimental data. Use of the values $n=1/8$–$1/10$ somewhat improves agreement of Eq. (2.85) with experimental data for $Re_{\varphi}\geq3.0\times10^6$ (Fig. 3.3).

Logarithmic model of velocity profiles allowed obtaining the following relations for C_M:

$$C_M^{-1/2} = 1.97\,\lg\left(Re_{\varphi}\sqrt{c_M}\right) + 0.03, \tag{3.23}$$

$$C_M = 0.982\left(\lg Re_{\varphi}\right)^{-2.58}. \tag{3.24}$$

Relation (3.23) [52] can serve as the upper limit for experimental data in Fig. 3.3, while the approximation of Dorfman, Eq. (3.24) [41], provides the best agreement with experiments in comparison with the other relations.

The author of the work [140] used also the logarithmic model for the velocity profiles and approximated (in analogy to [41]) his solution for the coefficient C_M over the range $Re_{\varphi}\approx4.0\times10^5$–$2.0\times10^6$ by the following relation:

$$C_M = 0.13Re_{\varphi}^{-0.185}. \tag{3.25}$$

Equation (3.25) agrees well with experimental data [47, 111] reprocessed in [140]; however, it exceeds the data of other authors by 5–10% for $Re_\varphi \geq 2.0 \times 10^6$ (Fig. 3.3).

Logarithmic approximations of the velocity profiles significantly complicate subsequent mathematical derivations; therefore they are almost never used in integral methods.

According to the work [72], for a boundary layer developing from a location $r=a$, the moment coefficient can be calculated by the following relation:

$$C_M = 0.142[(b/a)^{5.75} - 1]^{0.8}(b/a)^{-4.6}Re_\varphi^{-0.2}. \tag{3.26}$$

3.3.2 Surface Heat Transfer: Experimental and Theoretical Data of Different Authors

Using an integral method, Dorfman [41] obtained the following relations for the coefficients K_1 and K_2 in Eqs. (3.4) at the boundary condition (2.35) and $n_R=0.8$:

$$K_1 = 0.0197\,(n_* + 2.6)^{0.2}\,Pr^{0.6}, \tag{3.27}$$

$$K_2 = K_1\,(n_* + 2)\,/\,(n_* + 2.6)\,. \tag{3.28}$$

The value 0.6 of the exponent at the Prandtl number was suggested in works [138, 139].

Values of the constants K_1 and K_2 of Eqs. (3.27) and (3.28) for different values of n_* are given in Table 3.5 together with experimental data of different authors. As evident from Table 3.5, results of the authors [18, 32, 36, 98, 132] for an isothermal disk ($n_*=0$) obtained mainly in 50th and 60th years of the twentieth century agree well with Eqs. (3.27) and (3.28). However, the data obtained from more recent and accurate investigations are noticeably lower than the calculations given in Table 3.5: by 4.4% [146] and 9.5% [122].

For the case $q_w=const$ (which corresponds to the value $n_*=-0.6$), the value of K_1 according to experiments [19, 20, 46] is less than that predicted by Eq. (3.27)

Table 3.5 Values of the constants K_1 and K_2 for $n_R=0.8$ and $Pr=0.72$ (different authors)

Coefficient	Source or equation	$n_*=-0.6$	$n_*=0$	$n_*=2$	$n_*=6$
K_2	[18, 32, 36, 98]		0.015		
K_1	[132]		0.0194		
K_1	[122]		0.0179		
K_2	[122]		0.0138		
K_1	[11, 146]		0.0188		
K_2	[146]		0.0145		
K_1	[19, 20, 46]	0.0163			
K_1	Dorfman, Eq. (3.27)	0.0186	0.0196	0.022	0.0249
K_2	Dorfman, Eq. (3.28)	0.013	0.0151	0.0191	0.0231

by 14.1% (see Table 3.5). Results of numerical simulations [205] for the boundary condition q_w=*const* are generalized by Eq. (3.4) at n_R=0.83 and K_1=0.0111, K_2=0.0086.

In the light of these results, an assertion of the authors [18] that the constant K_2 for the cases q_w=*const* and T_w=*const* has the same value 0.015 looks implausible.

In experiments [133, 135], the Nusselt numbers for the case n_*=–0.2 were also 8–10% lower than data according to Eqs. (3.27) and (3.28). Unfortunately, the authors of the works [133, 135] did not suggest empirical values of constants K_1 and K_2; also the thermal boundary condition over the disk in these experiments did not strictly follow Eq. (2.35).

As a result of experiments [24, 59] on naphthalene sublimation in a rotating disk, the following relation for the coefficient K_1 by the Sherwood number was obtained

$$K_1 = Sh \cdot Re_\omega^{-0.8} = const, \tag{3.29}$$

where K_1=0.0512 [24] or K_1=0.0518 [59]. Authors of the works [24, 59] recommended to recalculate their data with the help of the methodology [28], which established an interrelation $Nu = Sh(Pr/Sc)^{0.4}$ between the Nusselt and the Sherwood numbers. The Schmidt number for naphthalene sublimation in air is equal to Sc=2.28, while the Prandtl number for heat transfer in air is equal to Pr=0.72. Mass transfer at constant concentration of a substance on a surface is analogous to heat transfer at a constant temperature on a surface. In view of this, a recalculation of Eq. (3.29) results in the value of K_1=0.0323 for the condition T_w=*const*, which is significantly different from the data in Table 3.5. This is an evidence of significant inaccuracy of the recalculation technique suggested in the work [28]. Turbulent heat and mass transfer of rotating disks for the Prandtl and Schmidt numbers larger than unity is analysed in detail in Chap. 8. There one can also find an improved methodology developed by the author of this monograph for turbulent flow regime, which establishes a relation between the Sherwood numbers for naphthalene sublimation in air and the Nusselt numbers for heat transfer in air.

A solution for the Nusselt number obtained in works [41, 68, 138, 139] based on a theory of local modelling for the cases where the boundary condition (2.35) does not hold is insufficiently accurate and therefore has not been widely used.

For a case where the boundary layer develops from a non-zero radial location r=a at the boundary condition (2.35), the authors of the work [72] based on the approach of Dorfman [41] obtained the following solution:

$$Nu = 0.0189(n_* + 2.6)^{0.2} \cdot Re_\omega^{0.8} \cdot \bar{r}^{0.25n_*+0.65}(\bar{r}^{1.25n_*+3.25} - 1)^{-0.2} \cdot Pr^{0.6}, \tag{3.30}$$

$$Nu_{av} = 0.0226 \cdot \frac{(n_* + 2)\, Re_\varphi^{0.8}}{(1.25n_* + 3.25)^{0.8}} \cdot \frac{\left[(b/a)^{1.25n_*+3.25} - 1\right]^{0.8}}{(b/a)^{0.6}\left[(b/a)^{n_*+2} - 1\right]}, \tag{3.31}$$

where $\bar{r} = r/a$. Because of the aforementioned inaccuracy in the choice of the constant α in the work [70], Eqs. (3.30) and (3.31) predict heat transfer rates underestimated by $\approx 4\%$.

Average Nusselt number for an entire disk. In a number of technical applications, one needs to calculate average Nusselt numbers Nu_{av} of an entire disk where regions of laminar and transitional flow or laminar, transitional and turbulent flow co-exist simultaneously.

The definition of Nu_{av} (see Nomenclature) can be transformed as follows:

$$Nu_{\mathrm{av}} = \frac{b\int_0^b Nu(T_w - T_\infty)dr}{\int_0^b (T_w - T_\infty)rdr}. \tag{3.32}$$

Assuming, following the work [36], that transition from laminar and turbulent flow occurs instantly at a radial location r_{tr} for the Reynolds number $Re_{\omega,\mathrm{tr}}$, one can present Eq. (3.32) in the following form:

$$Nu_{\mathrm{av}} = \frac{b\left[\int_0^{r_{\mathrm{tr}}} Nu_{\mathrm{lam}}(T_w - T_\infty)dr + \int_{r_{\mathrm{tr}}}^b Nu_{\mathrm{turb}}(T_w - T_\infty)dr\right]}{\int_{r_{\mathrm{tr}}}^b (T_w - T_\infty)rdr}. \tag{3.33}$$

The Nusselt numbers in Eq. (3.33) are specified by the first of Eq. (3.4) with allowance for the values $K_{1,\mathrm{lam}}$ and n_R=1/2 for laminar flow and the values $K_{1,\mathrm{turb}}$ and n_R for turbulent flow.

Assuming further that the temperature of the disk surface varies according to Eq. (2.35), one can obtain as a result of integration of Eq. (3.33) that

$$Nu_{\mathrm{av}} = K_{1,\mathrm{lam}} Re_{\omega,\mathrm{tr}}^{1/2}\left(\frac{Re_{\omega,\mathrm{tr}}}{Re_\varphi}\right)^{n_*/2+1/2} + \frac{2+n_*}{2n_R+1+n_*}K_{1,\mathrm{turb}}Re_\varphi^{n_R}\left[1-\left(\frac{Re_{\omega,\mathrm{tr}}}{Re_\varphi}\right)^{n_*/2+n_R+1/2}\right]. \tag{3.34}$$

For n_*=0 (an isothermal disk at T_w=*const*), Eq. (3.34) takes the following form:

$$Nu_{\mathrm{av}} = K_{1,\mathrm{lam}} Re_{\omega,\mathrm{tr}}^{1/2}\left(\frac{Re_{\omega,\mathrm{tr}}}{Re_\varphi}\right)^{1/2} + \frac{2}{2n_R+1}K_{1,\mathrm{turb}}Re_\varphi^{n_R}\left[1-\left(\frac{Re_{\omega,\mathrm{tr}}}{Re_\varphi}\right)^{n_R+1/2}\right]. \tag{3.35}$$

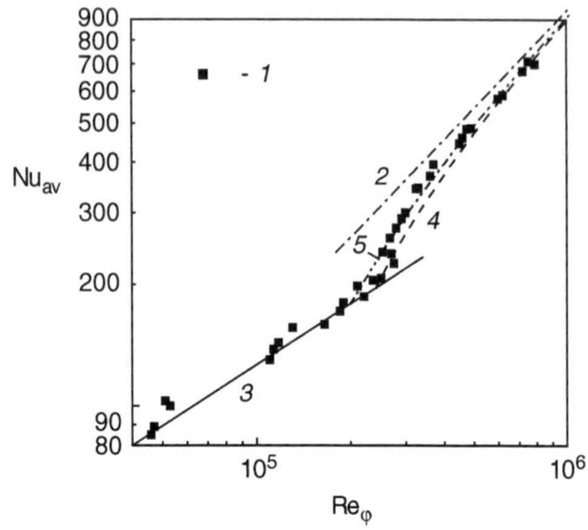

Fig. 3.7 Average Nusselt numbers of an *entire* disk rotating in still air. Experiments: *1* – [36]. Calculations by Eq. (8.4): *2* – region of developed turbulent flow, n_R=0.8, K_2=0.015 (Table 3.5) [36]; *3* – laminar flow, n_R=1/2, K_1=0.4 [36]. Calculation of Nu_{av} for an *entire* disk: *4* – Eq. (3.35) at $Re_{\omega,\mathrm{tr}}=2.4\times10^5$; *5* – Eq. (3.35) at $Re_{\omega,\mathrm{tr}}=2.0\times10^5$

Authors of the work [36] presented only a particular form of Eq. (3.35) at $n_R=0.8$, $2K_{1,\mathrm{turb}}/(2n_R+1)=0.015$, $K_{1,\mathrm{lam}}=0.4$.

Relation (3.34) is valid, if $Re_\varphi \geq Re_{\omega,\mathrm{tr}}$. When $Re_\varphi < Re_{\omega,\mathrm{tr}}$, then the second term in Eq. (3.34) should be neglected. In the limiting case $Re_\varphi >> Re_{\omega,\mathrm{tr}}$, Eq. (3.34) reduces to the second of Eq. (3.4) for developed turbulent flow over the entire disk with

$$K_{2,\mathrm{turb}} = \frac{2+n_*}{2n_R+1+n_*} K_{1,\mathrm{turb}}. \tag{3.36}$$

Relation (3.36) coincides with Eq. (3.28) for n_R=0.8.

The aforementioned particular form of Eq. (3.35) for the case $T_w=const$ was used in the work [36] to approximate original experimental data for the average Nusselt number of an *entire* disk rotating in still air (Fig. 3.7). Given a fixed value $Re_{\omega,\mathrm{tr}}=2.4\times10^5$, the approximation curve *4* lies 15% lower than experimental data over the range $Re_\omega\leq6.5\times10^5$. For higher Reynolds numbers, curve *4* agrees well with the experimental data. Agreement of Eq. (3.35) with experimental data can be improved assuming a somewhat lower value of $Re_{\omega,\mathrm{tr}}=2.0\times10^5$ (curve *5*).

3.3.3 Effect of Approximation of the Radial Velocity Profile on Parameters of Momentum and Thermal Boundary Layers

The integral method [41, 138, 139] is based on the von Karman's linear approximation (2.46) of the tangent of the flow swirl angle [80] jointly with the model (2.56) for the Stanton number. This sub-section is devoted to an investigation into the effect of different approximations of the tangent of the flow swirl angle, used jointly with the power-law model for the temperature profile in the boundary layer, on parameters of fluid flow and heat transfer of a free rotating disk [163, 165, 171, 172].

To attain this goal, the tangent of the flow swirl angle was approximated by the following relation:

$$\tan\varphi = \alpha(1-\xi)^{\sigma}, \tag{3.37}$$

where exponent σ accepted discrete values σ=2, 1 and 0 (as shown above in Sect. 2.4, the value σ=2 was accepted as the main one in the present integral method).

Resultant solutions for constants in Eqs. (2.82)–(2.92) take the following form:

$$\alpha = \left[\frac{C_1}{(3+m)B_0+(4+m)D_0}\right]^{1/2}, \tag{3.38}$$

$$H_9 = \alpha D_0(4+m), \tag{3.39}$$

$$\varepsilon_M = 8\pi\alpha\gamma D_0, \tag{3.40}$$

where B_0=B_1 and D_0=D_1 for σ=2; D_0=1/(n+1)–1/(n+2)–1/(2n+1)+1/(2n+2) and B_0=D_2 for σ=1; B_0=1/(2n+1) and D_0=1/(n+1)–1/(2n+1) for σ=0. Solution for σ=2 is a particular case of the solutions (2.88), (2.89), (2.90), (2.91) and (2.92) at κ=0, β=0.

Numerical values of the constants α, γ and ε_M obtained by Eqs. (3.38), (3.39) and (3.40) are documented in Table 3.6 (a part of the data for $\sigma = 1$ and 2 corresponds to the results in Table 3.4).

Profiles of the radial velocity component v_r for different values of σ are shown in Fig. 3.8. The profile for σ=0 differs significantly from the others. However, based on the data for the moment coefficient (Table 3.6), one can conclude that the errors in predicting the value of C_M in the case of a wrong choice of the value of σ (for instance, σ=0 instead of σ=2) do not exceed 5%, in spite of much larger errors in the data for the values of α, velocity profiles, etc.

An analytical solution of Eq. (2.75) with allowance for Eqs. (3.37), (3.38), (3.39) and (3.40) can be obtained only for the case $\Delta \geq 1$ [165, 171, 172]. It is obvious that at σ=2 this solution represents a particular case of Eqs. (2.93), (2.94) and (2.95) for N=0 and β=0.

Thus, relations for the constants in Eq. (3.4) are as follows:

$$n_R = (n+1)/(3n+1), \tag{3.41}$$

$$K_1 = K_3\Delta^{-n}Pr^{1-n_p}, \tag{3.42}$$

$$\Delta^{-n} = \left[\frac{4+m}{2+m+n_*}K_V Pr^{-n_p} + (1-K_V)\right]^{-1}, \tag{3.43}$$

$$K_1 = K_3 Pr\left[\frac{4+m}{2+m+n_*}K_V + (1-K_V)Pr^{n_p}\right]^{-1}, \tag{3.44}$$

$$K_2 = K_1\,(n_*+2)\,/\,(2+n_*+m)\,, \tag{3.45}$$

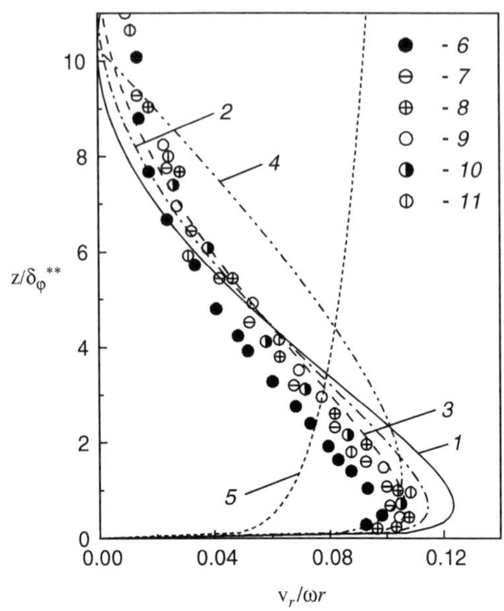

Fig. 3.8 Profiles of the radial velocity component in the turbulent boundary layer over a free rotating disk [163, 165, 171, 172]. For notations *1–4*, *6–11* see Fig. 2.3; *5* – σ=0, n=1/7

Table 3.6 Constants in Eqs. (2.82), (2.83), (2.84) and (2.85), (3.38), (3.40) and (3.41) for different values of σ [163, 165, 171]

Coefficient	n=1/7	n=1/8	n=1/9	n=1/10
$\alpha, \sigma=2$	0.2087	0.1842	0.1649	0.1493
$\alpha, \sigma=1$	0.162	0.143	0.128	0.116
$\alpha, \sigma=0$	0.0925	0.0818	0.0733	0.0664
$\gamma, \sigma=2$	0.530	0.4977	0.4773	0.4597
$\gamma, \sigma=1$	0.526	0.497	0.479	0.463
$\gamma, \sigma=0$	0.616	0.588	0.571	0.556
$\varepsilon_M, \sigma=2$	0.1466	0.1127	0.0901	0.0734
$\varepsilon_M, \sigma=1$	0.146	0.112	0.09	0.073
$\varepsilon_M, \sigma=0$	0.139	0.107	0.086	0.0704
$K_V, \sigma=2$	0.203	0.183	0.1661	0.1523
$K_V, \sigma=1$	0.167	0.15	0.1364	0.125
$K_V, \sigma=0$	0.111	0.10	0.0909	0.0833
$K_3, \sigma=2$	0.02683	0.02079	0.01673	0.0137
$K_3, \sigma=1$	0.0267	0.0207	0.0166	0.0136
$K_3, \sigma=0$	0.0255	0.0198	0.016	0.0131
n_R	0.8	0.8182	0.8333	0.8462
m	0.6	0.6363	0.6667	0.6923
$n_p, \sigma=2$	0.5018	0.4894	0.4797	0.4719
$n_p, \sigma=1$	0.48	0.471	0.463	0.457
$n_p, \sigma=0$	0.45	0.444	0.44	0.436

$$K_3 = A_c(1+\alpha^2)^{1/2} = C_n^{-2/(n+1)} \gamma^{-2n/(n+1)} (1+\alpha^2)^{0.5(1-n)/(n+1)}, \tag{3.46}$$

$$K_V = 1 - D_{2*}/A_{1*}. \tag{3.47}$$

Here D_{2*}=D_2 and A_{1*}=A_1 for σ=2; D_{2*}=1/(2n+1)–1/(2n+2) and A_{1*}=1/(n+1)–1/(n+2) for σ=1; D_{2*}=1/(2n+1) and A_{1*}=1/(n+1) for σ=0. Values K_3 and K_V independent of n_* are presented in Table 3.6. Equation (3.45) coincides with Eq. (3.36), if one takes into account the relation $2n_R = 1 + m$ following from Eqs. (2.83) and (3.41).

Denoting $G = Pr^{n_p}(2+m+n_*)/(4+m)$, expanding the power functions in Eq. (3.44) in the Taylor's series in the neighbourhood of the point G=1 (for $Pr \to 1$, $n_* \to 2$) and omitting negligible terms, one can derive [163, 165, 171, 172]

$$Nu = K_3 \left(\frac{2+m+n_*}{4+m} \right)^{K_V} Re_\omega^{n_R} Pr^{1-n_p(1-K_V)}. \tag{3.48}$$

Taking the value $n = 1/7$ and rounding down the value of K_V to 0.2 results in that relation (3.48), which coincides with the Dorfman's solution (3.27) for the Nusselt number, provided that the cumulative exponent at the Prandtl number is equal to 0.6 and remains invariable for the other values of n. Then the value of n_p, which still remains unknown, may be defined as

$$n_p = 0.4/(1 - K_V). \tag{3.49}$$

Values of n_p calculated by Eq. (3.49) and used further in all the other computations are documented in Table 3.6. Agreement of Eq. (3.48) with Eq. (3.44) takes place only for $Pr \to 1$ and $n_* \to 2$. Thus, from a purely mathematical point of view, Eq. (3.48) is a particular case of the solution (3.44).

Solution (2.93) for $\Delta \leq 1$ in the case of a free rotating disk (N=0 and β=0) remains a transcendental equation:

$$\Delta^{2n+1}\left(a_* - 2b_*\Delta + c_*\Delta^2\right) = \frac{4+m}{2+m+n_*}\left(a_* - 2b_* + c_*\right) Pr^{-n_p}. \tag{3.50}$$

Here, the subscript "T" of the coefficients a_*, b_* and c_* is neglected, which means equality n_T=n. Solution (3.50) is valid for $Pr \geq 1$ or $Sc \geq 1$ that takes place for fluids or naphthalene sublimation in air. Because of this, Eq. (3.50) is analysed in more detail in Chap. 8.

The effect of the value of σ in Eqs. (3.44) and (3.45) on the coefficient K_1 in comparison with Eq. (3.27) is illustrated in Table 3.7 for a series of discrete values of n_* at Pr=0.72 and 1, as well as in Fig. 3.9 for air at $Pr = 0.72$. It is obvious that the best agreement of Eq. (3.44) with experiments, in particular for $n_* \leq 0$, takes place at $\sigma = 2$, therefore the value $\sigma = 2$ was chosen as the baseline for further computations. According to Eq. (3.44), one can obtain the value $K_1 = 0.0115$ at $1/n = 8.7647$ that is only 3.6% different from the value $K_1 = 0.0111$ obtained in simulations of [205].

Table 3.7 Constants in Eqs. (3.44) and (3.45) for different values of σ [165, 171, 172]

Coefficient	Equations	n_*=–0.6	n_*=0	n_*=2	n_*=6
Calculation for Pr=0.72					
K_1	(3.44), σ=2	0.0169	0.0187	0.022	0.0246
K_2	(3.45), σ=2	0.0118	0.0144	0.0191	0.0229
K_1	(3.44), σ=1	0.0176	0.0191	0.0219	0.0240
K_1	(3.44), σ=0	0.0180	0.0191	0.0209	0.0222
Calculation for Pr=1					
K_1	Dorfman, Eq. (3.27)	0.0226	0.0238	0.0267	0.0303
K_2	Dorfman, Eq. (3.28)	0.0158	0.0183	0.0232	0.0282
K_1	Eq. (3.44), σ=2	0.0212	0.0232	0.0268	0.0296
K_2	Eq. (3.45), σ=2	0.0149	0.0178	0.0233	0.02755
K_1	Eq. (3.44), σ=1	0.0219	0.0237	0.0267	0.0289
K_1	Eq. (3.44), σ=0	0.0222	0.0235	0.0255	0.0269

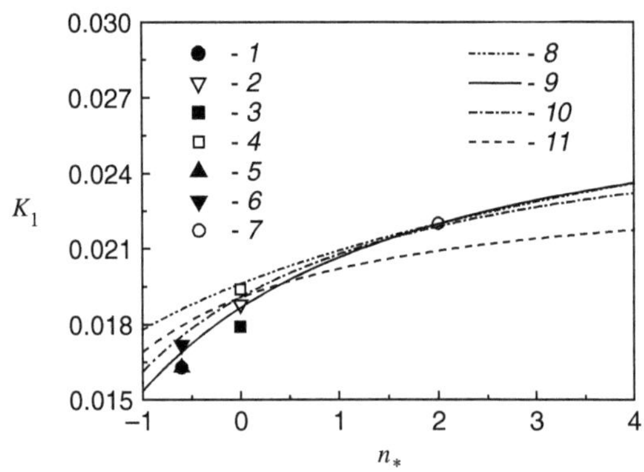

Fig. 3.9 Effect of the exponent n_* and value of σ on the constant K_1 in turbulent flow of air (Pr=0.72). Experiments: *1* – [19, 20, 46]; *2* – [11, 146]; *3* – [122]; *4* – [132]; *5* – [205], lower limit of variation of K_1; *6* – [205], upper limit of variation of K_1; *7* – [41, 77, 138, 139]. Calculations: *8* – Eq. (3.27) [41]; *9* – Eq. (3.44) for σ=2; *10* – Eq. (3.44) for σ=1; *11* – Eq. (3.44) for σ=0

Comparison of data in Tables 3.5 and 3.7 also shows that inaccuracies of the Dorfman's solutions (3.27) and (3.28) increase with increasing deviation of the Prandtl number from unity.

Figure 3.4 represents a comparison of the solution (3.44) for the Nusselt number at σ=2 with the experimental data on *local* heat transfer obtained in the work [46]. The value of the exponent n_* in Eq. (2.35) for the case q_w=*const* studied by the authors [46] can be found with the help of Eq. (3.4) and the definition of the Nusselt number [173]

$$\frac{T_w - T_\infty}{T_\infty} = \frac{q_w b}{\lambda T_\infty K_1} Re_\varphi^{-(n+1)/(3n+1)} x^{(n-1)/(3n+1)}. \tag{3.51}$$

Based on this, one can obtain that $n_*=(n-1)/(3n+1)=-m$ for the case q_w=*const* [173]. Values of m for different n are presented in Table 3.6.

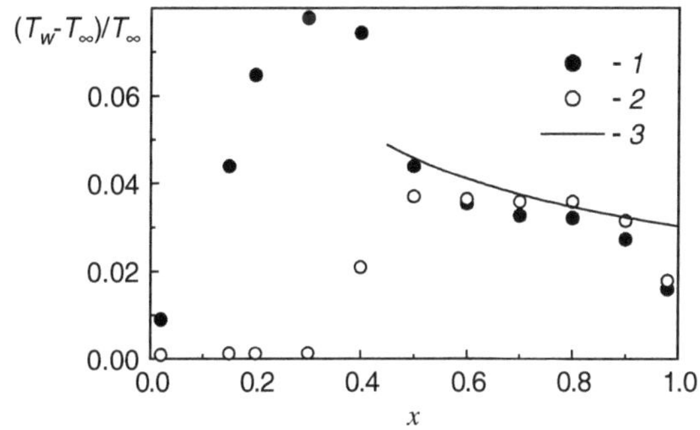

Fig. 3.10 Temperature distribution over the surface of a rotating disk in the experiments [46]. *1* – q_w=*const*; *2* – $T_w \approx$*const*. Calculation [173]: *3* – Eq. (3.51)

The disk temperature distributions obtained in the experiments [46] are shown in Fig. 3.10. In [46], the disk was divided into three annular regions. The region x =0–0.15 was left unheated; in the regions x = 0.15–0.4 and x = 0.4–0.96, the disk was heated by two separate heaters of the same power. Data *1* were obtained in the case where both heaters were switched on, and data *2* relate to the case where only the external heater was switched on.

In case *1*, the effect of the unheated section was weak; therefore, in the region $x > 0.5$, the power-law distribution of the disk temperature calculated by Eq. (3.51) at q_w=*const* is in fairly good agreement with the experimental data. The calculations involved use the following experimental values: K_1=0.0163, q_w=710 W/m^3, T_∞=298.9 K, $Re_\varphi=1.6 \times 10^6$, b=0.5 m; the value λ=0.02624 W/(m.K) was determined by the data of [22] for air at $T_\infty \approx 300$ K.

In the experiments of [46], fluid flow was turbulent at $Re_\omega > 3.6 \times 10^5$. For the case of $Re_\varphi=1.6 \times 10^6$ considered in Fig. 3.10 this means localization of the turbulent region at x=0.474–1.0.

In case *2* (the internal heater is switched off), the value T_w in the heated zone (turbulent flow) is approximately constant (Fig. 3.10).

The calculated values of the constants K_1 and K_2 obtained from Eqs. (3.44) and (3.45) at n=1/7 and σ=2 are equal to K_1=0.0169, K_2=0.0118 for n_*=–0.6, while the value obtained in experiments [19, 20, 46] is K_1=0.0163. Thus, Eq. (3.44) differs from experiments by only 3.7%.

A comparison with the experimental data of [46] for the local Nusselt number is given in Fig. 3.4. Calculations of Eq. (3.4) at $K_1 = 0.0169$ agree with experiments slightly worse than the results of calculations using the empirical value $K_1 = 0.0163$ [46]. Note the good agreement between the results of the calculation of Eq. (3.4) for $K_1 = 0.0187$ (T_w=*const*) and the experimental results of [46] with the internal heater switched off (data *2* for $T_w \approx$*const* in Fig. 3.10). The values $K_1 = 0.0187$, $K_2 = 0.0144$ at $n_* = 0$ predicted by the present integral method (see Table 3.7) agree with experiments [146] (deviation 0.5%) and [122] (deviation 4.5%) much better than those calculated by Dorfman's formula (3.27) (see Table 3.5).

3.3.4 Numerical Computation of Turbulent Flow and Heat Transfer for an Arbitrary Distribution of the Wall Temperature

In the real rotating-disk systems, the wall temperature distribution often cannot be described by the analytical relations (2.34) and (2.35). In these cases, a numerical version of the integral method can be used [179], and the distribution of T_w can be approximated by some other dependence. In works [133, 135], the Dorfman's method at a fixed value of $n=1/7$ was employed for numerical simulations of the conditions observed in experiments. As in the case of the analytical version of the Dorfman's method, calculations for the Nusselt number [133, 135] noticeably exceeded the experimental data at $dT_w/dr \approx 0$ and $dT_w/dr<0$. Calculations and experimental results agreed well at $dT_w/dr > 0$ with the exception for the cases of high Re_φ values, at which the calculated data were lower than the experimental results.

The authors of [136] simulated the experimental conditions of [133, 135] by solving numerically the differential equations of the boundary layer closed with the known Cebeci–Smith model of turbulent viscosity [22]. The agreement of the calculations with the experiments turned out to be good. This is indicative of the reliability of the experimental data of [133, 135]; therefore substantial inaccuracy of the Dorfman's method at $dT_w/dr \leq 0$ is responsible for the disagreement of the calculations and experiments performed in [133, 135].

Numerical simulations of turbulent heat transfer over a rotating disk have been performed by the author of this book with the help of the numerical version of the present integral method [179]. Boundary layer equations were used in the form of Eqs. (2.77) and (2.78). Consideration was given to the cases of positive, approximately constant and negative values of dT_w/dr in accordance with experimental conditions [133, 135].

In analogy to the work [136], experimental temperature distributions on the disk surface were chosen for further analysis that corresponds to determination of the Nusselt numbers from the measurements with the help of local heat flux meters

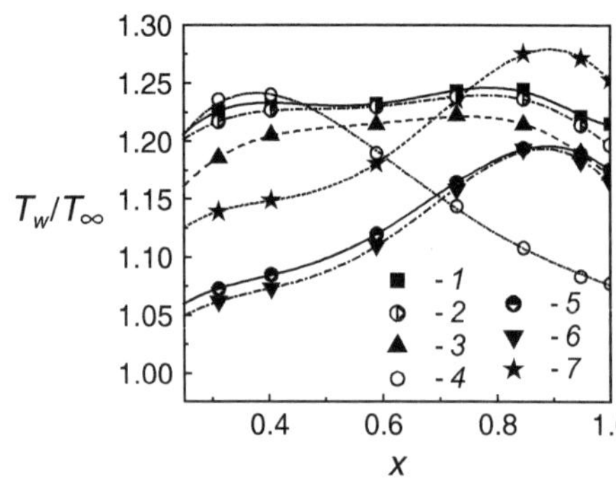

Fig. 3.11 Radial variation of the disk temperature T_w/T_∞: points are experiments [133, 135], *lines* are their polynomial approximations. *1* – case $n_*=0.1$, $Re_\varphi=1.135\times10^6$; *2* – 0.1 and 1.19×10^6; *3* – 0.1 and 3.2×10^6; *4* – –0.2 and 2.65×10^6; *5* – 0.4 and 2.67×10^6; *6* – 0.4 and 3.14×10^6; *7* – 0.6 and 1.59×10^6

[133, 135]. The Nu numbers obtained in such a way were corrected by subtracting the radiation component in accordance with the procedure suggested in [135].

The experimental surface temperature distributions of the disk obtained in [133, 135] were classified by their authors into four groups. These groups are characterized as those conventionally complying with the distribution (2.35) at the values of the power exponent $n_*=-0.2$, 0.1, 0.4 and 0.6. Examples of such distributions for a series of Re_φ values are given in Fig. 3.11. Inside each group, the T_w distributions differ from each other, maximum, by 10–15% for different Re_φ values, but at the same time they exhibit approximately the same general trend of the curves (see, e.g. curves *1–3* for the case with conventionally $n_*=0.1$). These curves are characterized mainly by the positive ($n_*=0.4$ and 0.6), approximately constant ($n_*=0.1$) and negative ($n_*=-0.2$) gradients of the wall temperature T_w.

Correspondence of relation (2.35) to the experiments at the n_* values indicated above is indeed rather conventional. For instance, Eq. (2.35) obviously does not assume maxima, minima and inflection points inside the region of determination of T_w over the disk radius. At the same time, all the dependences depicted in Fig. 3.11 have the aforementioned characteristic points. However, for the sake of convenience, the conventional subdivision of the curves into groups adopted in [133, 135] is retained in the present work. The discrete experimental values of the disk temperature (the tabular data given in work [135] are used here) were approximated in the present work in the form of the seventh-order polynomial to be used in calculations with the help of the integral method. Examples of such dependences partially published in the work [179] are shown in Fig. 3.11.

Results of computations of the local Nu number variation are given in Figs. 3.12 and 3.13. The common feature of the data in Figs. 3.12 and 3.13 is non-universality of the n and n_T values used in computations. However, comparisons of the calculated curves with the experimental data made it possible to develop some qualitative and quantitative recommendations for the choice of the n and n_T values for particular thermal and hydrodynamic conditions. It should also be noted that the calculated

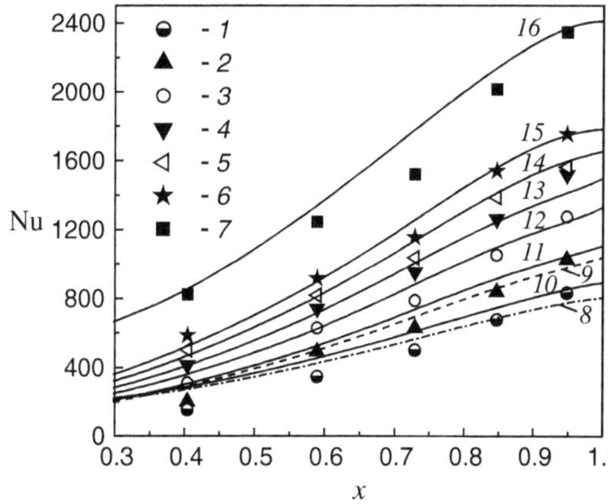

Fig. 3.12 Radial variation of the Nusselt number, case conventionally $n_*=0.1$: *1–7* – experiments [133, 135]. Calculations: *8* – $n=1/5$, $n_T=1/5$; *9* – $n=1/6$, $n_T=1/5$; *10–16* – $n=n_T=1/6$. *1, 8, 10* – $Re_\varphi=0.819\times10^6$; *2, 9, 11* – 1.08×10^6; *3, 12* – 1.35×10^6; *4, 13* – 1.6×10^6; *5, 14* – 1.88×10^6; *6, 15* – 2.14×10^6; *7, 16* – 3.2×10^6

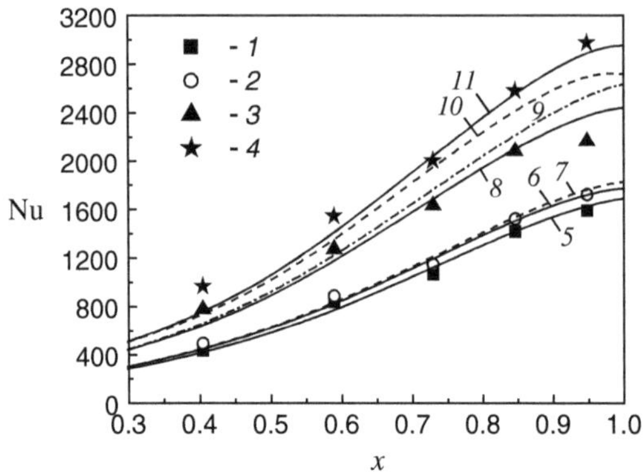

Fig. 3.13 Radial variation of the Nusselt number, case conventionally n_*=0.6: *1*, *2* – experiments [133, 135]; *5*, *6* – calculations at n=n_T=1/6.5; *7* – calculation at n=n_T=1/7. Case conventionally n_*=0.4: *3*, *4* – experiments [133, 135]; *8*, *10* – calculations at n=n_T=1/6; *9*, *11* – calculations at n=n_T=1/7. The Reynolds numbers Re_φ: *1*, *5* – 1.59 × 10^6; *2*, *6*, *7* – 1.71 × 10^6. *3*, *8*, *9* – 2.67 × 10^6; *4*, *10*, *11* – 3.14 × 10^6

curves in Figs. 3.12 and 3.13 agree well with the calculations [136] carried out, as indicated above, by a differential method.

Results of simulations of the case n_*=0.1 partially published in the work [179] are presented in Fig. 3.12. The calculations have been carried out mainly at n=n_T=1/6. Agreement of the calculated and experimental data is good, though for $x \le 0.85$ the values n=1/6 and n_T=1/5 allow slightly better agreement at Re_φ=1.08 × 10^6, and values n=n_T=1/5 provide better performance for Re_φ=0.819 × 10^6. This can be apparently attributed to the trend towards decreasing values of n and n_T at rather small Reynolds numbers Re_ω.

Figure 3.13 shows results of simulations of the rather similar cases n_*=0.4 and n_*=0.6, also partially published in the work [179]. At lower values of Re_φ=(1.59−1.71) × 10^6, good agreement with experiments can be attained by using the value n=n_T=1/6.5. For higher Reynolds numbers, one should use the value n=n_T=1/6 (at Re_φ=2.67 × 10^6) or n=n_T=1/7 (at Re_φ=3.14 × 10^6). It should be noted that the values n and n_T for the data given in Fig. 3.13 are, on the whole, slightly smaller as compared to the case n_*=0.1 (Fig. 3.12). At Re_φ=*idem*, the only difference in the boundary conditions for the computations being compared is the value of the radial temperature gradient of the disk. For Fig. 3.12, condition $dT_w/dr \approx 0$ is valid, while for Fig. 3.13 condition $dT_w/dr > 0$ is in force.

The assumption concerning the influence of the temperature gradient dT_w/dr on the exponents n and n_T is confirmed by calculations presented in Fig. 3.14 for the case n_*=–0.2. For lower values of the Reynolds number Re_φ=(0.548−1.08) × 10^6, one needed to choose the values of n_T=1/4 and n=1/6. At a higher value of Re_φ=2.65 × 10^6, the exponent n_T is smaller and equals n_T=1/5 at the same value of n=1/6. Thus, the influence of $dT_w/dr<0$ is evident again as compared to Fig. 3.12: for Re_φ=*idem*, the same value n=1/6 is retained and n_T increased.

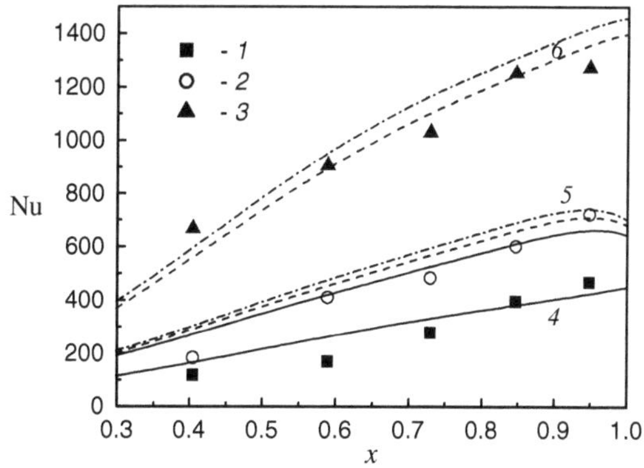

Fig. 3.14 Radial variation of the Nusselt number, case conventionally n_*=–0.2: *1–3* – experiments [133, 135]. Calculations [179]: *solid lines 4, 5* – n=1/6, n_T=1/4; *dashed lines 5, 6* – 1/6 and 1/5; *dash-dotted lines 5, 6* – 1/6 and 1/6. *1, 4* – $Re_\varphi=0.548\times10^6$; *2, 5* – 1.08×10^6; *3, 6* – 2.65×10^6

A characteristic feature of the data in Figs. 3.12, 3.13 and 3.14 is a dip of the experimental *Nu* numbers [133, 135] at the point $x\approx0.73$ as compared to the calculations. The same disagreement between the experiments and calculations at $x\approx0.73$ was noticed also in [136]; apparently, the reason lies in the systematic error of experimental measurements [133, 135] at this point.

It should be pointed out that velocity and temperature profiles have not been measured in the experiments [133, 135]; therefore, the above analysis of the effect of the exponents n and n_T is based on the indirect conclusions concerning behaviour of the Nusselt number. Nevertheless, taking into consideration the interrelation between the $n=n_T$ values and the exponent n_R at the Reynolds number in Eq. (3.4) for the Nusselt number under the condition (2.35), this analysis can be believed to be quite justified. For example, at $n=n_T=1/7$ we have $n_R=0.8$. It is evident that the rate of increase in the *Nu* numbers with the radius in Figs. 3.12 and 3.13 (the cases n_*=–0.2 and 0.1) corresponds to the smaller values of n_R, and, consequently, to the larger n and n_T, which is confirmed by numerical calculations. Besides, it is interesting to note that the authors of [46], who measured experimentally temperature profiles at a negative value of dT_w/dr in the case q_w=const (or $n_*\approx$–0.6), obtained values n_T=1/4–1/5 for $Re_\omega=10^6$ (see Fig. 2.5). This agrees with the data of the present investigations for n_*=–0.2.

Thus, the numerical simulations of turbulent heat transfer over a rotating disk with the help of the present integral method agree well with the known experimental data [133, 135] for arbitrary thermal boundary conditions different from the conditions (2.34) and (2.35). The calculations allow concluding that the thermal boundary conditions of the problem affect the exponents in the power-law approximation of the temperature profiles and suggest the optimum values of these parameters for the particular conditions considered.

3.4 Generalized Analytical Solution for Laminar and Turbulent Regimes Based on the Novel Model for the Enthalpy Thickness

As mentioned in Sect. 3.1, Dorfman's formula (3.6), which is the most general approximate analytical solution for laminar flow, exceeds the exact solution by 34–238% over the range of n_* from –1.5 to 0 at Pr=1–0.1, with the deviations increasing following decrease in the Prandtl number. Relations (3.7) and (3.8) offered in the works [26, 106] are much more accurate; however, they were obtained only for an isothermal disk, i.e. at n_*=0. Therefore, it was essential to obtain an approximate analytical solution with the help of the present integral method for the Prandtl numbers Pr=1–0.1 for any values of n_* varying from –2 to 4, which can provide much higher accuracy than Eq. (3.6) [174, 175, 177].

Tangential equation of the momentum boundary layer (2.23) and the thermal boundary layer equation (2.25) can be transformed as follows:

$$\frac{d}{dr}\left[r^4\delta K_V K_m\right] = \frac{c_f}{2}r^4(1+\alpha^2)^{1/2}, \tag{3.52}$$

$$\frac{d}{dr}\left[r^2\delta K_H K_m\left(T_w - T_\infty\right)\right] = \chi\frac{c_f}{2}r^2(1+\alpha^2)^{1/2}\left(T_w - T_\infty\right), \tag{3.53}$$

where the Reynolds analogy parameter χ is defined by Eq. (2.57); for the definition of the parameter K_m see Nomenclature.

In a generalized form that describes both laminar and turbulent flow, parameters of the boundary layer are specified by Eqs. (2.82), (2.83), (2.84), (2.85), (2.86) and (2.87), as well as by relations

$$K_V = const, \quad K_m = const. \tag{3.54}$$

Relations for the majority of the constants for turbulent flow are given above in Sect. 2.5, while K_m=αA_1, $K_V = 1 - D_2/A_1$.

Formally, Eqs. (2.82), (2.83), (2.84), (2.85), (2.86) and (2.87) transform to those for laminar flow at n=1 and m=0. The numerical solution of the self-similar Navier–Stokes equations was used in [138, 139] to obtain the following values for the constants for laminar flow:

$$\alpha = 0.8284, \quad K_V= 0.3482, \quad K_m = \frac{I_\infty \alpha}{\delta(\omega/\nu)^{1/2}} = \frac{I_\infty \alpha}{\gamma}, \tag{3.55}$$

$$I_\infty = \alpha^{-1}\int_0^\infty \frac{v_r}{\omega r}d(z\sqrt{\omega/\nu}) = 0.5338, \quad A_c = 0.6159(1+\alpha^2)^{-1/2}. \tag{3.56}$$

Under conditions where relation (2.35) is valid, the dimensionless temperature Θ and parameter K_H do not depend on the coordinate r. By substituting Eqs. (2.35), (2.82), (2.83), (2.84), (2.85), (2.86) and (2.87), (3.55) and (3.56) into Eqs. (3.52) and (3.53) one can obtain

$$(4+m)\gamma K_V K_m = A_c(1+\alpha^2)^{1/2}, \tag{3.57}$$

$$(2+m+n_*)\gamma K_H K_m = \chi A_c(1+\alpha^2)^{1/2}. \tag{3.58}$$

Equation (3.58) involves two unknown quantities: K_H and χ. Dorfman [41] proposed the model (2.56) that imposed an additional relation between K_H and χ. Drawbacks of the Dorfman's approach were demonstrated above. Results [173] for turbulent flow described in Sect. 3.3 make it possible to derive a significantly more accurate model, which is a consequence of the joint use of the power-law velocity and temperature profiles:

$$b_2 K_H = 1 - \chi Pr^{n_p}(1-K_V)b_1. \tag{3.59}$$

The correction factors b_1 and b_2 are introduced for laminar flow. For turbulent flow, b_1=1 and b_2=1, while the formula for the exponent n_p is given by Eq. (3.49).

Based on the definition of the Nusselt number and expressions (2.57), (3.58), it can be found that the Nu number still follows the first of Eq. (3.4) at $n_R = (n+1)/(3n+1)$ and

$$K_1 = (2+m+n_*)\gamma K_H K_m Pr. \tag{3.60}$$

Equating the values χ from Eqs. (3.58) and (3.59), expressing K_H from the resultant relation and substituting this value into Eq. (3.60), one can obtain

$$K_1 = A_c(1+\alpha^2)^{1/2} Pr\left[\frac{4+m}{2+m+n_*}K_V b_2 + (1-K_V)Pr^{n_p}b_1\right]^{-1}, \tag{3.61}$$

$$K_3 = A_c(1+\alpha^2)^{1/2}. \tag{3.62}$$

At b_1=1 and b_2=1, Eq. (3.61) becomes the same as solution (3.44) for turbulent flow. To remind, for laminar flow n=1, m=0, $n_R = 1/2$ and, as follows from the second of Eq. (3.56),

$$A_c(1+\alpha^2)^{1/2} = 0.6159. \tag{3.63}$$

Relation (3.61) for $Pr \to 0$ reduces to its asymptotic form, which for laminar flow and b_2=1 coincides with the solution obtained in the work [138]

$$K_1 = 0.6159 Pr(2+n_*)/(4K_V). \tag{3.64}$$

Constants b_1 and b_2 are found via agreeing Eq. (3.61) with the exact solution (Table 3.1). In order to satisfy the asymptotic solution (3.64), it is necessary that at Pr=0 the value b_1 be finite and b_2=1. Let us assume that

$$b_2 = 1 + b_3 Pr^{n_{p1}}, \quad b_1 = const, \quad b_3 = const. \tag{3.65}$$

Table 3.8 Values of n_p and n_{p1} depending on the Prandtl number [174, 175, 177]

Pr	0.9	0.8	0.72	0.71	0.6	0.5	0.4	0.3	0.2	0.1	0.01
n_p	0.7290	0.7349	0.7436	0.7435	0.7529	0.7608	0.7721	0.7860	0.8036	0.8338	0.9156
n_{p1}	0.9349	0.9354	0.9316	0.9334	0.9366	0.9439	0.9519	0.9602	0.9765	0.9893	0.999

The constants b_1 and b_3, which are independent of the Prandtl number, are found at Pr=1. Further, the exponents n_{p1} and n_p are determined for each value of the Pr number from Table 3.1. Because two unknown constants were determined each time in the matching procedure, there were two matching conditions as well: the values of the constant K_1 from the exact solution were substituted into Eq. (3.61) at n_*=0 and n_*=2. As a result, we have

$$b_1 = 0.6827, \quad b_3 = 0.5939, \tag{3.66}$$

and the values of the exponents are given in Table 3.8.

For convenience of computations, the exponents n_{p1} and n_p were approximated by the relation

$$n_{p1} = \sum_0^7 a_i Pr^i, \quad n_p = \sum_0^7 c_i Pr^i. \tag{3.67}$$

Here a_0=1, a_1=–0.0080733, a_2=–0.35578, a_3=0.54852, a_4=1.7994, a_5=–6.4323, a_6=7.3537, a_7=–2.9153; c_0=0.92485, c_1=–1.1705, c_2=3.5321, c_3=–3.3762, c_4=–5.7250, c_5=15.591, c_6=–12.282, c_7=3.2084.

The values of K_1 calculated by Eq. (3.61) in view of Eqs. (3.65), (3.66) and (3.67) are given in Table 3.9. Maximal deviation of the approximate K_1 values from those of the exact solution reach 3.1% in the worst case of n_*=–1.5 and Pr=1; the errors vanish at $Pr \le 0.1$. These errors are much less than those involved in the calculation of K_1 by Eq. (3.6) of Dorfman.

Let us show that the Nusselt number in the case of laminar flow at q_w=*const* is identically equal to that at T_w=*const*. The exponents at the r values in both sides of Eq. (3.53) must be equal. For q_w=*const*, this leads to the relation $1 + m + n_* = 1$. In laminar flow, m=0, therefore, n_*=0, just as for T_w=*const*. In turbulent flow, as shown above, the condition q_w=*const* is satisfied at n_*=–m.

Data of the exact solution (2.37), (2.38), (2.39), (2.40) and (2.41) for air at Pr=0.72 obtained by the author [174, 175, 177] and those of the approximate solution (3.61) are illustrated in Fig. 3.15 in comparison with the Dorfman's solution (3.6) and known experimental data.

Thus, in this section an improved formula for the parameter K_H (modified enthalpy thickness) was derived for laminar flow. The relations K_H and Nu suggested here extend the solution obtained in Sect. 3.3 for the case of turbulent flow onto the laminar flow conditions. The constants of the model determined via comparisons

Table 3.9 Values of the constant K_1 by Eqs. (3.61), (3.65), (3.66) and (3.67)

Pr	n_*=–2	n_*=–1.5	n_*=–1	n_*=–0.5	n_*=0	n_*=1	n_*=2	n_*=3	n_*=4
1.0	0.0	0.1261	0.2311	0.3199	0.3961	0.5197	0.6159	0.6928	0.7557
0.9	0.0	0.1180	0.2170	0.3012	0.3737	0.4922	0.5849	0.6594	0.7207
0.8	0.0	0.1093	0.2018	0.2809	0.3495	0.4623	0.5513	0.6233	0.6827
0.72	0.0	0.1019	0.1887	0.2635	0.3286	0.4365	0.5223	0.5921	0.6500
0.71	0.0	0.1009	0.187	0.2612	0.3259	0.4332	0.5185	0.5880	0.6457
0.6	0.0	0.0898	0.1673	0.2349	0.2943	0.3939	0.4742	0.5402	0.5955
0.5	0.0	0.0788	0.1477	0.2084	0.2623	0.3539	0.4287	0.4910	0.5437
0.4	0.0	0.0666	0.1258	0.1787	0.2263	0.3084	0.3767	0.4344	0.4839
0.3	0.0	0.0531	0.1012	0.1449	0.1849	0.2553	0.3153	0.3671	0.4122
0.2	0.0	0.0379	0.0730	0.1057	0.1362	0.1914	0.2400	0.2832	0.3217
0.1	0.0	0.0204	0.0399	0.0586	0.0766	0.1104	0.1417	0.1707	0.1977
0.01	0.0	0.00219	0.00438	0.00655	0.00871	0.01301	0.01726	0.02148	0.02565
10^{-3}	0.0	0.000221	0.000442	0.000662	0.000883	0.00132	0.00176	0.00220	0.002643

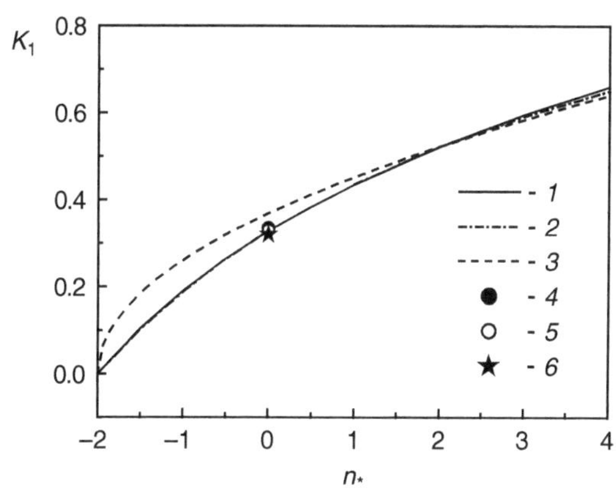

Fig. 3.15 Effect of the exponent n_* on the constant K_1 in Eq. (3.4) for laminar flow at Pr=0.72. *1* – exact solution [174, 175, 177], *2* – Eq. (3.61), *3* – Eq. (3.6). Experiments for n_*=0: *4* – 0.335 [46, 64, 143], *5* – 0.33 [11, 32, 146], *6* – 0.32 [24]

with the exact solution were approximated by polynomials. An improved approximate analytical solution for calculating the Nusselt number was obtained. The error of this formula with respect to the numerical exact solution does not exceed 3.1%.

3.5 Inverse Problem of Restoration of the Wall Temperature Distribution at a Specified Arbitrary Power Law for the Nusselt Number

3.5.1 Solution of the Problem

Most of the approximate analytical solutions of the thermal boundary layer equation are known to be obtained by way of solving a direct problem using a specified power-law distribution of the wall temperature (2.35) or, in a rewritten form,

$$\overline{\Delta T} = c_0 x^{n_*}. \tag{3.68}$$

The relation for the Nusselt number (3.4) can be presented as follows:

$$Nu = K_1 Re_\varphi^{(n+1)/(3n+1)} x^{2(n+1)/(3n+1)}, \quad Nu = K_1 Re_\varphi^{(n+1)/(3n+1)} x^{1+m}. \tag{3.69}$$

It is essential to restore a distribution for ΔT for the case where the Nusselt number is described by the following relation:

$$Nu = K_1 Re_\varphi^{(n+1)/(3n+1)} x^{m_x}, \quad Nu_b = K_1 Re_\varphi^{(n+1)/(3n+1)} x^{m_x - 1}, \tag{3.70}$$

where m_x=const is an arbitrary exponent, which is in the general case different from $2(n+1)/(3n+1) = 1+m$ in Eq. (3.69). One of the practical applications of such a solution is the problem of experimental determination of the stationary Nusselt number using a transient technique [154–156] (see Chap. 4). In this case, experimental distributions of ΔT are unsteady, while the steady-state function ΔT that corresponds to the obtained stationary Nusselt number may be found by means of solving an inverse problem.

A solution of the stated problem is performed with the help of the present integral method [176, 177]. The integral equation (3.53) of the thermal boundary layer can be reduced to the following form:

$$\frac{d}{dx}\left[Re_\omega \bar{\delta} K_H K_m \overline{\Delta T}\right] = \frac{Nu}{Pr}\overline{\Delta T}. \tag{3.71}$$

For laminar and turbulent boundary layers, Eq. (2.82) can be rewritten as follows:

$$\bar{\delta} = C_\delta^* x^m, \quad C_\delta^* = \gamma Re_\varphi^{-2n/(3n+1)}, \tag{3.72}$$

where m=1 and $K_m \bar{\delta} = I_\infty \alpha Re_\varphi^{-1/2}$ in laminar flow (see Eq. (3.55)). Choice of the values of C_δ^* or γ depends on the desired accuracy of their determination from the self-similar profiles of v_r and v_φ. Since parameters $\bar{\delta}$ and K_m are used in Eq. (3.71) as a product, then, in view of Eq. (3.72), values of C_δ^* or γ for laminar flow do not affect the subsequent transformations.

Taking a derivative of the left-hand side of Eq. (3.71) and dividing both sides of this equation by $Re_\omega \bar{\delta} K_H K_m \overline{\Delta T}$, one can deduce

$$\frac{d}{dx}\left[\ln\left(Re_\omega \bar{\delta} K_H K_m \overline{\Delta T}\right)\right] = \frac{Nu}{Pr}\frac{1}{Re_\omega \bar{\delta} K_H K_m}. \tag{3.73}$$

Taking into account Eq. (3.70) for Nu and (3.72) for $\bar{\delta}$, along with the relation $Re_\omega = Re_\varphi x^2$, one can obtain

$$\frac{d}{dx}\left[\ln\left(Re_\omega \bar{\delta} K_H K_m \overline{\Delta T}\right)\right] = \frac{K_1}{Pr K_m \gamma} \cdot \frac{x^{m_x - m}}{x^2 K_H}, \tag{3.74}$$

where the first of the multipliers in the right-hand side of Eq. (3.74) is constant, while the second one is a function of the radial coordinate x.

Equation (3.74) has to be integrated within the limits from x to 1 (provided that $x \neq 0$ to avoid uncertainty in both sides of this equation):

$$\left[\ln\left(Re_\omega \bar{\delta} K_H K_m \overline{\Delta T}\right)\right]_x^1 = \frac{K_1}{Pr K_m \chi} \cdot \int_x^1 \frac{x^{m_x - m} dx}{x^2 K_H}. \tag{3.75}$$

Parameter K_H is specified by Eq. (3.59). The Reynolds analogy parameter χ is given by transformed equation (2.57):

$$\chi = \frac{Nu}{\frac{C_f}{2} Re_\omega (1+\alpha^2)^{1/2} Pr} = C_\chi x^{m_x - m - 1}, \tag{3.76}$$

$$C_\chi = \frac{K_1}{A_c (1+\alpha^2)^{1/2} Pr}. \tag{3.77}$$

Let us denote

$$m_x^* = m_x - m. \tag{3.78}$$

In terms of Eq. (3.78), one can easily see that Eq. (3.69) for the Nusselt number is valid at $m_x^* = 1$, $m_x = 1 + m$.

Equation (3.59) can be expressed with the help of Eqs. (3.76), (3.77) and (3.78), as follows:

$$K_H = a_* + b_* x^{m_x^* - 1}, \tag{3.79}$$

$$a_* = \frac{1}{b_2}, \quad b_* = -Pr^{n_p}(1 - K_V)\frac{b_1}{b_2} C_\chi. \tag{3.80}$$

In view of Eqs. (3.79) and (3.80), the integral in the right-hand side of Eq. (3.75) takes the following form:

$$\int_x^1 \frac{x^{m_x - m} dx}{x^2 K_H} = \int_x^1 \frac{x^{m_x^*} dx}{a_* x^2 + b_* x^{m_x^* + 1}}. \tag{3.81}$$

Integrating Eq. (3.81) with the help of Mathematica software, one can obtain

$$\frac{1}{b_*(1 - m_x^*)} \cdot \ln \frac{a_* + b_* x^{m_x^* - 1}}{a_* + b_*} = \ln \left[\frac{a_* + b_* x^{m_x^* - 1}}{a_* + b_*}\right]^{\frac{1}{b_*(1 - m_x^*)}}. \tag{3.82}$$

With allowance for Eq. (3.72), the left-hand side of Eq. (3.75) takes the following form:

$$\left[\ln\left(Re_\omega \bar{\delta} K_H K_m \overline{\Delta T}\right)\right]_x^1 = \ln\left[\overline{\Delta T}\frac{K_H x^{2+m}}{K_H x = 1}\right]^{-1}. \tag{3.83}$$

Substituting Eqs. (3.82), (3.83) into Eq. (3.74), one can finally obtain

$$\overline{\Delta T} = \frac{a_* + b_*}{a_* + b_* x^{m_x^*-1}} x^{-2-m} \left[\left(\frac{a_* x^{1-m_x^*} + b_*}{a_* + b_*}\right)^{\frac{1}{1-m_x^*}} x^{-1}\right]^{-\frac{K_1}{PrK_m\gamma\, b_*}} \tag{3.84}$$

or

$$\overline{\Delta T} = \left[\frac{K_H}{K_{H_{x=1}}}\right]^{-\frac{K_1}{PrK_m\gamma\, b_*(1-m_x^*)}-1} x^{-2-m}. \tag{3.85}$$

3.5.2 Limiting Case of the Solution

Let us consider a limit of Eq. (3.85) at $m_x^* \to 1$ or, which is the same, at $y = 1 - m_x^* \to 0$. Utilizing the rule of L'Hospital, one can obtain

$$\lim_{y\to 0}\frac{\ln\frac{a_*+b_*x^{-y}}{a_*+b_*}}{b_* y} = -\frac{\ln x}{a_* + b_*} = \ln x^{-\frac{1}{a_*+b_*}}. \tag{3.86}$$

Arguments of logarithms in the right-hand sides of Eqs. (3.82) and (3.86) at $m_x^* \to 1$ become equal to each other:

$$\left[\frac{a_* + b_* x^{m_x^*-1}}{a_* + b_*}\right]_{m_x^*\to 1}^{\frac{1}{b_*(1-m_x^*)}} = \left[\frac{K_H}{K_{H_{x=1}}}\right]_{m_x^*\to 1}^{\frac{1}{b_*(1-m_x^*)}} = x^{-\frac{1}{a_*+b_*}}. \tag{3.87}$$

At $m_x^* \to 1$, solution (3.85) with allowance for relation (3.87) reduces to

$$\overline{\Delta T} = x^{\frac{K_1}{PrK_m\gamma\,(a_*+b_*)}-2-m}. \tag{3.88}$$

Comparing Eqs. (3.68) and (3.88), one can obtain

$$n_* = \frac{K_1}{PrK_m\gamma\,(a_* + b_*)} - 2 - m. \tag{3.89}$$

If the value of n_* is specified (as has been done elsewhere), then, taking into account that $a_* + b_* = K_H$ at m_x^*=1, one can derive a formula for K_1, which coincides with Eq. (3.61) obtained earlier for the boundary conditions (2.35) or (3.68). Thus, distribution (3.68) can be interpreted as a particular case of solution (3.85) at m_x^*=1.

3.5.3 Properties of the Solution for Temperature Head

Taking a derivative of Eq. (3.85) with respect to the coordinate x and equating the resulting expression to zero, one can find the abscissa of the extreme point x_{ext} for the temperature head [177, 178]:

$$x_{ext} = \left[\frac{\frac{K_1}{PrK_m\gamma} - b_*(m_x + 1)}{a_*(2 + m)}\right]^{\frac{1}{1-m_x^*}}. \qquad (3.90)$$

Relative thickness of the thermal boundary layer Δ for the turbulent flow can be found from Eq. (3.76) for χ and relation $\chi = \Delta^{-n}Pr^{-n_p}$. Consequently, the value of Δ varies according to the law

$$\Delta = (\chi Pr^{n_p})^{-1/n} = (C_\chi x^{m_x^*-1} Pr^{n_p})^{-1/n} = (C_\chi Pr^{n_p})^{-1/n} x^{(1-m_x^*)/n}. \qquad (3.91)$$

For the boundary condition (3.68), where $m_x^* = 1$, one can obtain from Eq. (3.91) that, as expected, Δ=*const*. At $m_x^* < 1$, the value of Δ increases with x. At $m_x^* > 1$, the dependence of Δ on x is decreasing.

Relation (3.79) for the value of K_H has a critical point K_H=0. Based on the physical sense of the model used, the parameter K_H can be only positive. Thus, parameters at the critical point can be calculated from the following relations:

$$\chi_{crit} = \frac{1}{b_1 Pr^{n_p}(1 - K_V)}, \qquad (3.92)$$

$$\Delta_{crit} = (\chi_{crit} Pr^{n_p})^{-1/n} = \left[\frac{1}{b_1(1 - K_V)}\right]^{-1/n}, \qquad (3.93)$$

$$x_{crit} = \left[\frac{\chi_{crit}}{C_\chi}\right]^{1/(m_x^*-1)}. \qquad (3.94)$$

Equations (3.90), (3.91), (3.92), (3.93) and (3.94) are important in calculations of distributions of $\overline{\Delta T}$ and their analysis.

3.5.4 Analysis of the Solution

The analytical solution of the heat transfer problem of a free rotating disk obtained in Sect. 3.5 belongs to a new class of solutions that generalizes the class of solutions known before [176, 177]. This novel solution can describe a much wider range of thermal boundary conditions at the wall and includes the previously known solution as a particular case, where one of the determining parameters of the problem becomes unity.

Indeed, the family of curves describing distributions of the temperature head (3.84) or (3.85), as well as the Nusselt number (3.70), is characterized by two independent parameters K_1 and m_x. At $m_x = 1 + m$ (or $m_x^* = 1$), the novel solutions (3.84) or (3.85) for $\overline{\Delta T}$ and Eq. (3.70) for the Nusselt number reduce to Eq. (3.68) for $\overline{\Delta T}$ and Eq. (3.69) (or (3.4)) for Nu, which have only one independent parameter K_1 or n_*. The relation between parameters K_1 and n_* for $m_x^* = 1$ is described by Eqs. (3.89) or (3.61) (most often, the exponent n_* is given, while the constant K_1 can be determined from the solution (3.61)). The rest of the parameters of the considered solutions are not free, as they are determined either by the physical properties of fluid (the Prandtl number) or by the models of the boundary layer (and thermal boundary layer) themselves with allowance for the flow regime given. Both of the compared distributions (3.68) and (3.84) (or (3.85)) for $\overline{\Delta T}$ are independent of the Reynolds number Re_φ.

Both one-parameter family of dependencies for the Nusselt number (3.69) (or Eq. (3.4)) and two-parameter family (3.70) allow obtaining monotonic distributions of the Nusselt number for all values of x, i.e. the sign of the derivative dNu/dx is constant for any x. At $m_x^* = 1$, the sign of the derivative $d\overline{\Delta T}/dx$ in the distribution (3.68) is also constant. However, at $m_x^* \neq 1$, the distributions of $\overline{\Delta T}$ described by Eqs. (3.84) or (3.85) are non-monotonic and allow obtaining (under certain conditions) curves of $\overline{\Delta T}$ with points of minimum or maximum.

Application to laminar flow. Let us consider laminar flow of air at $Re_\varphi = 53{,}500$, $Pr = 0.71$ (and $m = 0$, $m_x^* = m_x$). This corresponds to the conditions of experiments in the works [154–156]. Let us further analyse properties of the novel solution for the Nusselt number and temperature head [176, 177].

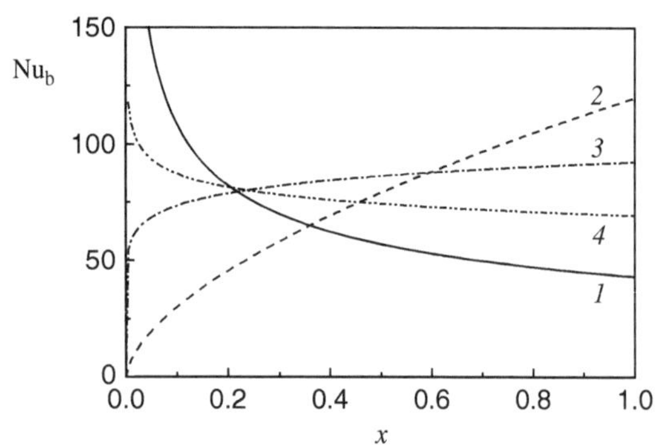

Fig. 3.16 Radial distribution of the Nusselt number Nu_b by Eq. (3.70) for laminar flow ($m = 0$) at $Re_\varphi = 53{,}500$, $Pr = 0.71$. *1* – $m_x = 0.6$, $K_1 = 0.187$; *2* – $m_x = 1.6$, $K_1 = 0.5185$; *3* – $m_x = 1.1$, $K_1 = 0.4$; *4* – $m_x = 0.9$, $K_1 = 0.3$

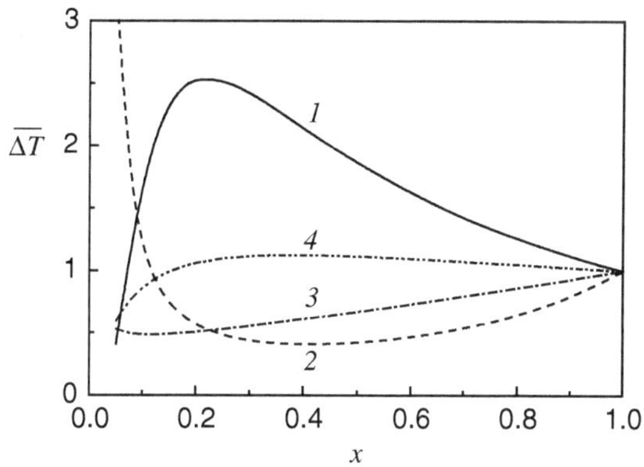

Fig. 3.17 Radial distribution of the temperature head $\overline{\Delta T} = (T_w - T_\infty)/(T_w - T_\infty)_{x=1}$ by Eq. (3.85) for laminar flow ($m = 0$) at $Re_\varphi = 53{,}500$, $Pr = 0.71$. *1* – $m_x = 0.6$, $K_1 = 0.187$; *2* – $m_x = 1.6$, $K_1 = 0.5185$; *3* – $m_x = 1.1$, $K_1 = 0.4$; *4* – $m_x = 0.9$, $K_1 = 0.3$

When the temperature head is determined by Eq. (3.68) at $n_* = -1$, the radial distribution of $\overline{\Delta T}$ is decreasing, while the Nusselt number Nu_b in accordance with Eq. (3.69) is constant with $K_1 = 0.187$; this case is also described by Eq. (3.70) at $m_x = 1$.

Taking the same value $K_1 = 0.187$ and assuming $m_x = 0.6$ instead of $m_x = 1$ in Eq. (3.70), one can obtain a function for the Nu_b numbers steadily decreasing in the radial direction from infinite values at $x \to 0$ (see curve *1* in Fig. 3.16) instead of being constant as at $m_x = 1$. For such set of parameters K_1 and m_x, the function $\overline{\Delta T}$ according to Eq. (3.85) increases from zero value at $x = 0$ and reaches its maximum $\overline{\Delta T} = 2.55$ at the point $x \approx 0.2$, see curve *1* in Fig. 3.17. Such a behaviour is a response to the sharp decrease in the Nu_b numbers in this region. Parameters at the point of maximum can be calculated by Eqs. (3.92), (3.93) and (3.94). Having come through its maximum, function $\overline{\Delta T}$ decreases with the increasing x according to the dependence close to x^{-1}. In all cases, $\overline{\Delta T} = 1$ at $x = 1$.

In the case where temperature head is specified by relation (3.68) at $n_* = 2$, value $\overline{\Delta T}$ grows up in the radial direction, while the Nu_b value is constant at $K_1 = 0.5185$, and m_x again equals to unity.

Taking the same value $K_1 = 0.5185$ and assuming $m_x = 1.6$ in Eq. (3.70), one can obtain a radially increasing function of Nu_b (instead of being constant as at $m_x = 1$). For such values of K_1 and m_x, the temperature head $\overline{\Delta T}$ falls down from infinity to the value $\overline{\Delta T} = 0.4075$ at the point of minimum $x \approx 0.4$, which is caused by the fast increase in the Nu_b values for small x. For $x > 0.4$, the function $\overline{\Delta T}$ of x becomes increasing according to a relation close to x^2 up to the value $\overline{\Delta T} = 1$ at $x = 1$.

Curves *3* (for $m_x = 1.1$, $K_1 = 0.4$) and *4* (for $m_x = 0.9$, $K_1 = 0.3$) in Figs. 3.16 and 3.17 characterize the cases of variation of the Nusselt numbers Nu_b and temperature heads $\overline{\Delta T}$ according to dependencies intermediate in comparison with cases *1* and *2* considered above.

Additional comparisons with experiments [154–156] are presented in Chap. 4.

Thus, Eq. (3.84) or (3.85) for $\overline{\Delta T}$ jointly with Eq. (3.70) for the Nu_b number provides much stronger potential of modelling different thermal boundary conditions

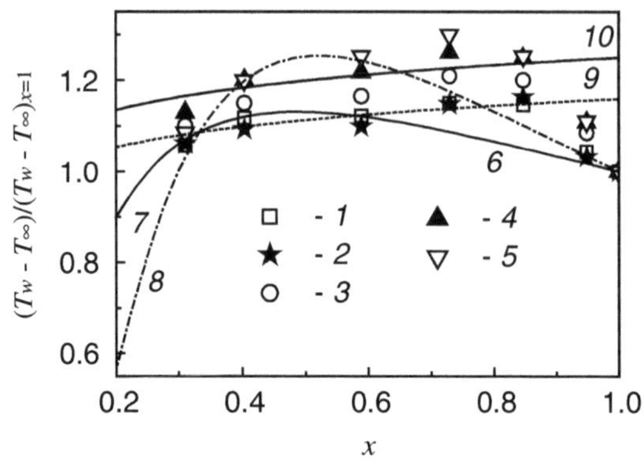

Fig. 3.18 Radial variation of the temperature head $\overline{\Delta T} = (T_w - T_\infty)/(T_w - T_\infty)_{x=1}$ for $Pr = 0.72$, case conventionally $n_* = 0.1$. Experiments [133, 135]: *1* – $Re_\varphi = 1.08 \times 10^6$; *2* – 1.6×10^6; *3* – 1.88×10^6; *4* – 2.14×10^6; *5* – 3.2×10^6. Calculations by Eq. (3.85) for $n = 1/6$: *6* – $K_1 = 0.0232$ and $m_x = 1.48$; *7* – 0.0229 and 1.43; *8* – 0.0224 and 1.38. Calculations by Eq. (3.68): *9* – $c_0 = 1.16$ and $n_* = 0.06$; *10* – $c_0 = 1.25$ and $n_* = 0.06$

for heat transfer of a rotating disk than Eqs. (2.35) and (3.4) (or Eqs. (3.68) and (3.69)) known earlier.

Application to turbulent flow. Experimental data [133, 135] were chosen for the analysis here [178, 189]. Results of the numerical modelling done above in Sect. 3.3.4 allowed, in particular, selecting values of the exponents n and n_T in the power-law approximation of the temperature and velocity profiles corresponding to experimental conditions of the works [133, 135]. These values will be used below in modelling conditions [133, 135] with the help of the solutions (3.70) and (3.85), with the condition $n = n_T$ being always held.

Results of modelling the case with, conventionally, $n_* = 0.1$ are shown in Figs. 3.18 and 3.19. Computations were performed at $n = n_T = 1/6$.

Experimental distributions of the temperature head $\overline{\Delta T}$ correlate well with Eq. (3.68) at $n_* = 0.06$, $c_0^* = 1.16$ for $Re_\varphi = (1.08\text{–}1.35) \times 10^6$ and at $n_* = 0.06$, $c_0^* = 1.26$ for $Re_\varphi = (2.14\text{–}3.2) \times 10^6$ over the range $x \approx 0.3$–0.85. Resulting from this fact is fair consistency of the experimental values for the Nusselt number and those calculated by Eq. (3.69) (with K_1 being computed from Eq. (3.44)) for $x \approx 0.3$–0.85.

However, for $x > (0.7$–$0.85)$, decreasing radial variation of experimental values of the wall temperature is observed contrary to the still increasing predicted values of $\overline{\Delta T}$. This phenomenon results in the tendency to exceed by the predicted Nusselt numbers over the experimental values in the area where signs of the derivatives $d\overline{\Delta T}/dx$ are in disagreement. This tendency is amplified with the increasing Reynolds numbers Re_φ. It is also necessary to note that the calculation of the *Nu* number from Eq. (3.69) with the use of Eq. (3.44) for K_1 is still better than the use of Dorfman's equation (3.27) at $n_* = 0.1$, $n = n_T = 1/7$, like it was done by the authors of the works [133, 135] (see curves *11–13* in Fig. 3.19 and explanations to them).

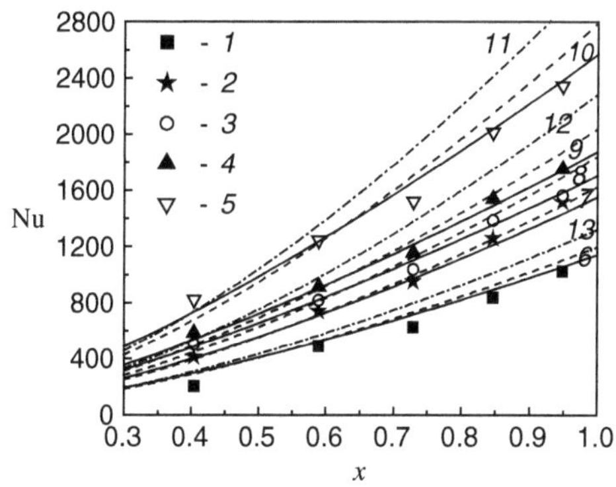

Fig. 3.19 Radial variation of the Nusselt number, case conventionally $n_* = 0.1$: *1–5* – experiments [133, 135]; *6–11* – calculations by Eqs. (3.69), (3.70) for $n = n_T = 1/6$. *Solid lines*, Eq. (3.70): *6, 7* – $K_1 = 0.0232$, $m_x = 1.48$; *8–10* – $K_1 = 0.0224$, $m_x = 1.38$. *Dashed lines 6–10* – Eq. (3.69), $K_1 = 0.0232$, $n_R = 0.778$, $m_x = 1.556$. *Lines 11–13* – Dorfman's equation (3.27) for $n_* = 0.1$. *1, 6, 13* – $Re_\varphi = 1.08 \times 10^6$; *2, 7* – 1.6×10^6; *3, 8* – 1.88×10^6; *4, 9, 12* – 2.14×10^6; *5, 10, 11* – 3.2×10^6

An improvement in the agreement with the experimental data for the Nusselt number can be attained using Eq. (3.70) (Fig. 3.19) and simulating distributions of $\overline{\Delta T}$ by Eq. (3.85) (Fig. 3.18). As a whole, the absolute errors of Eq. (3.85) with respect to the experimental data are not lower than the errors of Eq. (3.68). However, Eq. (3.68) predicts a constant sign of the derivative $d\overline{\Delta T}/dx$ for any x. On the contrary, Eq. (3.85) allows quite a good simulation also of the sign of the derivative $d\overline{\Delta T}/dx$, which in the considered case changes from positive to negative values with increasing x. Namely this is the main reason for the improvement of the consistency between the predicted and experimental values of the *Nu* number (Fig. 3.19).

Results of simulations of the case conventional $n_* = -0.2$ are shown in Figs. 3.20 and 3.21. Predictions using analytical formulas were done for $n = n_T = 1/6$. Numerical modelling with the help of polynomial approximations of the real distributions of $\overline{\Delta T}$ showed that somewhat better agreement with the experiments can be attained at a larger value $n_T = 1/5$ and the same value $n = 1/6$ (see Fig. 3.14). However, analytical formulas do not allow using values n_T and n different from each other.

As is evident from Fig. 3.20, a fair agreement of the predicted and experimental data for $\overline{\Delta T}$ is observed at $x \geq 0.6$. Only curve *4* exhibits qualitative consistency of the sign of the predicted derivative $d\overline{\Delta T}/dx$ with the experiments at $x = 0.35$–0.45.

It should be mentioned also that the agreement of Eq. (3.68) with the experiments is observed only at $n_* = -1.5$ (and $c_0 = 1.14$), but not at $n_* = -0.2$ (curve *6*), as suggested in the work [133, 135].

At first, differences in the distributions of $\overline{\Delta T}$ in Fig. 3.20 at $x \geq 0.6$ obtained from Eqs. (3.68) and (3.85) seem to be not too essential. However, it is evident from Fig. 3.21 that the Nusselt numbers are sensitive to the choice of the parameters K_1 and m_x.

Values of the Nu number calculated by Eq. (3.69) for $Re_\varphi = 1.08 \times 10^6$ show quite a good agreement with the experiments (the lower curve *4* in Fig. 3.21). However, the experimental point for the Nu number at $x \approx 0.4$ corresponds to laminar flow, and an agreement with it means that the predicted dependence *4* should be considered underestimated. Confirming this fact is the upper curve *4* for $Re_\varphi = 2.65 \times 10^6$, where the inadequate use of the model (3.68), (3.69) results in an essential underestimation of the calculations at $x \leq 0.7$ as compared to experiments, with this underestimation being not "compensated" with an insufficient development of the turbulent flow in the experiments (i.e. the turbulent flow is already developed at $x \approx 0.4$).

Again, an improvement of the agreement with the experimental data for the Nusselt number can be achieved due to use of Eq. (3.70), with the distributions $\overline{\Delta T}$ being modelled by Eq. (3.85). Parameters K_1 and m_x were chosen here based on the necessity to agree the predicted and experimental values of the Nu numbers.

A calculation of the Nusselt number by the Dorfman's equation (3.27) at $n_* = -0.2$ leads to obtaining the values of Nu noticeably overestimated in comparison with the experiments (curve *7* in Fig. 3.21).

Results for the quite close cases at conventionally $n_* = 0.4$ and $n_* = 0.6$ are shown in Figs. 3.22 and 3.23. For $Re_\varphi = 1.59 \times 10^6$, a good agreement with the experiments is attained at $n = n_T = 1/6.5$; for $Re_\varphi = 2.67 \times 10^6$, it is necessary to employ the values $n = n_T = 1/6$ (see line 3 in Fig. 3.23). Analysis of the distributions of the temperature head $\overline{\Delta T}$ in Fig. 3.22 at $x = 0.3$–0.6 shows that indeed the value $n_* = 0.6$ at $c_0^* = 1$ in Eq. (3.68) agrees well with the experiments for $Re_\varphi = 1.59 \times 10^6$. At the same time, the experiments for $Re_\varphi = 2.67 \times 10^6$ can be simulated better with the value $n_* = 0.7$ for $c_0 = 1$, but not with $n_* = 0.4$.

Equation (3.85) allows obtaining distributions of $\overline{\Delta T}$ with an error not exceeding the error of Eq. (3.68) (see Fig. 3.22). However, Eq. (3.85) provides an essential radial variation of the absolute value of the derivative $d\overline{\Delta T}/dx$ and thus more opportunities to select the parameters K_1 and m_x.

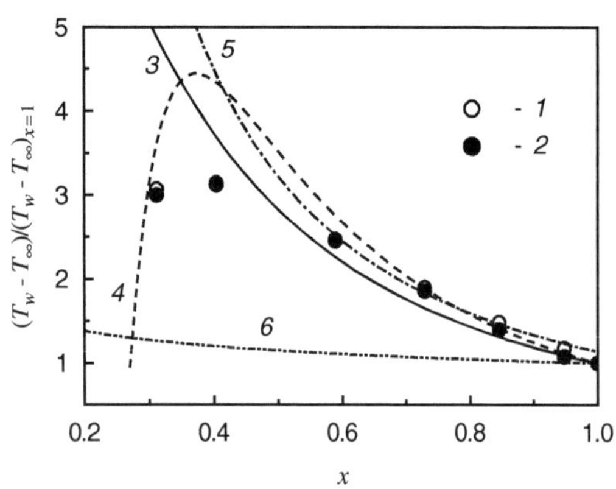

Fig. 3.20 Radial variation of the temperature head $\overline{\Delta T} = (T_w - T_\infty)/(T_w - T_\infty)_{x=1}$ for $Pr = 0.72$, case conventionally $n_* = -0.2$. Experiments [133, 135]: *1* – $Re_\varphi = 1.08 \times 10^6$; *2* – 2.65×10^6. Calculations by Eq. (3.85) for $n = 1/6$: *3* – $K_1 = 0.0157$, $m_x = 1.3$; *4* – $K_1 = 0.0137$, $m_x = 0.775$. Calculations by Eq. (3.68): *5* – $c_0 = 1.14$, $n_* = -1.5$; *6* – $c_0 = 1.0$, $n_* = -0.2$

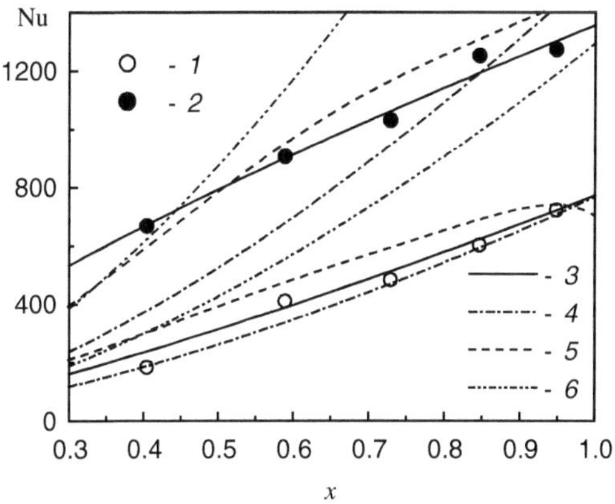

Fig. 3.21 Variation of the Nusselt number, case conventionally $n_* = -0.2$: *1, 2* – experiments [133, 135]; *3–5* – Eq. (3.69) for $n = n_T = 1/6$; *6* – Dorfman's equation (3.27) for $n_* = -0.2$. *Lower lines 3–6* and *1* – $Re_\varphi = 1.08 \times 10^6$; *upper lines 3–6* and *2* – 2.65×10^6. *Lower line 3* – Eq. (3.70), $K_1 = 0.0157$, $m_x = 1.3$; *upper line 3* – Eq. (3.70), $K_1 = 0.0137$, $m_x = 0.775$. *Lines 4* – Eq. (3.69), $K_1 = 0.0156$, $n_R = 0.778$, $m_x = 1.556$, $n_* = -1.5$. *Lines 5* – numerical modelling (see Fig. 3.14)

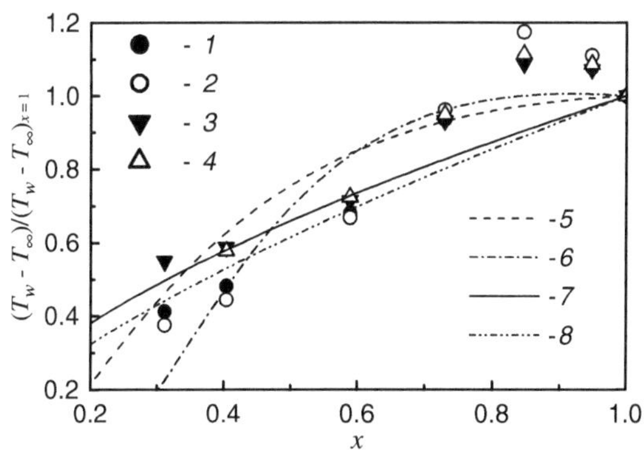

Fig. 3.22 Radial variation of the temperature head $\overline{\Delta T} = (T_w - T_\infty)/(T_w - T_\infty)_{x=1}$ for $Pr = 0.72$, cases conventionally $n_* = 0.4$ (data *1, 2, 6, 8*) and $n_* = 0.6$ (data *3, 4, 5, 7*). Experiments [133, 135]: *1* – $Re_\varphi = 2.67 \times 10^6$; *2* – 3.14×10^6; *3* – 0.615×10^6; *4* – 1.71×10^6. Calculations by Eq. (3.85): *5* – $K_1 = 0.0219$, $m_x = 1.48$, $n = 1/6.5$; *6* – $K_1 = 0.0249$, $m_x = 1.34$, $n = 1/6$. Calculations by Eq. (3.68): *7* – $c_0 = 1$, $n_* = 0.6$, *8* – $c_0 = 1$, $n_* = 0.7$

Analysis of the distributions of the local Nusselt numbers in Fig. 3.23 shows that Eq. (3.70) together with Eq. (3.85) at specially fitted values K_1 and m_x provides better agreement with the experiments than Eq. (3.69) together with Eq. (3.68). This is especially obvious for the higher value $Re_\varphi = 2.67 \times 10^6$. Inexactitudes of Eq. (3.69) for the K_1 values computed by the Dorfman's equation (3.27) are not so essential (see curves *6* and *7* in Fig. 3.23). This fact agrees with the earlier conclusions that the errors of the Dorfman's method become essential at $n_* \leq 0$.

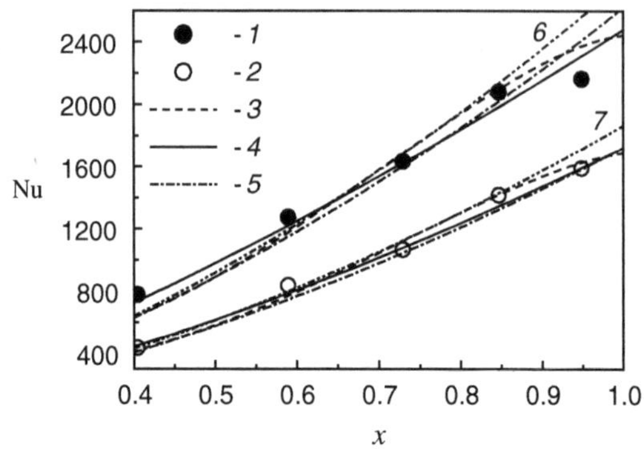

Fig. 3.23 Radial variation of the Nusselt number. *1*, *2* – experiments [133, 135]. *3* – numerical modelling (Fig. 3.13). Case conventionally $n_* = 0.4$ (*upper group of lines*, $Re_\varphi = 2.67 \times 10^6$), calculations at $n = n_T = 1/6$: *4* – Eq. (3.70), $K_1 = 0.0249$, $m_x = 1.34$; *5* – Eq. (3.69), $K_1 = 0.0262$, $n_R = 0.778$, $m_x = 1.556$, $n_* = 0.7$. Case conventionally $n_* = 0.6$ (*lower group of lines*, $Re_\varphi = 1.59 \times 10^6$), calculations at $n = n_T = 1/6.5$: *4* – Eq. (3.70), $K_1 = 0.0219$, $m_x = 1.48$; *5* – Eq. (3.69), $K_1 = 0.02265$, $n_R = 0.7896$, $m_x = 1.579$, $n_* = 0.6$. *Line 6* – Dorfman's equation (3.27), $n_* = 0.4$; *line 7* – Eq. (3.27), $n_* = 0.6$

Thus, the new solution for the temperature head $\overline{\Delta T}$ and the Nusselt number at a given value of the Prandtl number represents a two-parameter family of curves, i.e. depends on two free parameters K_1 and m_x^*. The new analytical solution given by Eqs. (3.70) and (3.85) provides much better accuracy of the agreement with the experiments [133, 135] than the previously known solutions (3.68) and (3.69). This significantly widens the potential for analytical modelling of different thermal boundary conditions for heat transfer of a rotating disk.

3.6 Theory of Local Modelling

3.6.1 Solution of the Problem

The basic assumption of the theory of local modelling is Eq. (2.56) at a constant value of M_s. Dorfman's solutions (3.6) and (3.27) were obtained namely at the condition $M_s = const$. An improvement of the agreement of the theory of local modelling with the experimental data for turbulent flow and exact solution for laminar flow can be attained only via assuming the value of M_s to be variable depending on the exponent n_*. Such an investigation has been performed in the work [89].

Thermal boundary layer equation (3.71) can be presented in the following form:

$$\frac{1}{\Delta T}\frac{d}{dr}\left[Re_T^{**} r \Delta T\right] = Nu/Pr. \tag{3.95}$$

Substituting the law of heat transfer, Eq. (2.56), into the right-hand side of Eq. (3.95) and using the boundary condition (2.35), one can obtain a relation for the Nusselt number:

$$Nu = M_s^{\frac{1}{1+\sigma}} (2n_R + n_* + 1)^{\frac{\sigma}{1+\sigma}} (1 + \alpha^2)^{\frac{1}{2(1+\sigma)}} Re_\omega^{\frac{1}{1+\sigma}} \; Pr^{\frac{1+\sigma-n_s}{1+\sigma}} . \tag{3.96}$$

For the constant value of the coefficient M_s found in the reference point $n_* = 2$ (see explanations to Eq. (2.56)) one can obtain the Dorfman's solutions (3.6) and (3.27).

For the sake of simplicity, the derivations presented below will be performed at the constant value of the Prandtl number $Pr = 0.72$. In this case, Eq. (3.44) for turbulent flow at $n = 1/7$, and relations for M_s and St have the following form:

$$\frac{1}{K_1} = 34.99 + \frac{48.33}{2.6 + n_*}, \tag{3.97}$$

$$M_s = \frac{1.252}{(34.99 + 48.33/(2.6 + n_*))^{1.25} (2.6 + n_*)^{0.25}}, \tag{3.98}$$

$$St = \frac{1.475 \cdot Re_T^{**-0.25}}{(34.99 + 48.33/(2.6 + n_*))^{1.25} (2.6 + n_*)^{0.25}}. \tag{3.99}$$

In the work [89], another relation for the constant K_1 was obtained based on the mathematical form of Eq. (3.97) and fitting of the coefficients to experimental data [46, 146]

$$\frac{1}{K_1} = 33.27 + \frac{56.16}{2.6 + n_*}, \tag{3.100}$$

with Eqs. (3.98) and (3.99) being modified accordingly. Equation (3.100) can be in fact treated as a somewhat modified Eq. (3.97), with the accuracy of Eq. (3.100) being somewhat lower than that of Eq. (3.97) (see Fig. 3.24).

For laminar flow at $Pr = 0.72$, Eq. (3.61) and the relation for the Stanton number take the following form:

$$K_1 = \frac{0.4435}{0.3486 + 2.002/(2 + n_*)}, \tag{3.101}$$

$$St = \frac{0.2922 \cdot Re_T^{**-1.0}}{(2 + n_*) (0.3486 + 2.002/(2 + n_*))^2}. \tag{3.102}$$

Thus, the use of the laws (3.99) and (3.102), which significantly improve agreement with experimental data, required cancellation of the fundamental statement of the theory of local modelling, namely, that the value of the coefficient M_s is constant. Finding the mathematical form of Eqs. (3.99) and (3.102) required the use of solution (3.44) obtained on the base of a principally different theoretical approach, which significantly devalues the fundamental law (2.56) of the theory of local modelling.

The theory of local modelling assumes using Eqs. (3.99) and (3.102) as the "baseline" relations in which additional effects of streamwise pressure gradient, porous injection/suction, physical properties variation with temperature, etc. can be taken into account in the form of multipliers. However, justification of such assumptions has never been performed.

In general, the theory of local modelling, even being used together with a variable relation for the coefficient M_s, seems to be a less perfect modelling tool, which requires purely empirical approaches in those cases, where the present integral method operates with clear model assumptions.

3.6.2 Other Interpretations

The Dorfman's approach [41] developed further by the authors of [138, 139] has been rather extensively used by different authors, and its results have been widely cited for several decades. Therefore, it makes sense to describe in more detail this approach in the form employed by the authors of works [41, 138, 139].

Equation (2.56) connecting the Stanton number and the enthalpy Reynolds number Re_T^{**} was rewritten by Dorfman [41] in the following form:

$$Nu = F(Pr)Re_{\omega}\left[2\pi Re_T^{**}\right]^{-\sigma} = \left[2\pi K_H K_m Re_{\omega}\delta/r\right]^{-\sigma}. \tag{3.103}$$

Expressing the Nusselt number with the help of the Reynolds analogy parameter χ and its definition, Eq. (2.57), one can obtain

$$F(Pr)\left[2\pi K_H K_m Re_{\omega}\delta/r\right]^{-\sigma} = \chi\frac{c_f}{2}(1+\alpha^2)^{1/2}Pr. \tag{3.104}$$

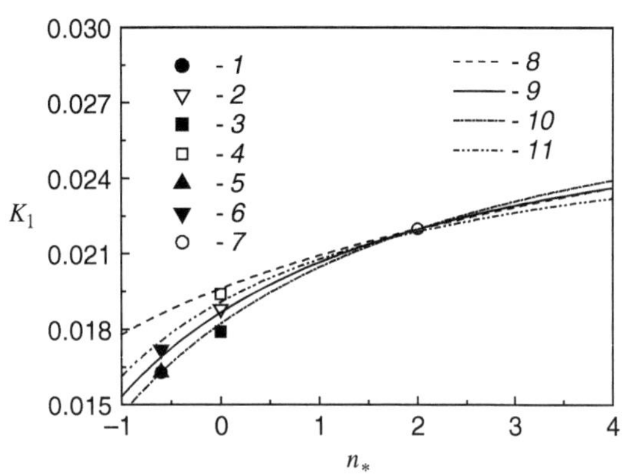

Fig. 3.24 Effect of the exponent n_* on the constant K_1 in turbulent flow of air ($Pr = 0.72$). For notations for experimental data *1–7* see Fig. 3.9. Calculations: *8* – Eq. (3.27); *9* – Eq. (3.97); *10* – Eq. (3.100); *11* – Eq. (3.44) for $\sigma = 1$ (see Fig. 3.10)

The unknown function $F(Pr)$ is excluded via substituting it with the value $\chi_{\rm quad}$, which is actually the Reynolds analogy parameter χ for $n_* = 2$ and $Pr \neq 1$. This is done in the following way. Let us assume that $K_H = K_{H,\rm quad}$ and $\chi = \chi_{\rm quad}$ for $n_* = 2$ and $Pr \neq 1$. Then, on the basis of this,

$$F(Pr)\left[2\pi K_{H,\rm quad} K_m Re_\omega \delta/r\right]^{-\sigma} = \chi_{\rm quad}\frac{c_f}{2}(1+\alpha^2)^{1/2}Pr. \tag{3.105}$$

Dividing Eq. (3.104) by Eq. (3.105), one can obtain

$$\left[K_H/K_{H,\rm quad}\right]^{-\sigma} = \chi/\chi_{\rm quad}. \tag{3.106}$$

Based on Eqs. (3.52) and (3.53), one can further obtain for $n_* = 2$, $Pr \neq 1$ and $\chi = \chi_{\rm quad}$ that

$$K_{H,\rm quad} = K_V \chi_{\rm quad}, \tag{3.107}$$

$$\left[K_H/K_V\chi_{\rm quad}\right]^{-\sigma} = \chi/\chi_{\rm quad}. \tag{3.108}$$

According to the works [138, 139]

$$\begin{aligned} \chi_{\rm quad} &= Pr^{-1/2} \quad \text{for laminar flow,} \\ \chi_{\rm quad} &= Pr^{-2/5} \quad \text{for turbulent flow.} \end{aligned} \tag{3.109}$$

The exponent σ should have such a value that the exponents at the coordinate r in both sides of Eq. (3.104) are equal to each other (so that conditions $\chi = const$, $K_H = const$ hold). Then

$$\sigma = 2n/(n+1). \tag{3.110}$$

Substituting the value of χ from Eq. (3.108) into Eq. (3.53) and using the accessory equation (3.52), one can obtain with allowance for Eq. (3.110)

$$K_H = \chi_{\rm quad} K_V \left(\frac{2+m+n_*}{4+m}\right)^{(n+1)/(3n+1)}, \tag{3.111}$$

$$\chi = \chi_{\rm quad}\left(\frac{2+m+n_*}{4+m}\right)^{2n/(3n+1)}. \tag{3.112}$$

As a result, the Nusselt number can be expressed by the following relation:

$$Nu = A_c(1+\alpha^2)^{1/2}\left(\frac{2+m+n_*}{4+m}\right)^{2n/(3n+1)} Re_\omega^{(n+1)/(3n+1)}\chi_{\rm quad}\,Pr. \tag{3.113}$$

For laminar flow, one has $n = 1$ and $m = 0$, which results in Eq. (3.113) with allowance for Eqs. (2.35) and (3.109) becoming identical to Eq. (3.6). For turbulent flow, Eq. (3.113) at $n = 1/7$ with allowance for Eq. (3.109) transforms to Eq. (3.27).

To our mind, the derivations of the authors [41, 138, 139] given above rather obscure the essence of the theory of local modelling described in the most strict form in Sect. 3.6.1.

Thus, Chap. 3 describes in detail solutions of different problems of convective heat transfer of a free rotating disk obtained using the present integral method (developed in Chap. 2), which provide noticeably higher accuracy than those known earlier. These solutions serve also as validations of the present integral method confirming its superior accuracy in comparison with the previously known methods. Also, Chap. 3 contains an overview of the most important finding of the other authors regarding transitional flow, average heat transfer of an *entire* disk, etc.

Chapter 4
Unsteady Laminar Heat Transfer of a Free Rotating Disk

4.1 Transient Experimental Technique for Measuring Heat Transfer over Rotating Disks

Transient heat transfer investigations are a matter of great importance both from the fundamental and applied points of view, primarily, in turbomachinery. One of the most widely used applications comprises various transient experimental techniques of determining surface heat transfer of a body in stationary hydrodynamic conditions. Such techniques, currently employing thermochromic liquid crystals, are based on the known fact that after a certain period of time from the beginning of the cooling process, the surface heat transfer coefficient becomes a time-independent function equal to its value for steady-state heat transfer under the same boundary conditions. The heat flux in this situation can be computed from more or less simple analytical solutions for unsteady heat conduction inside solid bodies for known surface temperatures. Experimentally measured dependencies of the surface temperature on time are substituted into the mentioned analytical solutions, and the resultant algebraic equations are further solved to find the heat transfer coefficients [91, 112, 121, 154–156, 176, 215].

Most often, the experimental data analysis is based on the theory of one-dimensional heat conduction in a semi-infinite slab with a convective boundary condition at the interface between the slab and the cooling/heating medium with a step change in the coolant's temperature T_∞ [9, 60, 91, 115, 154–156, 215]:

$$F_t(t) = \frac{T_w(t) - T_\infty}{T_{w,i} - T_\infty} = \exp(\gamma^2) \cdot \mathrm{erfc}(\gamma), \quad \gamma = \alpha\sqrt{a_w t}/\lambda_w. \tag{4.1}$$

Here $T_w(t)$ is the instantaneous disk surface temperature; subscript "w" denotes physical properties of the wall (i.e. of the disk); α, $T_{w,i}$ and T_∞ are time-independent. Having measured the temporal curve of the local surface temperature $T_w(t)$, one can solve Eq. (4.1) for the heat transfer coefficient. The use of this technique is restricted by an obvious consideration that heat conduction must involve only a small fraction of the real wall thickness for the semi-infinite slab assumption to hold. Therefore, the measurement time is strictly limited. In accordance with the classical theory

I.V. Shevchuk, *Convective Heat and Mass Transfer in Rotating Disk Systems*, Lecture Notes in Applied and Computational Mechanics 45, DOI 10.1007/978-3-642-00718-7_4,
© Springer-Verlag Berlin Heidelberg 2009

[160], the time of measurements expressed in terms of the Fourier number should not exceed the value of $Fo = 1/4$; according to the data of [214], the limiting value of the Fourier number is $Fo = 1$. The reason for this restriction is that at $Fo>1/4$ [160] (or $Fo>1$ [214]) the temperature of the body of the entire disk diverts from the initial value $T_{w,i}$, and thus the entire disk takes part in the heat transfer process.

Experimental techniques for determining heat transfer coefficients based on the solution of Eq. (4.1) have been widely used in experimental investigations of heat transfer of rotating disks [91, 123, 151, 154–156]. However, not all the researchers who used the transient techniques to determine heat transfer coefficients obtained experimental data agreeing with theoretical solutions and data of steady-state experimental methods for the same boundary conditions.

For example, in work [156] a Plexiglas® rotating disk ($b = 0.123$ m, $s = 0.01$ m) was placed in a thermally insulated box and heated up to a constant temperature of $T_{w,i} = 40°$C (Fig. 4.1). Then the box cover was suddenly removed and the disk started cooling down, due to convective heat transfer, to room temperature of $T_\infty = 24°$C without additional input of heat. Unsteady surface temperatures were measured by means of liquid crystals applied onto the disk surface as ring-like bands. The heat transfer coefficient was computed by Eq. (4.1). In laminar flow regime ($Re_\varphi = 53{,}500$), instantaneous distributions $T_w(t)$ obtained in [156] increased weakly according to dependence agreeing rather well with Eq. (2.35) for $n_* = 0.06$ (see Fig. 4.2), and heat transfer coefficients decreased abruptly with an increasing local radius r, with the values of Nu_b at $r \to 0$ and $r \to b$ differing from each other by a factor of approximately two (Fig. 4.2) [176]. This contradicts the known theoretical and experimental data discussed above in Chap. 3, in accordance with which the Nusselt number Nu_b under conditions of laminar flow is constant (curve *4*, Fig. 4.2). Experimental data for Nu_b [156] agree with the value $Nu_b = 77.6$ (or $K_1 = 0.336$), obtained via the solution of Eqs. (2.37), (2.38), (2.39), (2.40) and (2.41) for $n_* = 0.06$, only at $r \to b$ (or $x \to 1$).

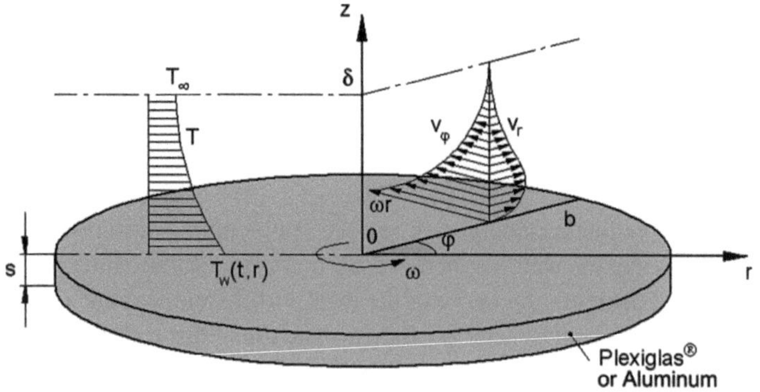

Fig. 4.1 Geometrical arrangement and main parameters of the problem of fluid flow and unsteady heat transfer over a rotating disk in still air

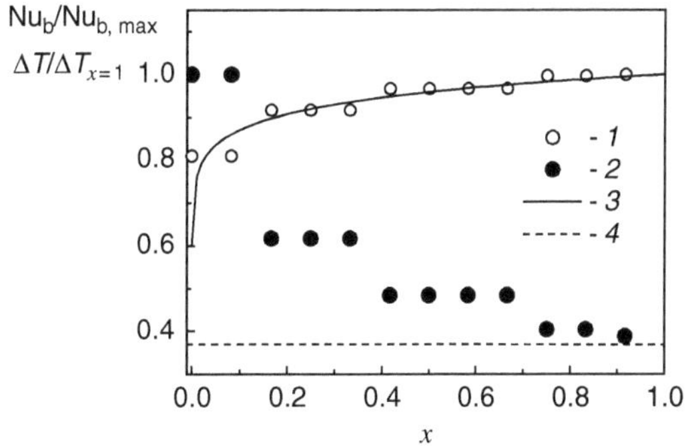

Fig. 4.2 Radial variations of the relative Nusselt number and temperature head over a rotating disk for $Re_{\varphi} = 53{,}500$. Experiments [156, 176], transient technique based on Eq. (4.1): *1* – $\Delta T/\Delta T_{x=1}$, *2* – $Nu_b/Nu_{b,\max}$. Theoretical predictions for $n_* = 0.06$: *3* – $\Delta T/\Delta T_{x=1}$ by Eq. (2.35), *4* – $Nu_b/Nu_{b,\max}$ by Eq. (2.37), (2.38), (2.39), (2.40) and (2.41). Here $Nu_{b,\max} = 209.9$, $\Delta T = T_w - T_\infty$

Obviously, some factors affecting quantitative characteristics of unsteady heat transfer of a rotating disk that lead to the aforementioned deviations of the heat transfer coefficients measured in [156] from theoretical and experimental data of other authors should be additionally taken into consideration.

An advantage of solution (4.1) is its simplicity, which however disappears once one has to take into consideration curvature of the surface at which the measurements are to be made [215]. An alternative to Eq. (4.1) is to use the solution for unsteady heat transfer of a slab of a finite thickness $s/2$, where the back face is insulated [45], or a slab of a finite thickness s (Fig. 4.1) with identical heat transfer coefficients on both faces [115]:

$$F_t(t) = \vartheta(t,y=1), \quad \vartheta(t,y) = \sum_{m=1}^{\infty} E_m \cos(\mu_m y) \exp(-\mu_m^2 Fo), \tag{4.2}$$

$$E_m = \frac{2\sin(\mu_m)}{\mu_m + \sin(\mu_m)\cos(\mu_m)}, \quad \cot(\mu_m) = \mu_m/Bi, \tag{4.3}$$

where eigenvalues μ_n are defined by Eq. (4.3); $\vartheta(t,y) = (T(t,y) - T_\infty)/(T_{w,i} - T_\infty)$; $Bi = 0.5\alpha s/\lambda_w$; $y = z/(0.5s)$.

4.2 Self-Similar Navier–Stokes and Energy Equations

As mentioned above, the transient technique for determining heat transfer coefficients is based on the fact that after a certain period of time since the unsteady process of cooling or heating begins, a regime of heat transfer arises in which the heat transfer coefficient of a body under unsteady-state conditions acquires a value equal to that in steady-state heat transfer under the very same thermal boundary conditions. However, this technique may be used only when the following conditions are satisfied:

(a) heat transfer must occur in a regime in which the heat transfer coefficient ceases to vary in time; therefore, time limitations of this regime must be found;
(b) one must know a correlation between variation of the disk surface temperature and the heat transfer coefficient; this correlation, which can be found from a solution of the unsteady heat conduction problem, is different for different geometries of the system and for different initial and thermal boundary conditions.

This sub-section is devoted to a solution of the non-stationary equation of a laminar thermal boundary layer jointly with stationary Navier–Stokes equations in a self-similar form for a specified law of temporal variation of the wall temperature, as well as for determination of variation of the heat transfer coefficient with time [187, 192].

The thermal boundary layer equation for an unsteady axisymmetric problem in cylindrical polar system of coordinates has the form of Eq. (2.16).

Assuming that the non-dimensional temperature $\theta(t,z) = (T(t,r,z) - T_\infty)/(T_w(r,t) - T_\infty)$ is a self-similar function independent of r, one can transform Eq. (2.16) as follows:

$$\frac{\partial\theta}{\partial t} + \frac{\theta}{\Delta T_t}\frac{\partial\Delta T_t}{\partial t} + \mathrm{v}_r\theta\,\frac{1}{\Delta T_t}\frac{\partial\Delta T_t}{\partial r} + \mathrm{v}_z\frac{\partial\theta}{\partial z} = a\frac{\partial^2\theta}{\partial z^2}, \tag{4.4}$$

where $\Delta T_t(t,r) = T_w(t,r) - T_\infty$; $T_w(r,t)$ is the instantaneous temperature of the disk surface (Fig. 4.1); $T_{w,i}$ is the initial disk temperature.

For the instantaneous heat transfer coefficient to be equal to its stationary value existing during the steady-state conditions before the onset of the unsteady cooling/heating of the disk, the instantaneous temperature head ΔT_t should behave in such a way that

$$\Delta T_t(t,r) = \Delta T(r)F_t(t), \tag{4.5}$$

$$\frac{1}{\Delta T_t}\frac{\partial\Delta T_t}{\partial r} = \frac{1}{\Delta T}\frac{d\Delta T}{dr}, \tag{4.6}$$

where $\Delta T(r) = T_{w,i}(r) - T_\infty$, and $F_t(t) = \Delta T_t(t,r)/\Delta T(r)$ a function of the only variable t. This is an indispensable condition, because the first two terms in Eq. (4.4) become negligible very fast with an increasing process time t, and namely the third term determines the effect of the radial disk's surface distribution on heat transfer.

In doing so, Eq. (4.4) takes the following form:

$$\frac{\partial\theta}{\partial t} + \frac{\theta}{F_t}\frac{\partial F_t}{\partial t} + \mathrm{v}_r\theta\,\frac{1}{\Delta T}\frac{d\Delta T}{dr} + \mathrm{v}_z\frac{\partial\theta}{\partial z} = a\frac{\partial^2\theta}{\partial z^2}. \tag{4.7}$$

In particular, in view of the boundary condition (2.35), Eq. (4.7) transforms to

$$\frac{\partial\theta}{\partial t} + \frac{\theta}{F_t}\frac{\partial F_t}{\partial t} + n_*\theta\frac{\mathrm{v}_r}{r} + \mathrm{v}_z\frac{\partial\theta}{\partial z} = a\frac{\partial^2\theta}{\partial z^2}. \tag{4.8}$$

If the disk is isothermal at the initial moment of time, then $\Delta T(r)=const$, $d\Delta T/dr=0$, $n_*=0$. If, additionally to the above condition, the external cylindrical surface of the disk is thermally insulated and the heat transfer coefficient is constant over the surface (which takes place in laminar flow), then the disk remains further isothermal during the process of unsteady cooling/heating. This is also valid for the case where the external radius b of the disk is rather large, and the boundary conditions on the external cylindrical surface have no effect on the inner region of the disk. As a result, Eq. (4.8) can be simplified to

$$\frac{\partial\theta}{\partial t} + \frac{\theta}{F_t}\frac{\partial F_t}{\partial t} + \mathrm{v}_z\frac{\partial\theta}{\partial z} = a\frac{\partial^2\theta}{\partial z^2}. \tag{4.9}$$

Using self-similar variables and functions, Eq. (4.8) can be reduced to an ordinary differential equation, which can be solved numerically using MathCAD software. The form of self-similar variables for this problem can be found using the method described in [22]. Let us introduce a linear transformation using the constants α_k $(k = 1,\ldots,4)$ and A:

$$t = A^{\alpha_1}\bar{t}, \quad z = A^{\alpha_2}\bar{z}, \quad \mathrm{v}_z = A^{\alpha_3}\bar{\mathrm{v}}_z, \quad \theta = A^{\alpha_4}\bar{\theta}. \tag{4.10}$$

Expressions (4.10) are substituted into Eq. (4.8). The initial and transformed forms of Eq. (4.8) are invariant, if the cumulative exponents of the constant A in terms of the transformed equation are equal to each other. As a result, one can obtain $\alpha_2 = \alpha_1/2$, $\alpha_3 = -\alpha_1/2$, $z^2/t = \bar{z}^2/\bar{t}$, $\mathrm{v}_z t^{1/2} = \bar{\mathrm{v}}_z\bar{t}^{1/2}$. The temperature θ transforms into itself.

As a result, the self-similar variables, the thermal boundary layer equation and the boundary conditions have the following form:

$$\theta'' = Pr\left[g_*\theta + \theta'(H - \eta/2) + n_* F\theta\right], \tag{4.11}$$

$$\eta = z/(\nu t)^{1/2}, \quad H(\eta) = \mathrm{v}_z(t/\nu)^{1/2}, \tag{4.12}$$

$$g_* = \frac{t}{F_t}\frac{dF_t}{dt}, \tag{4.13}$$

$$\theta = 1 \quad \text{for } \eta = 0 \quad \text{and} \quad \theta = 0 \quad \text{for } \eta \to \infty. \tag{4.14}$$

Here primes denote derivatives with respect to η.

The function H(η) in Eq. (4.11) is time-independent and found as a result of a solution of a steady-state self-similar system of the Navier–Stokes equations (2.37), (2.38), (2.39) and (2.40) (for $N=0$, $\beta=0$), in which, as applied to the problem at hand, the time t plays the role of a parametric non-dimensionalizing variable instead of the commonly used quantity $1/\omega$. As a dimensionless parametric variable in these equations, ωt is used, while the non-dimensional functions

$$F(\eta) = \mathrm{v}_r r/t, \quad G(\eta) = \mathrm{v}_\varphi t/r \text{ and } P(\eta) = -pt/(\rho\nu) \tag{4.15}$$

are also time-independent. The relevant boundary conditions look as follows:

$$\eta = 0: \quad F = H = 0, \quad G = \omega t, \tag{4.16}$$

$$\eta \to \infty: \quad G = F = 0. \tag{4.17}$$

The Nusselt number Nu_b is calculated by the formula

$$Nu_b = K_1 Re_\varphi^{1/2}, \quad K_1 = -\left(\frac{d\theta}{d\zeta}\right)_{\zeta=0} = \frac{1}{\sqrt{\omega t}}\left(\frac{d\theta}{d\eta}\right)_{\eta=0}, \tag{4.18}$$

where $\zeta = z\sqrt{\omega/\nu}$.

Authors of the work [26] introduced self-similar variables for a process of unsteady mass transfer in case of orthogonal uniform air flow impingement onto a rotating disk, which are analogous to Eq. (4.12) for the parameters η and $\mathrm{H}(\eta)$ within a constant used as additive term in the brackets:

$$\eta = z\left[\frac{1}{(4D_m t)^{1/2}} + \frac{Sc}{(1+Sc)^{2/3}}\frac{(Re_a^2 + Re_{\omega d}^2)^{1/4}}{d}\right], \tag{4.19}$$

$$H(\eta) = \frac{\mathrm{v}_z}{D_m}\left[\frac{1}{(4D_m t)^{1/2}} + \frac{Sc}{(1+Sc)^{2/3}}\frac{(Re_a^2 + Re_{\omega d}^2)^{1/4}}{d}\right]^{-1}. \tag{4.20}$$

Such a modification does not change the essence of the self-similar transformation performed. It should be noticed that the self-similar variable (4.19) and the function (4.20) were introduced in the work [26] without any justification of such a choice, while the above-mentioned transformations (4.10) are based on the formal and strict mathematical background of the group methods [22].

4.3 Exact Solution for Surface Heat Transfer of an Isothermal Rotating Disk

This section represents a theoretical solution [61, 187] of the problem studied experimentally in the work [156] and described in Sect. 4.1.

The following values of physical and geometric parameters were used in the calculations: for Plexiglas® [156] $\lambda_w = 0.19$ W/(m^2·K), $a_w = 1.086\times10^{-7}$ m^2/s;

for air [22] $\lambda = 0.02624$ W/(m^2·K), $a = 2.216 \times 10^{-5}$ m^2/s; $Pr = 0.71$; the disk thickness $s = 0.01$ m; $\omega = 52.36$ 1/s (500 r.p.m.), which corresponds to the case $Re_\phi = 5.35 \times 10^4$ [156]. The values of γ and Bi were calculated using a steady-state value of $K_1 = 0.326$.

Use of Eq. (4.1) for $F_t(t)$ allows obtaining that

$$g_* = \gamma^2 - \gamma/(\pi^{1/2} F_t), \quad \gamma = K_1 Pr^{-1/2} (a_w/a)^{1/2} (\lambda/\lambda_w) \sqrt{\omega t}. \tag{4.21}$$

Substituting numerical values of physical constants into Eq. (4.21), one can obtain $\gamma = 0.0768\sqrt{\omega t}$. The numerical solution of Eq. (4.11) (for $n_* = 0$) jointly with Eqs. (2.37), (2.38), (2.39) and (2.40) was obtained using the MathCAD software.

As one can see from the data in Fig. 4.3, the constant K_1 (and the Nusselt number Nu_b) accepts stationary values very rapidly.

It was mentioned in the work [187] that the constant K_1 accepts its stationary value for $\omega t \approx 1000$ or $t \approx 19$ s, with the function $F_t(t)$ being respectively equal to ≈ 0.876. In this case, deviation of the constant K_1 from its stationary value is 0.14%, which is a too strict criterion. If one assumes a standard 1% deviation as a criterion of the steady-state conditions, then the constant K_1 reaches its stationary value already for $\omega t \approx 130$ or $t \approx 2.5$ s, with the function $F_t(t)$ being respectively equal to ≈ 0.96. Unfortunately, the authors of [156] did not provide the reader with the experimentally obtained curves $F_t(t)$ that would be helpful for comparisons.

In the calculations of the authors [26] for unsteady mass transfer, the Sherwood number attained its stationary value even earlier, already for $\omega t \approx 10$. The reason for this was probably that the concentrations on the surface of the disk C_w and at the infinity C_∞ were time-independent, and at the initial moment of time the concentration on the disk changed abruptly from the value C_∞ to C_w. Thus, the data of the work [26] can serve as the lowest estimate of the time necessary for the unsteady heat or mass transfer coefficients to attain their stationary values, which can be obtained using self-similar solutions of the problem.

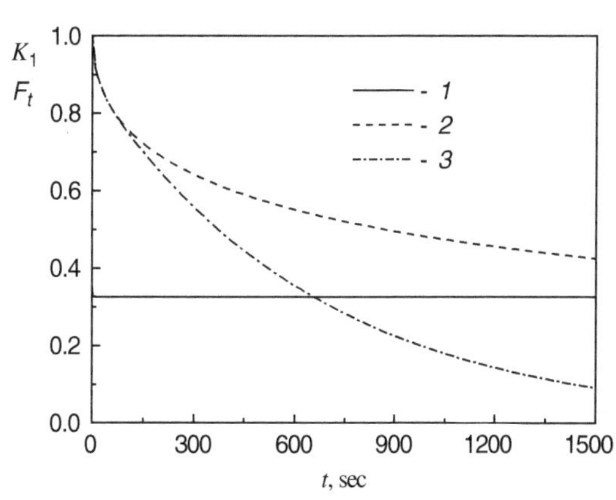

Fig. 4.3 Variation of the values K_1 and F_t with time t. *1* – K_1; *2* – $F_t(t)$, Eq. (4.1); *3* – $F_t(t)$, Eq. (4.2)

If one interprets the disk as an isothermal slab of finite thickness s (Fig. 4.1) with identical heat transfer coefficients at $z=0$ and $z=-s$, then the function $F_t(t)$ is described by Eq. (4.2). Using Eq. (4.2) for $F_t(t)$, one can obtain $Bi=0.395$. Already for $t=69$ s, the value of the Fourier number is equal to $Fo=0.3$, and the function $F_t(t)$ with an error of 0.37% may be calculated by Eq. (4.2) using only the first term of the series, i.e. the so-called regular regime of heat transfer begins [115]. The function $K_1(t)$ calculated by Eq. (4.18) based on a solution of the systems (2.37), (2.38), (2.39), (2.40), (4.11), (4.12), (4.13) and (4.14) and using Eq. (4.2) for $F_t(t)$ coincides with the curve for $K_1(t)$ obtained using Eq. (4.1) (see Fig. 4.3).

The data for variation of the constant K_1 with time obtained in calculations are in agreement with the known phenomenon that, beginning with some moment of time, the heat transfer coefficient of a given surface under non-stationary conditions is independent of the rate of heating/cooling, all other conditions being equal. In doing so, the values of $F_t(t)$ calculated by Eq. (4.2) are much (by a factor of 2–4 at $t>700$ s) lower than those of $F_t(t)$ calculated by Eq. (4.1) (see Fig. 4.3). The explanation of this fact is obvious: according to Eq. (4.2), a real disk of finite thickness (rather thin in the case being treated) cools down much faster than a semi-infinite body cooling down according to Eq. (4.1), because the amount of heat accumulated by a semi-infinite body is much larger than the amount of heat accumulated by a rather thin disk. In the range of short time intervals, both values of function $F_t(t)$ coincide until the process of cooling down extends up to the middle of the disk of finite thickness.

Some additional results for unsteady distributions of the surface temperature of an isothermal disk are given in Sect. 4.5.4.

Significant discrepancy between transient experimental measurements of the heat transfer coefficient α and the known experimental and theoretical data [26, 46, 138, 139] may be caused by the use of Eq. (4.1) to restore the values of the heat transfer coefficient in such a way that the time of measurements exceeded the limiting value mentioned in Sect. 4.1. The actual rate of cooling of the disk surface must correspond to Eq. (4.2); for instance, at the moment of time $t>700$ s, this rate must have been 2–4 times higher than that predicted by Eq. (4.1). For Eq. (4.1) at $t>700$s to agree with Eq. (4.2), one must substitute into Eq. (4.1) the values of α, which are more than twice higher than those calculated by Eq. (3.4) with the steady-state value of $K_1=0.326$. Therefore, use of Eq. (4.1) to restore experimental values of the heat transfer coefficients at $t>700$ s results in obtaining values of α, which are two or more times higher than its actual values.

As applied to experimental techniques for a disk of finite thickness, one can also recommend using Eq. (4.2) in the regular regime of heat transfer. The conditions in which this regime takes place for different values of the Biot number that depends, in particular, on both the flow regime and the Prandtl number (via the heat transfer coefficient) are discussed below in detail in Sect. 4.5. It is, however, obvious that the heat transfer coefficient accepts its steady-state value a fortiori before the regular regime of heat transfer sets in.

Heat transfer coefficients in more complex conditions may be restored using numerical solutions of two-dimensional problems of heat conduction in a disk taking into account the thermal boundary conditions on the reverse side of the disk and on its rim.

4.4 Numerical Solution of an Unsteady Conjugate Problem of Hydrodynamics and Heat Transfer of an Initially Isothermal Disk

4.4.1 Computational Domain and Grid

The commercially available CFD code CFX-5 by ANSYS Inc. was employed for the numerical simulations by the authors [61, 191], whereas a case-sensitive mesh generated manually by the Patran volume mesher was used instead of that provided by the standard automatic mesher implemented in CFX-5. Because of the axisymmetric character of the problem, the overall computational domain comprised only a sector of 45 degree-angle with a periodic boundary condition specified on each flat vertical side of the sector (Fig. 4.4). This domain consisted of two parts: a stationary fluid domain with the radius and height of 0.5 m and a rotating fluid domain that contained a solid sub-domain, namely the disk itself (heated by an energy source) of radius 0.123 m and height of 0.005 m in accordance with the experimental setup of [156]. The rotating and stationary domains are connected to each other by a so-called frozen rotor domain interface. The outer flat and top cylindrical boundaries were defined as walls with a free slip condition and the fixed (ambient) temperature of $T_\infty = 297.15$ K. All lower surfaces of the computational domain were symmetry planes. These pre-conditions allowed the simulation of conjugate heat transfer of a free rotating heated disk with finite dimensions in a stationary fluid domain where the outer boundaries did not affect fluid flow and heat transfer intensity in the near-wall region.

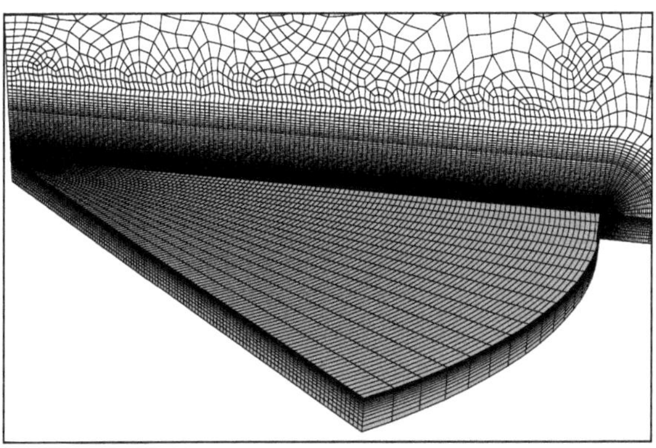

Fig. 4.4 Computational mesh for simulations of fluid flow and heat transfer of a rotating disk in cylindrical coordinates

For meshing this domain, mainly hexahedrons and a few prisms with 18 elements in angular direction were used. The overall 333,520 nodes were partitioned in such a way that 11% were located inside the disk zone, 54% were used to model the rotating fluid domain with an extension of 0.01 m around the solid disk and the remaining 35% were necessary to realize the stationary fluid region. Preliminary investigations of the mesh structure and necessary density in the rotating domain showed that the manually created structured mesh was necessary for obtaining really smooth distributions of heat transfer values in the disk surface area according to the self-similar solution [41, 138, 139, 174]. In order to reduce the overall number of nodes, an unstructured coarser mesh was used in the outer stationary part of the computational domain. This was sufficient to simulate the ambience of still air.

4.4.2 Validation for Steady-State Fluid Flow and Heat Transfer

In order to validate the numerical model with its boundary conditions described in the previous sub-section, several runs of simulations at $\omega = 52.36$ 1/s (or 500 r.p.m.) were performed to obtain a well converging solution for fluid flow and heat transfer parameters. Physical properties of Plexiglas® and air coincide with those given in Sect. 4.3. Target (normalized log) residual for all items was set to 1×10^{-6}. To reach this value, a timestep of $10 \cdot 1/\omega$ has been used during the start-up of the simulation, which finally needed to be reduced to $0.1 \cdot 1/\omega$.

Results for fluid flow are presented in Fig. 4.5, whereby dimensionless functions F, G and H, Eqs. (2.31), have been chosen for v_r, v_φ and v_z, respectively. For comparison purposes, numerical solution of the self-similar ordinary differential equations (2.31)–(2.40) for $N=0$, $\beta=0$ is also presented in Fig. 4.5 (dotted lines). Data of numerical simulations taken at domain centre ($x=0.3$) were chosen for the analysis. In this area, undisturbed laminar fluid flow existed without any influence of flow around the disk's rim, whilst such disturbed flow took place already at $x \geq 0.6$.

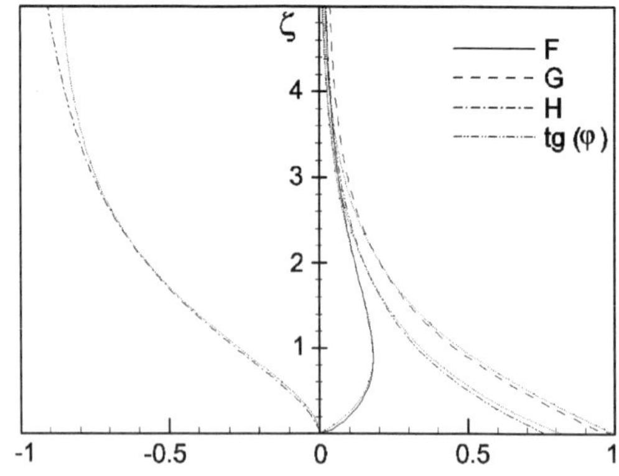

Fig. 4.5 Profiles of the velocity components F, G, H and function $tan\varphi$ for steady-state laminar flow: comparisons of the simulations using CFX-5 (at $x=0.3$) and self-similar solution (*dotted lines*)

In addition to the velocity components, the tangent of the flow swirl angle $tan\varphi$ has been calculated and plotted. Both solutions, numerical and self-similar, are matching very well over the whole boundary layer. Only in the near-wall region slight deviations of the CFX-5 data from the self-similar solution occur, among which the most remarkable are in the distributions of $tan\varphi$. Its wall value at $\zeta = 0$ is 0.8284 for the self-similar solution and 0.7645 for the numerical simulations using CFX-5.

Temperature profiles within the boundary layer and inside the disk taken at several radial locations are presented in Fig. 4.6. Again, numerical simulations and self-similar solution correspond to each other very well for regions without influence of the flow around the disk's rim. Deviations of the temperature profiles computed using the CFX-5 from the self-similar solution are noticeable only above $\zeta \approx 1.3$. Also the average value of the constant $K_1 = 0.341$ in the relation (3.4) for the Nusselt number computed from results of CFX-5 simulations for undisturbed laminar flow at $x \leq 0.6$ is in close consistence with the value $K_1 = 0.326$ by the self-similar solution for $Pr = 0.71$ and $T_w = const.$

The model provides an adequate and exact prediction of conduction heat transfer also inside the disk.

For the unsteady temperature distribution inside the disk written as the function $\theta(t,y)$, one can obtain an analytical solution

$$\theta(t,y) = \frac{1}{F_t(t)} \sum_{m=1}^{\infty} E_m \cos\left[\mu_m(1+y)\right] \exp\left(-\mu_m^2 \mathrm{Fo}\right), \tag{4.22}$$

which is a modification of Eq. (4.2), with the variable $y = z/(0.5s)$ varying over the range $y = -1$ to 0 (or $z = -s/2$ to 0).

For the steady-state one-dimensional conduction (in z-direction) with a heat source inside the disk, one can easily deduce an analytical solution (valid for the same range of y)

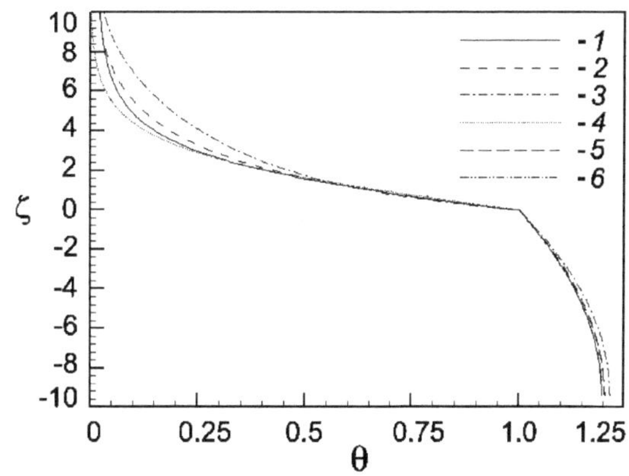

Fig. 4.6 Profiles of the dimensionless temperature θ for $t = 0$ s [191]. *1* – $x = 0.5$; *2* – 0.7; *3* – 0.9; *4* – boundary layer, self-similar solution; *5* – disk, analytical solution (4.22) for $t = 400$ s (merges with line *2* inside the disk); *6* – solution (4.23) (merges with line *1* inside the disk). Here $\zeta = 0$ is the disk surface; $\zeta = -9.37$ is the symmetry plane of the disk that corresponds to $z = -s/2$

$$\theta = -Bi(0.5y^2 + y - 1/Bi). \quad (4.23)$$

Profiles of θ computed using CFX-5 exhibit good agreement with analytical solutions (4.22) and (4.23) (see Fig. 4.6). It is worth noticing adequate physical behaviour of the CFX-5 results demonstrating zero derivatives at the symmetry planes and only minor dependence on the radial location.

Thus, results of simulations of steady-state fluid flow and conjugate heat transfer parameters of a heated rotating disk are in good agreement with the self-similar solution for fluid flow and convective heat transfer and the analytical solution for heat conduction inside the disk. Hence, further two conclusions can be drawn. First, the model itself and the computational domain can be used for further correct and adequate predictions of the transient problem of heat transfer. Second, the steady-state solution is suitable for providing initial velocity, pressure and temperature distributions for the transient simulations.

4.4.3 Unsteady Fluid Flow and Heat Transfer

The temperature, velocity and pressure fields resulted from the steady-state simulations described above were used as initial conditions for the transient run. Switching off the energy source used to heat the disk during the steady-state run allowed for analysing unsteady laminar heat transfer of a rotating disk in still air at the same angular speed of the rotating domain $\omega = 52.36$ 1/s. The calculation has been carried out during 1500 s of physical time. Within this period the average disk surface temperature decreased from the initial value $T_w = 312.60$ K to $T_w = 298.80$ K. The surrounding air temperature has been kept constant at $T_\infty = 297.15$ K.

Expressed in terms of the non-dimensional function $F_t(t)$, a value of 0.106 was reached up to the time of the simulation abort. Variation of this function versus time is shown in Fig. 4.7 and compared with Eqs. (4.1) and (4.2). Function $F_t(t)$ computed using CFX-5 is in very good agreement with Eq. (4.2). Thus, the real cooling rate of the relatively thin ($s = 0.01$ m) Plexiglas® disk indeed very significantly exceeds the cooling rate predicted by Eq. (4.1) when the time of the cooling process exceeds a certain limiting value (see Sect. 4.1). The reason for this discrepancy is that Eq. (4.1) is valid only for high Biot numbers that means a disk of a semi-infinite thickness or made of a material with low thermal conductivity k_w. Such a disk indeed should cool down much slower than a thin Plexiglas® disk used in experiments [154–156]. (It is easy to ascertain that Eqs. (4.1) and (4.2) can provide the same numerical results at $\mathrm{Bi} \to \infty$.)

Thus, again, once the time of measurements exceeds the aforementioned limiting value, the choice of Eq. (4.1) to process the transient experimental data may be the reason of significant deviation of the experimental results for the Nusselt number from Eq. (3.4) at $K_1 = 0.326$, like for instance in works [154–156].

The high quality of simulations can also be confirmed by the excellent agreement of the constant K_1 with the self-similar solution also presented in Fig. 4.7.

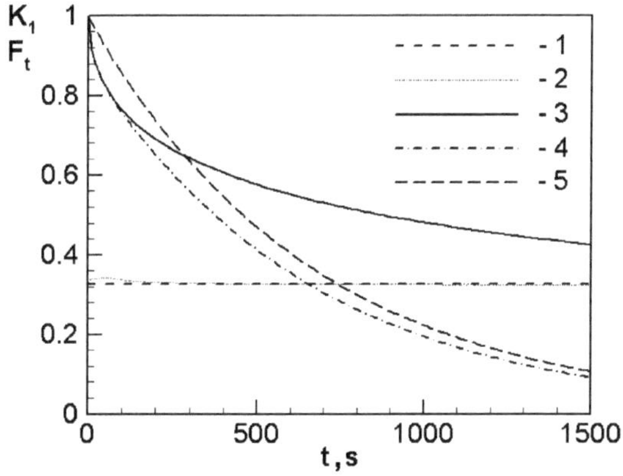

Fig. 4.7 Variation of the parameters K_1 and F_t with time according to the self-similar solution and simulations using the CFX-5 [191]. *1* – K_1, self-similar solution; *2* – K_1, CFX-5; *3* – $F_t(t)$, Eq. (4.1); *4* – $F_t(t)$, Eq. (4.2); *5* – $F_t(t)$, CFX-5

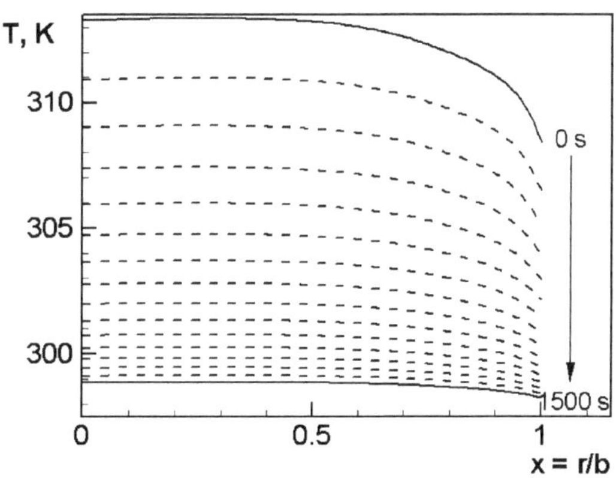

Fig. 4.8 Simulations using CFX-5 for the cooling rate of the disk surface versus time starting from with an average temperature of $T_w = 312.6$ K at the initial moment of time $t = 0$ s [191]. *Dashed lines* are representing intermediate distributions of $T_w(t,x)$ taken every 100 s

In consistency with physical expectations, the disk cooled down in a very homogenous manner. As obvious from Fig. 4.8, the surface temperature at each moment of time is nearly constant up to $x = 0.5$ and decreases subsequently to the value at the disk's edge. The decreased value of T_w at $x = 1$ is of course a result of cooling down of the outer cylindrical surface of the disk. The profile of T_w flattens more and more until the ambient temperature T_∞ is almost reached at $t = 1500$ s.

The main point of interest is behaviour of the Nusselt number (Fig. 4.9). Flow in the immediate vicinity of the rotation axis occurs under conditions of a kind of orthogonal flow impingement onto the disk and at the onset of the boundary layer development, which results in a time-independent moderate bump in the profile of Nu_b up to approximately $x = 0.2$. At each timestep, the Nusselt number is approximately constant between $x = 0.2$ and $x = 0.6$. The significant peak of the Nu_b number distributions in the outer part of the disk surface at $t = 0$ s gradually vanishing with time is caused by the non-uniform initial distribution of the disk surface temperature $T_{w,i}$. Noticeably decreasing radial distribution of $T_{w,i}$ results in radial heat conduction from the region of $x \leq 0.6$ towards the outer part of the disk. After a while

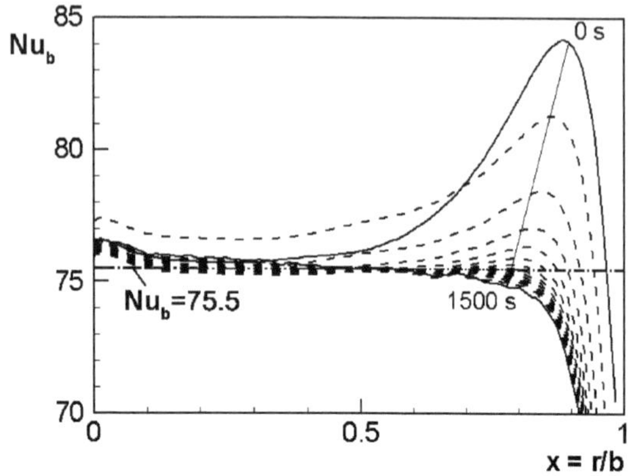

Fig. 4.9 Transitional behaviour of the Nusselt number on the disk: CFX-5 simulations from t=0 s to t=1500 s [61]. *Dashed lines* are intermediate distributions, taken every 100 s

($t \geq 500$ s), surface temperatures become more uniform, because a certain part of the heat accumulated by the disk is transferred to the ambience by the outer cylindrical surface of the disk. As a result, heat conduction to the region $r \to b$ becomes insignificant, and the aforementioned peak of the Nusselt number at $x \approx 0.8$ vanishes.

Average values of Nusselt number for the undisturbed region $0.2 \leq x \leq 0.6$ disregarding effects near rotation axis and outer radius of the disk has been calculated and plotted versus time in Fig. 4.10. As evident from Fig. 4.10, the Nusselt number becomes practically time-independent after about 300 s of physical time. After this period of time, the standard deviation from the mean value of $Nu_b = 75.497$ is only 0.103. This mean value of Nu_b is in excellent agreement with Eq. (3.4), which is based on the self-similar solution for laminar flow at $K_1 = 0.326$ and $n_R = 0.5$.

Thus, numerical simulations of the problem of unsteady conjugate laminar heat transfer of a rotating disk (at steady-state fluid flow conditions) using the CFD code CFX-5 are in good agreement with the self-similar solution applicable to this

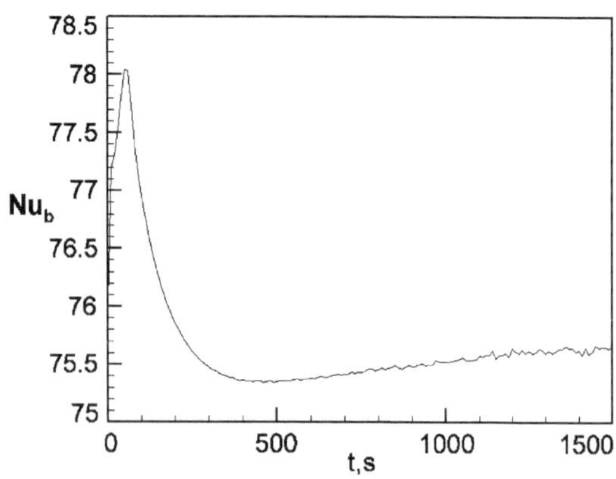

Fig. 4.10 Variation of the average Nusselt number with time in the region of $0.2 \leq x \leq 0.6$

problem. CFX-5 simulations are also in good consistency with analytical solutions for heat conduction inside the disk.

Both methods showed that the heat transfer coefficient very quickly becomes time-independent and equal to its value for steady-state conditions determined earlier theoretically and experimentally by the authors of works [19, 20, 41, 46, 146, 174].

It was shown that the solution (4.1) for unsteady one-dimensional heat conduction in a semi-infinite plate may (if the requirements limiting its use are not satisfied) significantly underestimate the real cooling rate of the disk surface and consequently overestimate the heat transfer coefficient calculated from this solution using known experimental instantaneous disk surface temperature distributions. The alternative solution for unsteady one-dimensional heat conduction in a finite-thickness plate given by Eq. (4.2), which for $Fo \geq 0.3$ degenerates just to the first term of the Fourier series, is analysed below.

4.5 Unsteady Conjugate Laminar Heat Transfer of a Rotating Non-uniformly Heated Disk

4.5.1 Problem Statement

Strictly saying, both solutions (4.1) and (4.2) are valid only for the initially isothermal surfaces. In many industrial applications, the surface temperature varies at least in the streamwise direction. A remedy for such a situation was suggested via dividing the experimental surface on a rather narrow quasi-isothermal zones [154–156]. Within each zone, the surface temperature was assumed to be constant and undergoing a spatial step change at the boundary between the neighbouring zones instead of the monotonic temperature variation that takes place in practice (see Fig. 4.2). However, validity of this approach is under question, as the experimental data [154–156] themselves require reasonable physical explanation. An attempt to find a quantitative estimate of the 2D heat conduction effects was made in [12, 13, 91, 223]. Conjugate unsteady heat transfer between parallel co-rotating disks was studied also in [107].

This section generalizes the solution (4.2) for it to be valid for the case of non-uniform initial surface temperature distribution [192]. The fact that Eq. (4.2) degenerates just to the first term of the Fourier series for $Fo \geq 0.3$ reveals one of the important fundamental properties of the transient heat transfer such as existence of the so-called regular regime of heating or cooling of a body when the temporal dependence of the local temperature in any location of the body follows a simple exponential function [115].

The objective of this investigation consists in finding a quasi-conjugate heat transfer solution for a disk initially pre-heated non-uniformly, which includes (a) a self-similar solution of the transient laminar convective heat transfer, (b) a solution of the unsteady two-dimensional heat conduction problem with a non-uniform initial temperature distribution, (c) answer to the question whether the shape of the initial temperature non-uniformity holds with time and (d) validation of a transient technique for the experimental determination of the heat transfer coefficient based on the regular heat transfer regime theory, which is free of the restriction in the time of measurements mentioned in Sects. 4.1, 4.2, 4.3 and 4.4.

The following problem is studied in this section. A disk rotating in still air and being identical to that studied in Sects. 4.1, 4.2, 4.3 and 4.4 (see Fig. 4.1) is pre-heated in such a way that its surface temperature $T_{w,i}(r)$ at the initial moment of time $t=0$ s follows the power-law distribution given by Eq. (2.35) at arbitrary value of the exponent n_*. Immediately for $t>0$, the disk starts cooling down up to the temperature T_∞ without additional input of heat.

Of course, Eq. (2.35) does not generally hold at $r=0$, because it does not provide axial symmetry of the disk's temperature at this point. Besides, choice of a thermally isolated outer cylindrical disk's surface invalidates Eq. (2.35) also at $r=b$. None of these restrictions is in force as applied to the solution of the thermal boundary layer equation for an infinite-radius disk using boundary condition (2.35). Since such a solution is also an integral part of the present research, the boundary condition (2.35) is still used in this section with an intention also to figure out the magnitude of the possible numerical inaccuracies in the heat conduction solution caused by Eq. (2.35). More detailed insight into this particular feature of the obtained solution is presented in the sub-section discussing the results.

For comparison purposes, unsteady cooling of a disk made of aluminium was also simulated, whose physical properties are [10]: $\lambda_w=204$ W/(m^2·K), $a_w=0.842\times10^{-4}$ m^2/s.

4.5.2 *Self-Similar Solution of the Transient Laminar Convective Heat Transfer Problem*

In order to further denote the heat transfer coefficients and the Biot numbers, subscript "1" is used for the outer cylindrical surface of the disk and subscript "2" for its flat surface.

In analogy to Sects. 4.1, 4.2, 4.3 and 4.4, the value $Bi_2=0.395$ was hold constant throughout the numerical computations. Because the constant K_1 varies with n_*, the condition $Bi_2=const$ means a different value of the Reynolds number Re_φ for every particular n_* provided that $Re_\varphi=5.35\times10^4$ at $n_*=0$.

The system of Eqs. (4.11), (4.12), (4.13) and (4.14) was solved with the help of the MathCAD software. Consequently, in similarity to the results of Sect. 4.3, it was obtained that the values K_1 and Nu_b in the unsteady conditions accept their stationary values very quickly. Based on the very same standard 1% deviation, the constant K_1 reaches its stationary value at $\omega t\approx40$ for $n_*=2$ and at $\omega t\approx300$ for $n_*=-1$. In both cases $t\approx2$ s, while, to remind, the Reynolds number Re_φ and angular speed of rotation ω decrease with increasing n_* to keep the Biot number constant at $Bi_2=0.395$. For these small process times, Eqs. (4.1) and (4.2) coincide and predict vales of F_t equal to $F_t\approx0.96$. Assuming that $Re_\varphi=5.35\times10^4$ is invariant,

K_1 anyway becomes stationary very quickly: at $t \approx 5.5$ s for $n_* = -1$ and at $t \approx 0.7$ s for $n_* = 2$.

4.5.3 Solution of the Unsteady Two-Dimensional Problem of Heat Conduction in a Disk

Equations of unsteady 2D heat conduction in a disk and boundary conditions are [192]

$$\frac{\partial \vartheta}{\partial Fo} = \frac{1}{H^2}\left(\frac{\partial^2 \vartheta}{\partial x^2} + \frac{1}{x}\frac{\partial \vartheta}{\partial x}\right) + \frac{\partial^2 \vartheta}{\partial y^2}, \tag{4.24}$$

$$Fo = 0: \quad \vartheta = x^{n_*}, \tag{4.25}$$

$$x = 0: \quad \frac{\partial \vartheta}{\partial x} = 0, \quad x = 1: \quad \left(\frac{\partial \vartheta}{\partial x}\right)_{x=1} = -Bi_1 \vartheta_{x=1}, \tag{4.26}$$

$$y = 0: \quad \frac{\partial \vartheta}{\partial y} = 0, \quad y = 1: \quad \left(\frac{\partial \vartheta}{\partial y}\right)_{y=1} = -Bi_2 \vartheta_{y=1}, \tag{4.27}$$

where, to remind, $y = z/(0.5s)$; $Bi_1 = \alpha_1 b/\lambda_w$; $Bi_2 = 0.5\alpha_2 s/\lambda_w$; $Fo = 4a_w t/s^2$; $H = b/(0.5s)$. It is important to point out that the temperature field is specified in the form of the function $\vartheta(t,y) = (T(t,y) - T_\infty)/c_{o*}$, where $c_{0*} = (T_{w,i} - T_\infty)_{n_*=0} = const$ has a sense of the radially constant difference between the disk surface temperature at the initial moment of time $T_{w,i}$ for $n_* = 0$ (or, which is the same, for $T_{w,i} = const$) and fluid temperature at the infinity T_∞.

Boundary condition (4.25) is in fact Eq. (2.35) which is rewritten in view of the new variables.

The system of Eqs. (4.24), (4.25), (4.26) and (4.27) is solved using the method of separation of variables, whose fundamentals as applied to the problems of the type considered in this section are given in the monograph [115]. The resulting solution has the following form:

$$\vartheta(Fo,x,y) = \sum_{n=1}^{\infty}\sum_{m=1}^{\infty} D_n E_m J_0(\mu_{xn} x)\cos(\mu_{ym} y)\exp\left[-\left(\mu_{xn}^2/H^2 + \mu_{ym}^2\right)Fo\right], \tag{4.28}$$

$$D_n = \frac{{}_1F_2(1 + n_*/2;\, 1,\, 2 + n_*/2;\, -\,\mu_{xn}^2/4)/(2 + n_*)}{0.5[J_0^2(\mu_{xn}) + J_1^2(\mu_{xn})]}, \tag{4.29}$$

$$\frac{J_1(\mu_{xn})}{J_0(\mu_{xn})} = \frac{Bi_1}{\mu_{xn}}. \tag{4.30}$$

Here, ${}_1F_2$ is a hypergeometric function of the argument $-\mu_{xn}^2/4$ [219]. Constant E_m and eigenvalues μ_{ym} are given by Eqs. (4.3). At $n_* = 0$ the numerator of

Eq. (4.29) is equal to $J_1(\mu_{xn})/\mu_{xn}$, and Eq. (4.28) coincides with the solution presented for such a case in [115]. It is obvious that by neglecting radial heat conduction effects one can reduce Eq. (4.28) to Eq. (4.2).

Using a thin-slab approximation, one can neglect temperature variation inside the disk in the y-direction and substitute the last term in Eq. (4.24) with the so-called source term $-\vartheta Bi_2$. In doing so, the final solution of Eq. (4.24) reduces to the following relation:

$$\vartheta(Fo,x) = \sum_{n=1}^{\infty} D_n J_0(\mu_{xn} x) \exp\left[-\left(\mu_{xn}^2/H^2 + Bi_2\right) Fo\right]. \qquad (4.31)$$

4.5.4 Analysis of the Solutions for Unsteady Heat Conduction in a Disk

In order to render the temporal dependencies of all the functions in the present paper, the most general appearance, a non-dimensional time in the form of the Fourier number was used. The interrelation between the real and non-dimensional time is: for the Plexiglas® disk $t = 230.17\ Fo$ and for the aluminium disk $t = 0.297\ Fo$.

Unsteady surface temperature distributions in an isothermal disk at $n_* = 0$. Detailed results for this case are documented in Sects. 4.1, 4.2, 4.3 and 4.4. Here, we present some new findings in comparison with what was found above. Variation of the non-dimensional disk surface temperature F_t with time is shown in Fig. 4.11. One-dimensional (4.2) and two-dimensional (4.28) solutions for a slab of a finite thickness coincide over the whole range of the variation of the Fourier number. Solution for a semi-infinite slab (4.1) agrees well with Eqs. (4.2) and (4.28) only up to a certain limiting value of Fo, which can be estimated as $Fo = 0.456$, 0.608 and 0.790, when the divergence between Eqs. (4.1) and (4.2) taken as a criterion is set

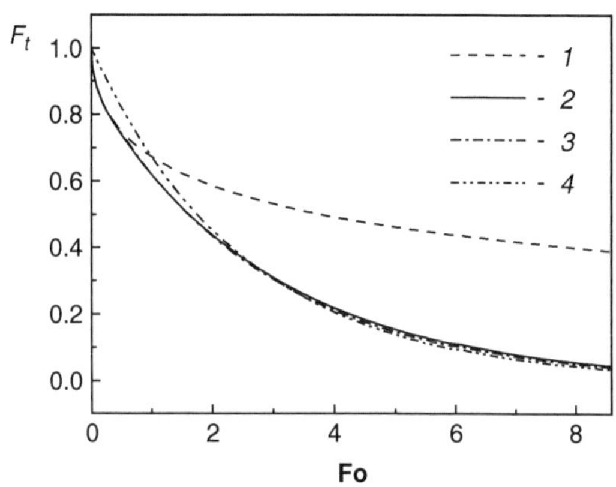

Fig. 4.11 Variation of F_t versus Fo at $n_* = 0$. *1* – Eq. (4.1); *2* – Eq. (4.2); *3* – Eq. (4.28); *4* – Eq. (4.31)

as 1, 2.5 and 5%, respectively. Having passed this threshold value of the argument *Fo*, data from Eq. (4.1) exceed more and more noticeably solutions (4.2) and (4.28). As said above, from the physical point of view this means that a finite-thickness disk cools down more and more rapidly than an infinite-thickness disk. Solution (4.31) for a thin disk deviates from Eq. (4.28) quite noticeably at $Fo<2$ and $Fo>4$. This means that the 0.01 m thick disk considered in this chapter as a basic geometry is insufficiently thin for the solution (4.31) to be valid. Results of additional computations showed that for numerical data from Eqs. (4.28) and (4.31) to coincide, the disk should be 0.001 m thick. Thus, solution (4.31) is of no use for the present research and will not be discussed further in this section.

Unsteady surface temperature distributions of a non-isothermal disk. At the beginning we will discuss the results obtained for the disk made of Plexiglas®. The results obtained based on solution (4.28) differ for the cases where n_* is moderately or strongly different from zero. Cases with $n_* = -1$ and 2 illustrated in Figs. 4.12 and 4.13 should be classified as those strongly different from zero. During the cooling process, curves of the normalized wall temperature $\vartheta_w/\vartheta_w(x = 1)$ seem to repeat the initial power-law distributions (4.25) for the most part of the disk. Plots of $\vartheta_w/\vartheta_w(x = 1)$ at $n_* = -0.25$, -0.5 and 0.5 and 1 (not shown here) behave in the same way.

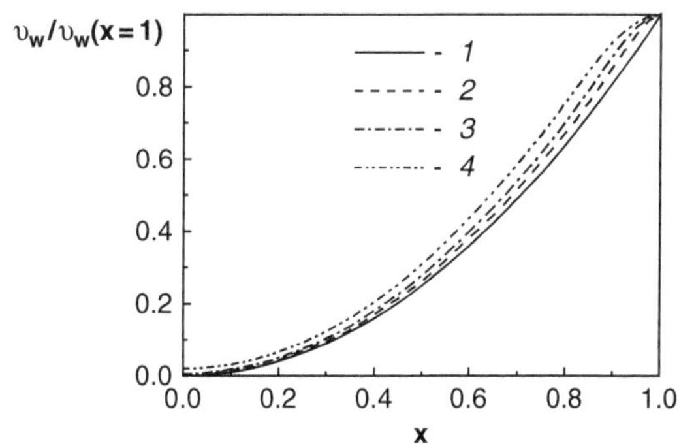

Fig. 4.12 Variation of the function $\vartheta_w/\vartheta_w(x = 1) = \vartheta(Fo,x,y = 1)/\vartheta(Fo,x = 1, y = 1)$ versus x according to Eq. (4.28) at $n_* = 2$. *1* – Eq. (4.25); *2* – $Fo = 0.261$; *3* – $Fo = 0.869$; *4* – $Fo = 2.607$

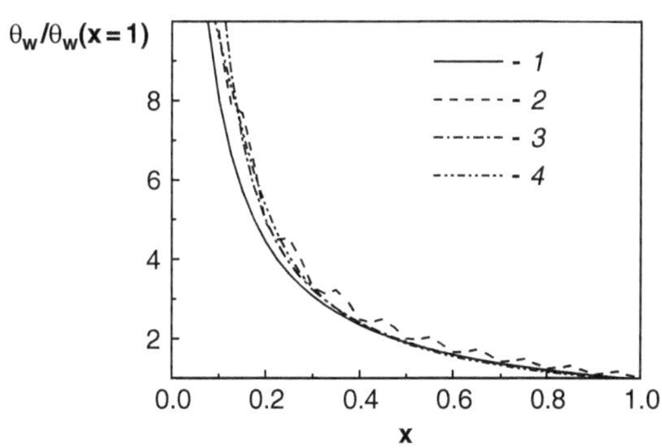

Fig. 4.13 Variation of the function $\vartheta_w/\vartheta_w(x = 1) = \vartheta(Fo,x,y = 1)/\vartheta(Fo,x = 1, y = 1)$ versus x according to Eq. (4.28) at $n_* = -1$. *1* – Eq. (4.25); *2* – $Fo = 0.00652$; *3* – $Fo = 0.869$; *5* – $Fo = 2.607$

There are two visible features of the plots of $\vartheta_w/\vartheta_w(x = 1)$ worth mentioning here.

First, strictly speaking, Eq. (4.25) does not agree with the boundary conditions (4.26) at the points $x = 0$ and $x = 1$. Hence, solution (4.28) adjusts itself to Eqs. (4.26) as *Fo* increases, thus creating instantaneous distributions of $\vartheta_w/\vartheta_w(x = 1)$ distorted in the neighbourhood of the points $x = 0$ and $x = 1$ as compared to Eq. (4.25). Hence, the temporal behaviour of the heat transfer coefficient $\alpha_{2,t}$ should be studied within a narrower region, say, at $x = 0.2$–0.9.

Second, at small Fourier numbers the aforementioned mathematical contradiction of Eqs. (4.25) and (4.26) causes oscillations of the temperature profiles visible in Figs. 4.13 and 4.14. This is a modest price to be paid for the possibility to operate with the boundary condition (4.25) as applied to the problem under investigation.

An opportunity to discern the differences between the cases with moderate and strong deviations of n_* from zero is provided in Fig. 4.14, where the instantaneous surface temperature profiles ϑ_w are plotted as divided by x^{n_*}. The enlarged scale of Fig. 4.14 shows that for the initial radial surface temperature distributions (4.25) moderately different from the isothermal case (e.g. $n_* = 0.5$) the non-stationary plots of ϑ_w/x^{n_*} look like horizontal straight lines over the region $x = 0.2$–0.95 (and even the initial oscillations at $Fo = 0.0261$ are almost negligible). This in turn means that conditions (4.5) and (4.6), which are necessary for the transient experiment to be valid, hold at $x = 0.2$–0.95 for the whole duration of the cooling process. When the temperature distributions (4.25) strongly deviate from the isothermal case (e.g. $n_* = 2$), the plots of ϑ_w/x^{n_*} at relatively small values of *Fo* (and even at $Fo = 0.261$) oscillate around some horizontal lines. However, following the further increase in the *Fo* number, the curves of ϑ_w/x^{n_*} become more noticeably inclined towards the outer edge of the disk (i.e. towards the value $x = 1$). The reason lies in the redistribution of heat owing to the radial heat conduction from the more heated outer part of the disk into its less heated inner part. As a result, one obtains a distorted (as compared to Eq. (4.25)) distribution of ϑ_w/x^{n_*} at high values of *Fo*, which does not comply in full with conditions (4.5) and (4.6). This phenomenon is more noticeable

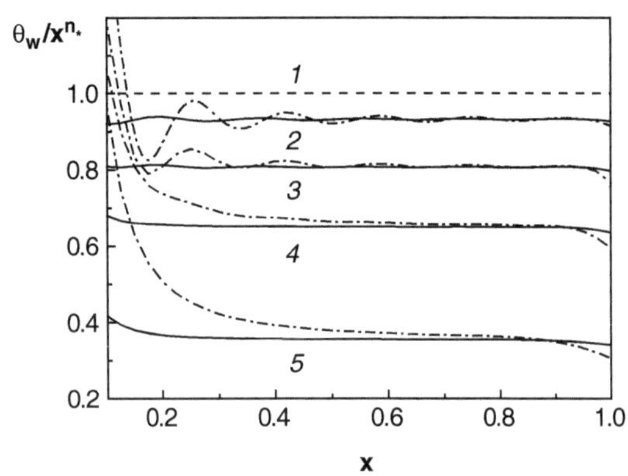

Fig. 4.14 Variation of the function $\vartheta_w/x^{n_*} = \vartheta(Fo,x,y = 1)/x^{n_*}$ versus x according to Eq. (4.28) at $n_* = 2$ (*dash-dotted lines*) and $n_* = 0.5$ (*solid lines*) for different values of *Fo*. *1* – Eq. (4.25); *2* – $Fo = 0.0261$; *3* – $Fo = 0.261$; *4* – $Fo = 0.869$; *5* – $Fo = 2.607$

for positive values of n_* than for negative ones, because for positive values of n_* the total amount of heat accumulated by the hotter part of the disk is larger (because the hotter part of the disk is volumetrically larger in this case).

Transient values of the heat transfer coefficient based on the semi-infinite slab solution. However, the target quantity in the present investigation is the heat transfer coefficient rather than the surface temperature. Values of the heat transfer coefficient computed with the help of Eqs. (4.1) and (4.2) for an isothermal disk at $n_* = 0$ are presented in Fig. 4.15.

Assuming Eq. (4.2) to be a true solution for unsteady heat transfer and specifying a constant value of α_2 in the boundary condition (4.27), one can equate Eqs. (4.1) and (4.2) and compute "transient" values of the heat transfer coefficient $\alpha_{2,t}$ from Eq. (4.1) depending on the Fourier number. In doing so, one can imitate the conditions that take place during transient experiments, with Eq. (4.2) playing a role of the "experimental" data source. The data obtained in such a way and plotted in Fig. 4.15 as curve *1* testify that the experimental technique based on the use of Eq. (4.1) can produce valid experimental data for $\alpha_{2,t}$ again up to a certain limiting value of the *Fo* number. The limiting value can be estimated as $Fo = 0.3$, 0.391 and 0.487 in order for the deviation of $\alpha_{2,t}$ from α_2 not to exceed 1, 2.5 and 5%, respectively. These threshold values are lower than those obtained above at the comparison of deviation of the values F_t computed by Eq. (4.1) from those obtained by Eq. (4.2). The explanation lies in the fact that an increase in the $\alpha_{2,t}$ values leads to a corresponding increase in the function $\exp(\gamma^2)$ and a decrease in the function $\mathrm{erfc}(\gamma)$, which form a product in Eq. (4.1) and thus mutually compensate in part the deviation of $\alpha_{2,t}$ from the pre-determined value α_2. Thus, based on the 1%-restriction in the error of $\alpha_{2,t}$, the limiting measurement time should be less than $Fo = 0.3$ or 69 s, which agree rather with the classical value $Fo = 1/4$ [160] than with the revised suggestion $Fo = 1$ [214].

Data for the instantaneous heat transfer coefficient $\alpha_{2,t}$, obtained for all input values of n_* from –1 to 2 investigated using the semi-infinite-slab solution and averaged over the region $x = 0.2$–0.9 to avoid the oscillations of the obtained solution at low

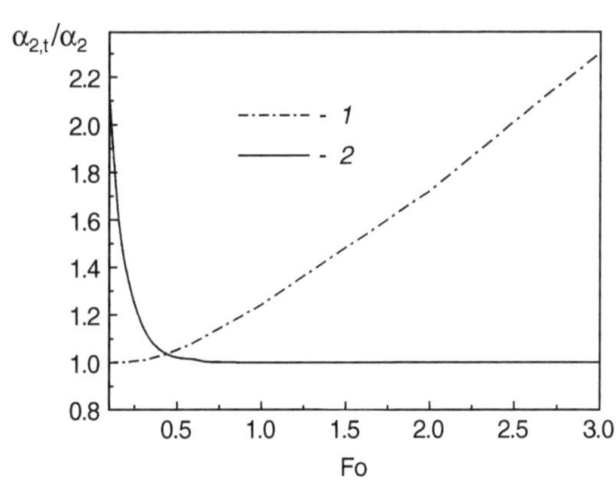

Fig. 4.15 Variation of the ratio $\alpha_{2,t}/\alpha_2$ depending on the Fourier number for $n_* = 0$. *1* – Eq. (4.1); *2* – Eq. (4.2)

Fo values, agree with maximal discrepancy of 1% with line *1* in Fig. 4.15. Thus, the fundamental conclusion from this finding is that the semi-infinite slab approach (4.1) is still valid as an experimental transient technique also for the cases with strong initial radial (i.e. streamwise) temperature gradients in the surface under investigation provided that the measurement time is restricted with the aforementioned limiting Fourier number values. Of course, this conclusion is valid for the disks made of Plexiglas®.

Transient values of the heat transfer coefficient based on the regular heat transfer regime theory. An alternative technique of determining heat transfer coefficients from the experimentally measured instantaneous distributions of the surface temperature is based on the theory of regular heat transfer regime [115]. The key statement of this theory is that with increasing Fourier number the series solutions (4.2) and (4.28) degenerate just to their first terms. Taking a derivative of logarithm of the first term of Eq. (4.28) with respect to time, one can obtain

$$-\frac{\partial\vartheta(Fo,x,1)}{\partial t} = \left(\mu_{xn}^2/H^2 + \mu_{y1}^2\right)\frac{a}{(s/2)^2} = m, \tag{4.32}$$

$$-\frac{\partial\vartheta(Fo,x,1)}{\partial t} = \mu_{y1}^2\frac{a}{(s/2)^2} = m \quad \text{for } Bi_1 = 0. \tag{4.33}$$

Solution (4.33) for the disk, whose outer rim is thermally insulated to avoid radial conduction effects in this location ($Bi_1 = 0$, $\mu_{x1} = 0$), is more convenient and will be used in further derivations. The regular regime of heat transfer takes place when the experimentally measured curve of the function $-\partial\vartheta(Fo,x,1)/\partial t$ becomes equal to a constant denoted further as m. Having experimentally found m and keeping in mind Eq. (4.3), one can easily deduce

$$\mu_{y1} = 0.5s\sqrt{m/a}, \quad \alpha_{2,t} = \frac{\lambda\sqrt{m/a}}{\mathrm{ctg}(0.5s\sqrt{m/a})}. \tag{4.34}$$

It is obvious that Eqs. (4.33) and (4.34) could have been obtained from the solutions (4.2) and (4.3) that disregard radial conduction effects. Assuming again that Eq. (4.2) is a true solution for the unsteady heat transfer at a constant value of α_2 in the boundary condition (4.27), instantaneous values of m and "transient" values of the heat transfer coefficient $\alpha_{2,\mathrm{t}}$ can be computed depending on the *Fo* number. Values of $\alpha_{2,\mathrm{t}}$ presented in Fig. 4.15 decrease quite rapidly. Already at $Fo = 0.543$ errors in determination of $\alpha_{2,\mathrm{t}}$ are equal to 1% and become negligible with further increase in the Fourier number.

The advantage of approach (4.34) consists in the fact that the regular heat transfer regime, once it is already established, imposes no restriction on the duration of measurements within the practical timescale providing visible differences between the surface and ambient temperatures. Within these time limits, the heat transfer coefficient $\alpha_{2,\mathrm{t}}$ remains constant.

At moderate radial surface temperature gradients, i.e. at n_* from –0.5 to 0.5, and for radial locations at $x = 0.25$–0.9, solution (4.34) as a basis for the experimental technique is still valid [192]. The curves of the radial distributions of the heat transfer coefficient $\alpha_{2,t}$ obtained using the regular heat transfer regime theory and plotted in Fig. 4.16 can prove this conclusion. For the values of n_* equal to –1 and 1, the region of the validity of Eq. (4.34) narrows down to $x = 0.45$–0.85, while for $n_* = 2$ approach Eq. (4.34) cannot be recommended at all. Thus, one can conclude that the regular heat transfer regime theory has restrictions caused by the radial heat conduction effects at high values of n_*, which distort the initial temperature distribution (4.25) and finally the computed radial curves for $\alpha_{2,t}$ at high values of *Fo*. A remedy may consist in use of the disks (or, in general, the objects of the experimental study) made of the materials with lower thermal conductivity in comparison with that of Plexiglas®.

Disk made of aluminium. In order to further illustrate the effect of the disk material, we have modelled unsteady cooling of a disk made of aluminium. From the very beginning of the cooling process even technique (4.1) is inapplicable, because already at $t \approx 1$ s the distribution of $\vartheta_w/\vartheta_w(x = 1)$ becomes radically distorted in comparison with that given by Eq. (4.25) because of the radial heat conduction (see Fig. 4.17).

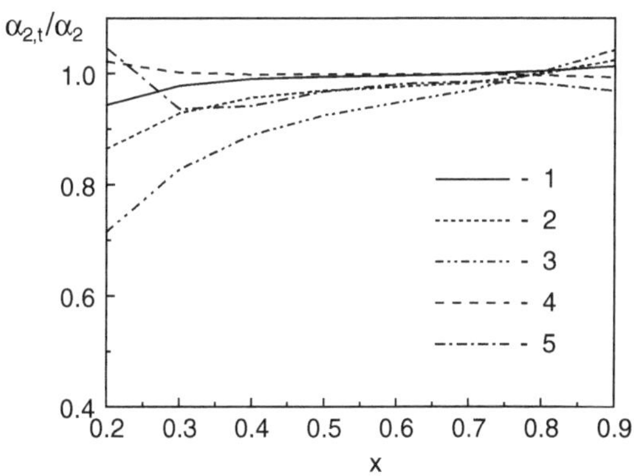

Fig. 4.16 Variation of the ratio $\alpha_2/\alpha_{2,t}$ versus x for $Fo = 5$. *1* – $n_* = 0.5$; *2* – $n_* = 1.1$; *3* – $n_* = 2$; *4* – $n_* = -0.25$; *4* – $n_* = -1$

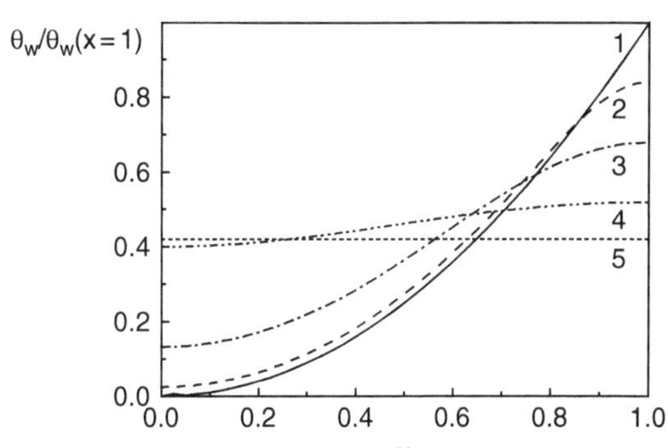

Fig. 4.17 Variation of the function $\vartheta_w/\vartheta_w(x = 1) = \vartheta(Fo,x,y = 1)/\vartheta(Fo,x = 1, y = 1)$ according to Eq. (4.28) versus x at $n_* = 2$ for an aluminium disk. *1* – Eq. (4.25); *2* – $Fo = 3.705$ ($t = 1.1$ s); *3* – $Fo = 20.208$ ($t = 6$ s); *4* – $Fo = 84.2$ ($t = 25$ s); *5* – $Fo = 471.52$ ($t = 140$ s)

As a result, one can obtain very high positive values of $\alpha_{2,t}$ at $x>0.85$ and very high negative values at $x<0.8$. A disk having initial temperature distribution with a value $n_* = 2$ becomes isothermal very rapidly.

As the most important findings of the research described in the present section, one can mention the following:

(1) A self-similar solution of transient laminar convective heat transfer obtained for an initially non-isothermal disk. Similarly to what was obtained earlier for an isothermal disk, the values of the heat transfer coefficient reached the steady-state values very rapidly (in maximum in 6 s).
(2) An analytical solution (4.28) of the unsteady two-dimensional heat conduction problem at the non-uniform initial temperature distribution (4.25).
(3) Confirmation that conditions (4.5) and (4.6) are necessary for the shape of the initial temperature non-uniformity to hold with time. These conditions are fulfilled in full for a disk made of Plexiglas® in the sense of the applicability of the semi-infinite-slab approach (4.1), which thus may be used as an experimental technique for determining the heat transfer coefficients $\alpha_{2,t}$ for all the studied values of n_* from –1 to 2, provided that the aforementioned limit of the measurement time is satisfied.
(4) Validation of a transient technique for the experimental determination of $\alpha_{2,t}$ based on the regular heat transfer regime theory. This technique is valid in full at n_* varying from –0.5 to 0.5, partially valid for n_* equal to –1 and 1, and cannot be recommended for $n_* \geq 2$.

Chapter 5
External Flow Imposed over a Rotating Disk

5.1 Rotation of a Disk in a Fluid Rotating as a Solid Body Without Imposed Radial Flow

5.1.1 Turbulent Flow

Geometrical arrangements with disks rotating with an angular velocity ω in a fluid, which rotates in the same direction with a velocity Ω different from ω, may be found in many technical applications such as parallel disks co-rotating with different angular velocities, disks rotating in swirling flows, etc. [114, 138, 139, 196]. If $\omega>\Omega$ (or $\beta=v_{\varphi,\infty}/(\omega r)<1$), then fluid flow on the disk surface caused by rotation is radially outward (centrifugal). A schematic diagram of such flow is outlined in Fig. 5.1. Physically, flow for $\beta<1$ looks like flow over a free disk for $\beta=0$, with quantitative relations for the velocity and temperature fields being dependent on a particular non-zero value of β.

If $\omega<\Omega$ (or $\beta>1$), fluid flow on the disk surface becomes radially inward (centripetal), and ultimately for $\beta\to\infty$ the problem reduces to the case of a rotating fluid over a stationary surface [138, 139, 158, 196]. However, this particular case falls out of the subject of the present research.

As well known, radial variation of the parameter β depends on a specific type of the problem under consideration. Two most widely occurring cases are the solid-body rotation with $\beta=\Omega/\omega=const$ (2.32) and free-vortex rotation with $\beta=const/r^2$ or $v_{\varphi,\infty}=(v_{\varphi,\infty}r)_i/r$ (this case is considered in detail in Chap. 6), though experimental and theoretical investigations have revealed some other forms of radial variations of the parameter β caused by specific flow patterns in different types of rotating-disk flows [138, 139, 188].

Finding an analytical solution of the given problem for $\beta=const$ is a problem of special interest and importance. As shown in [138, 139, 196], such an analytical solution can serve as a base for a development of an integral method, which in turn can be applied for cases with other arbitrary distributions of β.

For the considered case of solid-body rotation of fluid and thermal boundary conditions (2.34), (2.35) or (2.36), a self-similar solution of the Navier–Stokes and

I.V. Shevchuk, *Convective Heat and Mass Transfer in Rotating Disk Systems*, Lecture Notes in Applied and Computational Mechanics 45, DOI 10.1007/978-3-642-00718-7_5,
© Springer-Verlag Berlin Heidelberg 2009

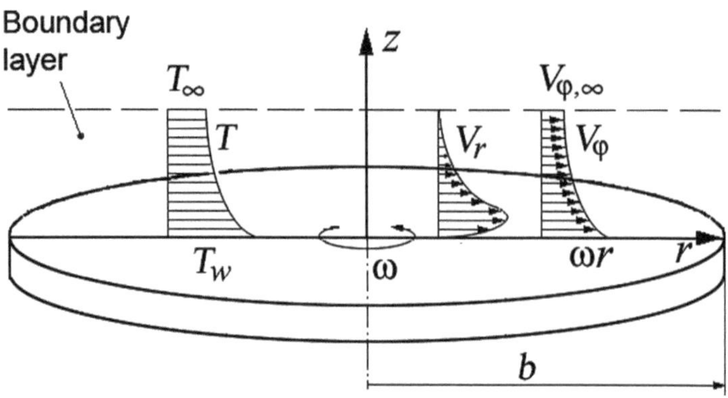

Fig. 5.1 Schematic outline of velocity and temperature profiles on a disk rotating in a rotating fluid for β<1. Tangential velocity component v_φ is turned 90 degree

energy equations can be found only for laminar flow (see Sect. 5.1.2). For turbulent flow modelled using an integral method for boundary layer equations under condition β=*const* with the help of approximations (2.45) and (2.46) for velocity profiles (with *n*=1/7), solutions (2.82), (2.83), (2.84), (2.85) and (2.86) remain in force. For the parameters α, γ, ε_m and ε_M to be substituted into Eqs. (2.82), (2.83), (2.84), (2.85) and (2.86), authors [130, 138] obtained the following solutions:

$$\alpha^2 = \frac{2300(1+8\beta)}{49(1789-409\beta)}, \quad \gamma_* = \left[\frac{81(1+\alpha^2)^{3/8}}{49(23+37\beta)\alpha}\right]^{4/5}, \tag{5.1}$$

$$\varepsilon_m = \frac{49\pi}{60}\mathrm{sgn}\,(1-\beta)\,|1-\beta|^{8/5}\,\alpha\gamma_*,$$
$$\varepsilon_M = \frac{49\pi}{4140}\mathrm{sgn}\,(1-\beta)\,|1-\beta|^{8/5}\,(23+73\beta)\,\alpha\gamma_*. \tag{5.2}$$

A similar integral method based on approximations (2.45) and (2.47) (also with *n*=1/7) allowed obtaining the following solutions for the same parameters in Eqs. (2.82), (2.83), (2.84), (2.85) and (2.86) for the condition β=*const*≤1 [196]:

$$\alpha^2 = \frac{0.2222(\beta+0.125)}{0.6374-0.1176}, \quad \gamma_* = 0.04809(1+\alpha^2)^{3/10}\,[\alpha\,(0.2425+0.2772\beta)]^{-4/5}, \tag{5.3}$$

$$\varepsilon_m = 1.6332\alpha(1-\beta)\gamma_*, \quad \varepsilon_M = 0.123(1+\alpha^2)^{0.375}\gamma_*^{-0.25}\,|\beta-1|^{1.75}. \tag{5.4}$$

The present integral method allowed obtaining an analytical solution of the boundary layer equations (2.73), (2.74) and (2.75) for boundary conditions (2.32), (2.33), (2.34), (2.35) and (2.36), which is analysed below. The solution for the momentum boundary layer has a form of Eqs. (2.82)–(2.92), while the solution for the thermal boundary layer is written in a form of Eqs. (2.93), (2.94) and (2.95). In these solutions, one should accept N=0 (or κ=0) for the case of pure solid-body rotation of fluid considered here. Solution (5.3), (5.4) is in fact a particular case of the solution (2.82)–(2.92) for N=0 and n=1/7.

As shown by the author of [163], according to Eq. (2.88) a solution for κ=0 exists only for $C_5 \leq \beta \leq H_8/H_7$. For n=1/7 these constants take the values C_5=–0.125 and H_8/H_7=5.412; for n=1/9 one can obtain C_5=–0.1 and H_8/H_7=6.026 (in [138] for n=1/7 these quantities take the values C_5=–0.125 and H_8/H_7=4.374). Constants α, γ, ε_m^* and ε_M computed according to Eqs. (2.88), (2.89), (2.90) and (2.91) [163, 196] are given in Table 5.1 in comparison with corresponding data by Eqs. (5.1) and (5.2) [138]. Like for a free rotating disk, values α computed with the help of the present integral method exceed data of [138], which however does not affect the moment coefficient (except for the region β>3, where solution [138] tends to infinity earlier).

For 0.6<β<3, constant ε_m^* (mass flowrate through the boundary layer) obtained by the two methods are close to each other. According to the definition of mass flowrate (see Nomenclature), this is obviously a result of two mutually compensating effects: smaller values of the definite integral of the profile of the radial velocity v_r in the boundary layer because of the smaller fullness of this profile (the present integral method, Fig. 2.3) are compensated by larger values of the boundary layer thickness (constant γ, Table 5.1).

For further development of a model of heat transfer for laminar flow, it is important to derive a relation for enthalpy thickness $\overline{\delta}_T^{**}$ based on the present integral method for turbulent flow (Sect. 2.4) for n=n_T. This relation has the following form [190]:

$$\overline{\delta}_T^{**} = -(1-\beta)\,\alpha\,(-A_1 + \chi Pr^{n_p} D_2). \tag{5.5}$$

For the currently considered case N=0, the analytical solution (2.93) for the thermal boundary layer for $\Delta \leq 1$ remains a transcendental equation, with correlations for constants E_1, E_2, C_4, and C_5 being simplified in comparison with the case $N \neq 0$.

Table 5.1 Constants in Eqs. (2.82), (2.83), (2.84), (2.85) and (2.86) for β=*const*, N=0

	Solution (2.88), (2.89), (2.90) and (2.91) [163, 196], n=1/7				Solution (5.1), (5.2) [138], n=1/7			
β	α	γ	ε_m^*	ε_M	α	γ	ε_m^*	ε_M
0	0.2087	0.5299	0.1806	0.1466	0.162	0.526	0.219	0.146
0.2	0.343	0.3084	0.1727	0.1205	0.267	0.286	0.196	0.1208
0.4	0.4445	0.2232	0.162	0.0845	0.348	0.197	0.1766	0.085
0.6	0.5331	0.1754	0.1527	0.0482	0.42	0.149	0.1603	0.0486
0.8	0.6151	0.1443	0.1449	0.0165	0.487	0.119	0.1446	0.0172
1	0.6936	0.1224	0.1386	0	0.553	0.0983	0.1395	0
2	1.084	0.0683	–0.1208	0.322	0.907	0.0498	0.116	0.326
3	1.564	0.046	–0.1175	1.28	1.445	0.031	–0.1151	1.356
4	2.348	0.0335	–0.1285	3.368	3.182	0.02	–0.1629	4.682

For the case $\Delta \geq 1$ that corresponds to the cases of air cooling in rotating-disk systems studied in the present work, the value of Δ can be expressed from Eq. (2.93) explicitly [163]

$$\Delta^{-n} = \left[\frac{(4+m)K_V Pr^{-n_p} + (2+m+n_*)(1-K_V) - Pr^{-n_p} 2\beta/(\beta-1)}{(2+m+n_*) - n_*\beta/(\beta-1)} \right]^{-1}, \tag{5.6}$$

where the parameter K_V is determined by Eq. (3.47).

For the conditions of the so-called Dorfman analogy [41], where $n_*=2$ and $Pr=1$, one can deduce from Eq. (5.6) that $\Delta=1$. Equations for the Stanton and Nusselt numbers, Eqs. (2.94) and (2.95), remain invariable. The form of Eq. (2.95) for the Nu number with allowance for Eq. (5.6) practically coincides for $n_*=2$ with the following solution:

$$Nu = C_n^{\frac{2}{n+1}} \left[\gamma\, |\beta - 1|^{(1-n)/(3n+1)} \right]^{\frac{2n}{n+1}} (1+\alpha^2)^{\frac{1-n}{2(n+1)}} |\beta - 1|^{(1-n)/(1+n)} Re_\omega^{(n+1)/(3n+1)} Pr, \tag{5.7}$$

obtained in [130] with the help of the integral method of Dorfman [41] (see Sect. 3.6.2) also for the case of $n_*=2$, $\beta=const$ and boundary equation (2.36). However, Eq. (5.7) does not take into account the effect of n_*, and the power exponent at the Prandtl number in this equation is obviously erroneous. Nevertheless, numerical values of the Nusselt number computed by Eqs. (5.7) and (2.95) for $n_*=2$, $Pr=1$, $\beta=idem$ and $n=idem$ differ by only 2–3% in spite of the differences in numerical values of the constants α and γ calculated for Eq. (5.7) by Eqs. (5.1), and for Eq. (2.95) by Eqs. (2.88) and (2.89).

Comparing relation (2.57) between the functions q_w and $\tau_{w\varphi}$ and relation (2.71) between the Stanton number and the friction coefficient, one can easily ascertain that the Reynolds analogy parameter χ in frames of the model used in the present work and its interrelation with the constant K_1 have the following form:

$$\chi = \Delta^{-n} Pr^{-n_p}, \quad K_1 = \chi A_c (1+\alpha^2)^{1/2} (1-\beta) Pr. \tag{5.8}$$

Then Eq. (5.6) valid for the boundary condition (2.36) can be transformed to the following expression for χ [190]:

$$\chi = \frac{(2+m+n_*) + n_*\beta/(1-\beta)}{(2+m+n_*)Pr^{n_p}(1-K_V) + (4+m)K_V + 2\beta/(1-\beta)}. \tag{5.9}$$

For the boundary condition (2.35), which is the main one used in the present work, Eq. (5.9) should be simplified as [190]

$$\chi = \frac{2+m+n_*}{(2+m+n_*)(1-K_V)Pr^{n_p} + (4+m)K_V + 2\beta/(1-\beta)}. \tag{5.10}$$

Effect of the swirl parameter β on the Reynolds analogy parameter χ in turbulent flow for four selected values of the power exponent n_* is illustrated in Fig. 5.2.

More detailed data for χ, along with the constant K_1, for Pr=0.71 and 1 and a series of discrete values of n_* over the range of n_*=–2.6–4 are given in Tables 5.2, 5.3, 5.4 and 5.5. As follows from these data, heat transfer rate decreases with an increasing β due to a decrease in the shear of the tangential velocity component, which is the driving force in convective transfer of momentum and heat in the currently considered physical problem. As usually, an increase in the exponent n_* leads to an increase in the heat transfer rate [190].

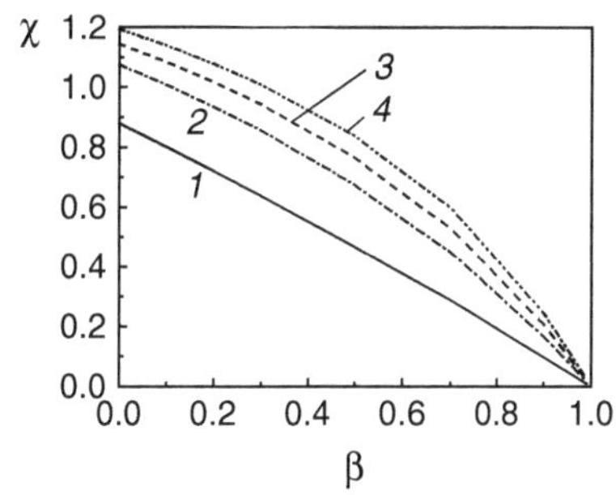

Fig. 5.2 Effect of the swirl parameter β on the Reynolds analogy parameter χ in turbulent flow for Pr=0.71. *1* – n_*=–0.6; *2* – n_*=1; *3* – n_*=2; *4* – n_*=3

Table 5.2 Constant K_1 in turbulent flow for Pr=1 and β=*const*

β	n_*=–2.6	n_*=–0.6	n_*=0	n_*=0.5	n_*=1	n_*=2	n_*=3	n_*=4
0.0	0.0	0.0212	0.0232	0.0244	0.0254	0.0268	0.0278	0.0286
0.1	0.0	0.0201	0.0222	0.0236	0.0247	0.0264	1.024	0.0284
0.2	0.0	0.0182	0.0204	0.0219	0.0231	0.0249	0.0262	0.0272
0.3	0.0	0.0159	0.0181	0.0196	0.0208	0.0227	0.0241	0.0252
0.5	0.0	0.0107	0.0126	0.0139	0.015	0.0169	0.0183	0.0195
0.7	0.0	0.0054	0.0065	0.00741	0.0082	0.0096	0.0107	0.0117
0.9	0.0	0.001	0.0013	0.00155	0.0018	0.0022	0.0026	0.0029
0.99	0.0	2.8×10^{-5}	3.6×10^{-5}	4.3×10^{-5}	5.0×10^{-5}	6.3×10^{-5}	8.3×10^{-5}	9.0×10^{-5}

Table 5.3 Parameter χ in turbulent flow for Pr=1 and β=*const*

β	n_*=–2.6	n_*=–0.6	n_*=0	n_*=0.5	n_*=1	n_*=2	n_*=3	n_*=4
0.0	0.0	0.7913	0.865	0.9106	0.9466	1	1.0376	1.0655
0.1	0.0	0.7273	0.8054	0.8548	0.8944	0.9539	0.9966	1.0286
0.2	0.0	0.6606	0.7416	0.794	0.8366	0.902	0.9496	0.9859
0.3	0.0	0.5909	0.6731	0.7274	0.7725	0.8429	0.8954	0.936
0.5	0.0	0.4417	0.5194	0.5736	0.6204	0.697	0.7571	0.8054
0.7	0.0	0.278	0.3389	0.3841	0.4251	0.4964	0.5565	0.6077
0.9	0.0	0.013	0.1238	0.1448	0.1651	0.2035	0.2393	0.2728
0.99	0.0	0.01	0.0129	0.0154	0.0178	0.0227	0.0275	0.0323

Table 5.4 Constant K_1 in turbulent flow for Pr=0.71 and β=*const*

β	n_*=–2.6	n_*=–0.6	n_*=0	n_*=0.5	n_*=1	n_*=2	n_*=3	n_*=4
0.0	0.0	0.0167	0.0185	0.0196	0.0218	0.0218	0.0227	0.0234
0.1	0.0	0.0157	0.0176	0.0188	0.0198	0.0213	0.0223	0.0232
0.2	0.0	0.0141	0.016	0.0173	0.0183	0.0199	0.0211	0.022
0.3	0.0	0.0122	0.0141	0.0153	0.0164	0.0180	0.0193	0.0203
0.5	0.0	0.008	0.0095	0.0106	0.0116	0.0131	0.0144	0.0154
0.7	0.0	0.0039	0.354	0.00553	0.0062	0.0073	0.0082	0.009
0.9	0.0	0.00075	0.00096	0.00112	0.0013	0.0016	0.0019	0.0022
0.99	0.0	2.0×10^{-5}	2.6×10^{-5}	3.0×10^{-5}	3.5×10^{-5}	4.5×10^{-5}	5.4×10^{-5}	6.4×10^{-5}

Table 5.5 Parameter χ in turbulent flow for Pr=0.71 and β=*const*

β	n_*=–2.6	n_*=–0.6	n_*=0	n_*=0.5	n_*=1	n_*=2	n_*=3	n_*=4
0.0	0.0	0.8788	0.9707	1.0285	1.0747	1.144	1.1935	1.2305
0.1	0.0	0.8006	0.8963	0.9579	1.0078	1.0841	1.1395	1.1816
0.2	0.0	0.7205	0.818	0.8821	0.9351	1.0175	1.0785	1.1256
0.3	0.0	0.6384	0.7354	0.8008	0.8557	0.943	1.0091	1.061
0.5	0.0	0.4677	0.5557	0.6183	0.6729	0.764	0.8368	0.8963
0.7	0.0	0.2881	0.354	0.4036	0.4491	0.5295	0.5984	0.658
0.9	0.0	0.0986	0.1257	0.1475	0.1686	0.2089	0.2468	0.2825
0.99	0.0	0.01	0.013	0.0154	0.0179	0.0228	0.0276	0.0325

5.1.2 Laminar Flow

Exact solution. Results of an exact solution of the problem of laminar heat transfer of a disk rotating in a co-rotating fluid for boundary conditions (2.32), (2.33), (2.34) and (2.35) and $\beta<1$ obtained in works [190] are discussed below. As said above, an exact solution of the considered problem is possible to obtain numerically with the help of the self-similar (with respect to r) functions (2.31) for a=0. As a result, the problem is reduced to the system of ordinary differential equations (2.37), (2.38), (2.39), (2.40) and (2.41) for N=0.

Respective boundary conditions take the following form:

$$\begin{aligned} &\zeta = 0: \quad F = H = 0, \quad G = 1, \quad \theta = 1, \\ &\zeta \to \infty: \quad G = \beta, \quad F = 0, \quad H \to const, \quad \theta = 0. \end{aligned} \tag{5.11}$$

The Nusselt numbers are calculated by Eqs. (3.4) for n_R=1/2, while the constant K_1 is given by Eq. (3.5). The numerical solution of Eqs. (2.37), (2.38), (2.39), (2.40) and (2.41) was performed using MathCAD software.

Profiles of the velocity components F, G, H and the dimensionless temperature θ are presented in Figs. 5.3, 5.4, 5.5 and 5.6 for different values of β. The radial velocity F has a form inherent to wall jets (Fig. 5.4), while the axial velocity H (Fig. 5.5) is negative, i.e. directed towards the disk surface. Temperature profiles θ (Fig. 5.6) look similar to the tangential velocity G [190].

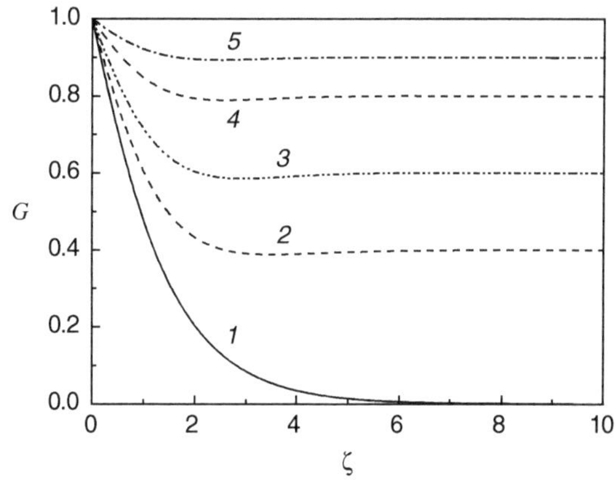

Fig. 5.3 Profiles of the tangential velocity component *G*: *1* – β=0; *2* – 0.4; *3* – 0.6; *4* – 0.8; *5* – 0.9

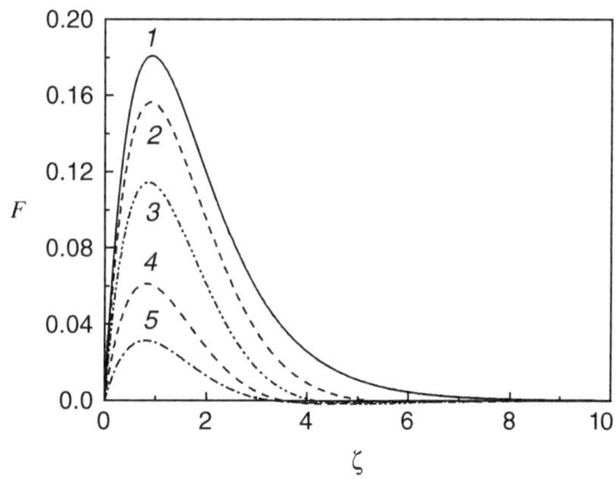

Fig. 5.4 Profiles of the radial velocity component *F*: *1* – β=0; *2* – 0.4; *3* – 0.6; *4* – 0.8; *5* – 0.9

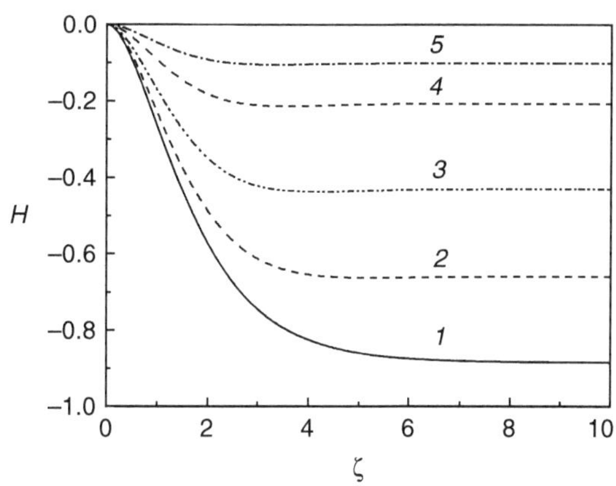

Fig. 5.5 Profiles of the axial velocity component *H*: *1* – β=0; *2* – 0.4; *3* – 0.6; *4* – 0.8; *5* – 0.9

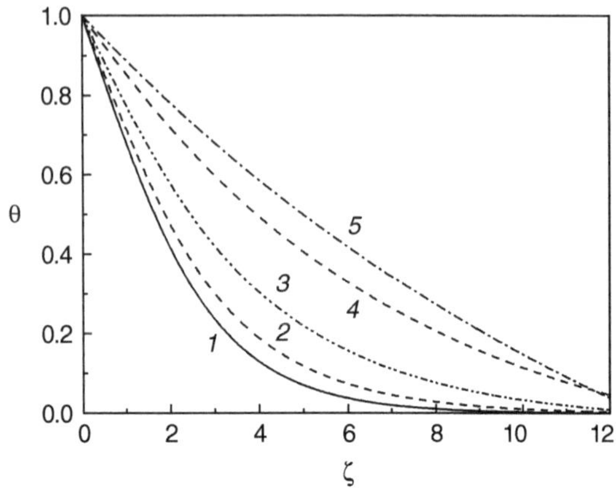

Fig. 5.6 Dimensionless temperature θ profiles: *1* – β=0; *2* – 0.4; *3* – 0.6; *4* – 0.8; *5* – 0.9

To remind, for the physical problem under consideration, shear of the tangential velocity G (Fig. 5.3) is the factor that drives flow. Increase in the swirl parameter β or, in other words, in the value of the angular velocity of the fluid rotation Ω reduces the shear of the tangential velocity G and, therefore, intensity of radial and axial flow over the disk as well as near-wall gradient of the temperature profile θ (Figs. 5.4, 5.5 and 5.6) [190].

In Tables 5.6 and 5.7, data for parameters of laminar flow are given, which are needed to calculate the following quantities:

$$\tau_{w\varphi}/(\rho\omega^2r^2) = -G'_w Re_\omega^{-1/2}, \quad \tau_{wr}/(\rho\omega^2r^2) = F'_w Re_\omega^{-1/2},$$
$$\dot{m}_d/(\mu r) = -\pi H_\infty Re_\omega^{1/2}, \quad C_M = -2\pi G'_w Re_\varphi^{-1/2}. \tag{5.12}$$

Thus, an increase in the swirl parameter β leads to a decrease in the wall values of the shear stresses and mass flowrate through the boundary layer. The data in

Table 5.6 Parameters of laminar flow for N=0 and β=$const$≤1

β	F'_w	$-G'_w$	α	$\tau_{w\varphi}/\tau_{w\varphi 0}$	$-H_\infty$
0.0	0.5102	0.6159	0.8284	1.0	0.8845
0.1	0.5134	0.6016	0.8535	0.9767	0.9176
0.2	0.5019	0.5721	0.8773	0.9289	0.8617
0.3	0.4769	0.5305	0.8989	0.8614	0.7687
0.4	0.4396	0.4787	0.9183	0.7772	0.6600
0.5	0.3908	0.4177	0.9357	0.6782	0.5457
0.6	0.3315	0.3484	0.9513	0.5657	0.4308
0.7	0.2622	0.2716	0.9654	0.4409	0.3178
0.8	0.1835	0.1876	0.9780	0.3046	0.2080
0.9	0.09594	0.09695	0.9895	0.1574	0.1020
1.0	0.0	0.0	1.0	0.0	0.0

Table 5.7 Parameters of laminar flow for N=0 and β=$const$≥1

β^{-1}	$-\beta^{-3/2}F'_w$	$\beta^{-3/2}G'_w$	α	$-\tau_{w\varphi}/\tau_{w\varphi 0}$	$\beta^{-3/2}H_\infty$
0.0	0.9420	0.7729	1.2188	1.2549	1.3696
0.1	0.8449	0.7184	1.1761	36.8861	0.1199
0.2	0.7517	0.6584	1.1417	11.952	0.2077
0.3	0.6602	0.5931	1.1132	5.8601	0.2633
0.4	0.5691	0.5225	1.0892	3.3533	0.2915
0.5	0.4776	0.4469	1.0688	2.0522	0.2935
0.6	0.3852	0.3664	1.0512	1.2801	0.2724
0.7	0.2915	0.2814	1.0359	0.7800	0.2307
0.8	0.1961	0.1918	1.0225	0.4352	0.1702
0.9	0.09901	0.09798	1.0106	0.1863	0.07778
1.0	0.0	0.0	1.0	0.0	0.0

Tables 5.6 and 5.7 agree with and complete analogous results obtained in works [138, 139] (in the first line of Table 5.7, factor $\beta^{-3/2}$ is omitted).

Surface heat transfer is also deteriorated with an increase in β. The problem of heat transfer solved in work [190] is considered below while describing the integral method.

Integral method: fluid flow. An integral method developed based on the exact self-similar solution allowed obtaining an approximate analytical solution of the problem considered in this section [190]. The integral method is based on the very same ideology as that described in Sect. 2.4.1, which means that a form of the solution already found for turbulent flow is extended onto the case of laminar flow by way of assigning few constants of the solution a status of free parameters to be found empirically via comparing and agreeing them with the self-similar solution [190].

Integral equations for the momentum boundary layer have a form of Eqs. (2.22) and (2.23). These equations are the same both for laminar and turbulent flow.

They are completed with models for the velocity profiles [190], which are in fact modifications of Eqs. (2.43):

$$\overline{\mathrm{v}}_\varphi = 1 - g, \quad \overline{\mathrm{v}}_r = \frac{\alpha}{\alpha_0} f. \tag{5.13}$$

Here g and f are the self-similar functions of the only variable ξ. Parameter α stands for the wall value of the tangent of the flow swirl angle; α_0=0.8284 (a free rotating disk). It follows from here that

$$g = \frac{\beta - G}{\beta - 1}, \quad f = \frac{\alpha_0 F}{\alpha\,(1-\beta)}. \tag{5.14}$$

Though being self-similar with respect to r, functions g and f (together with G and F) as well as integral parameters computed based on these functions depend parametrically on the value of β. An allowance for the effect of β on the functions

g and f is a principal difference of the present integral method from the method of authors [138, 139]. In the works [138, 139], the functions g and f were calculated using data for a free disk and were assumed to be independent of β. This assumption, as shown beneath, leads to noticeable inaccuracies compared to the exact solution.

It should also be emphasized that a fundamental assumption of the present integral method is that the velocity profiles g and f are universally the same when plotted versus the coordinate $\xi = z/\delta$ for β=*idem*. Using the coordinate ξ instead of ζ allows avoiding the effect of the boundary layer thickness δ on the profiles of g and f. Parameter δ thus becomes an unknown to be found, together with the unknown α, via a solution of the integral boundary layer equations.

After integration with account of model (5.13), Eqs. (2.22) and (2.23) accept a form still valid both for laminar and turbulent flow:

$$\frac{d}{dr}\left\{\delta\omega^2 r^3 (1-\beta)^2 \left(-B_1\alpha^2\right)\right\} + \delta\,(\omega r)^2 (\beta - 1)\; C_3\,(\beta - C_5) = r\tau_{wr}/\rho, \quad (5.15)$$

$$\frac{d}{dr}\left\{\delta\omega^2 r^4 (\beta-1)(1-\beta)\,\alpha D_1\right\} + \delta\omega r^2(\beta-1)\,\alpha A_1 \frac{d}{dr}\left(\beta\omega r^2\right) = r^2\tau_{w\varphi}/\rho. \quad (5.16)$$

Equations (5.15) and (5.16) are in fact somewhat transformed to Eqs. (2.73) and (2.74), respectively, valid for the case of $\kappa=0$ and $N=0$. Here

$$A_1 = \frac{1}{\alpha_0}\int_0^1 f\,d\xi, \quad A_3 = \int_0^1 g\,d\xi, \quad B_1 = \frac{1}{\alpha_0^2}\int_0^1 f^2\,d\xi, \quad (5.17)$$

$$C_1 = \int_0^1 g^2\,d\xi, \quad D_1 = \frac{1}{\alpha_0}\int_0^1 f g\,d\xi, \quad (5.18)$$

$$C_3 = C_1 - 2A_3, \quad C_5 = C_1/C_3. \quad (5.19)$$

For turbulent flow, functions

$$g = 1 - \xi^n, \quad f = \alpha_0\xi^n(1-\xi)^2, \quad (5.20)$$

the first of which coincides with Eq. (2.45), while the second is a modified equation (2.47), are obviously independent of both r and β. Relations for constants as well as for shear stress components in Eqs. (5.15) and (5.16) for turbulent flow are given in Sect. 2.4.

For the further use in the laminar flow case, all above constants are reduced to free disk conditions by way of multiplying each of them with $\zeta_{\delta 0} = \delta_0\sqrt{\omega/\nu}$, so that

$$a_1 = A_1\zeta_{\delta 0}, \quad a_3 = A_3\zeta_{\delta 0}, \quad b_1 = B_1\zeta_{\delta 0}, \quad c_1 = C_1\zeta_{\delta 0}, \quad d_1 = D_1\zeta_{\delta 0}, \quad (5.21)$$

$$c_3 = C_3\zeta_{\delta 0} = c_1 - 2a_3, \quad c_5 = C_5\zeta_{\delta 0} = c_1/c_3. \quad (5.22)$$

Here δ_0 is the boundary layer thickness over a free disk (β=0).

In the numerical solution of Eqs. (2.37), (2.38), (2.39) and (2.40), finding of the boundary layer thickness δ depends on accepted inaccuracy of determination of the values F and G at the outer edge of the boundary layer in comparison with the values specified in boundary conditions (5.11). Therefore, integration of functions F and G in the present work is performed using the variable $\zeta = z\sqrt{\omega/\nu}$ instead of ξ. The upper limit of the integration $\zeta_\delta = \delta\sqrt{\omega/\nu}$ is chosen large enough to provide invariable integral values within the limits of four meaningful digits. It is obvious that here $\xi = z/\delta = \zeta/\zeta_\delta$ and $\zeta_\delta/\zeta_{\delta 0} = \delta/\delta_0$.

Values of the constants defined in Eqs. (5.17), (5.18), (5.19), (5.21) and (5.22) are computed based on the exact solution of the system (2.37), (2.38), (2.39) and (2.40) in view of relations (5.14)

$$\begin{aligned}
&\tfrac{a_1\delta}{\delta_0} = A_1\zeta_\delta = \int_0^{\zeta_\delta} \tfrac{F}{(1-\beta)\alpha}\,d\zeta, \quad \tfrac{b_1\delta}{\delta_0} = B_1\zeta_\delta = \int_0^{\zeta_\delta} \tfrac{F^2}{(1-\beta)^2\alpha^2}\,d\zeta,\\
&\tfrac{a_3\delta}{\delta_0} = A_3\zeta_\delta = \int_0^{\zeta_\delta} \tfrac{\beta-G}{\beta-1}\,d\zeta, \quad \tfrac{c_1\delta}{\delta_0} = C_1\zeta_\delta = \int_0^{\zeta_\delta} \tfrac{(\beta-G)^2}{(\beta-1)^2}\,d\zeta,\\
&\tfrac{d_1\delta}{\delta_0} = D_1\zeta_\delta = \int_0^{\zeta_\delta} \tfrac{F(G-\beta)}{(\beta-1)^2\alpha}\,d\zeta.
\end{aligned} \tag{5.23}$$

The right-hand sides of Eqs. (5.23) may be calculated with the highest desirable precision and therefore they are constant for any value of β.

Shear stress components can be expressed in the following way:

$$\tau_{w\varphi} = \frac{\mu\omega r}{\delta}\cdot\left(\frac{dg}{d\xi}\right)_{\xi=0} = \mu\omega r\left(\frac{\omega}{\nu}\right)^{1/2}\cdot\left(\frac{dG}{d\zeta}\right)_{\zeta=0}, \tag{5.24}$$

$$\tau_{wr} = \frac{\mu\omega r}{\delta}\cdot\left(\frac{df}{d\xi}\right)_{\xi=0} = \mu\omega r\left(\frac{\omega}{\nu}\right)^{1/2}\cdot\left(\frac{dF}{d\zeta}\right)_{\zeta=0}, \tag{5.25}$$

$$\frac{\tau_{wr}}{\tau_{w\varphi}} = \left(\frac{df}{d\xi}\right)_{\xi=0}\Big/\left(\frac{dg}{d\xi}\right)_{\xi=0} = \left(\frac{dF}{d\zeta}\right)_{\zeta=0}\Big/\left(\frac{dG}{d\zeta}\right)_{\zeta=0} = -\,\alpha. \tag{5.26}$$

It is accepted here that $(dg/d\xi)_{\xi=0} = const$ does not depend on β. It follows then from Eqs. (5.24) and (5.25) that

$$(dF/d\zeta)_{\zeta=0} = 0.5102\tau_{wr}/\tau_{wr0} = 0.5102(\alpha/\alpha_0)(\tau_{w\varphi}/\tau_{w\varphi 0}), \tag{5.27}$$

$$(dG/d\zeta)_{\zeta=0} = -0.6159\tau_{w\varphi}/\tau_{w\varphi 0} = -0.6159\delta_0/\delta, \tag{5.28}$$

in view of the fact that, according to Eq. (3.1), the following is true: $(dG/d\zeta)_{\zeta=0} = -0.6159$ and $(dF/d\zeta)_{\zeta=0} = 0.5102$ for β=0.

Accepting as usually that α=*const*, δ=*const* and solving Eqs. (5.15) and (5.16), one can finally obtain

$$\alpha = \left[-c_3(\beta - c_5)/(\beta h_7 + h_8)\right]^{1/2}, \tag{5.29}$$

$$h_7 = -3b_1 - 4d_1 + 2a_1, \quad h_8 = 3b_1 + 4d_1, \tag{5.30}$$

$$\delta/\delta_0 = \tau_{w\varphi 0}/\tau_{w\varphi}, \tag{5.31}$$

$$\frac{\tau_{w\varphi}}{\tau_{w\varphi 0}} = \frac{\delta_0}{\delta} = \left[(1-\beta)\frac{\alpha}{\alpha_0} \cdot \frac{d_1 + \beta(0.5a_1 - d_1)}{d_{1,0}}\right]^{1/2}. \tag{5.32}$$

Values of the main parameters of this problem computed based on the self-similar solution are presented in Table 5.8.

Values of α and $\tau_{w\varphi}/\tau_{w\varphi 0}$ were determined by Eqs. (5.26) and (5.28), while constants a_1, a_3, b_1, c_1 and d_1 were found via numerical integration of the velocity profiles using Eqs. (5.23) with allowance for Eq. (5.31) for δ/δ_0 [190]. It may be concluded from Table 5.8 that constants a_1, a_3, b_1, c_1 and d_1 vary significantly with β, and taking into account this variation is obligatory for accurate predictions of α and $\tau_{w\varphi}/\tau_{w\varphi 0}$ and other parameters of the problem.

In order to provide convenience in use of the integral method, constants a_1, a_3, b_1, c_1 and d_1 were approximated by the fourth-order polynomials with respect to β:

Table 5.8 Values α, $\tau_{w\varphi}/\tau_{w\varphi 0}$ a_1, a_3, b_1, c_1 and d_1 based on the exact solution and the integral method (bold font)

β	α	$\tau_{w\varphi}/\tau_{w\varphi 0}$	a_1	a_3	b_1	c_1	d_1	d_2	K_V
0.0	0.8284	1	0.5338	1.2715	0.07887	0.6726	0.18587	0.34793	0.3482
	0.8284	**1**	**0.5338**	**1.2715**	**0.07887**	**0.6726**	**0.18587**	**0.34793**	**0.3482**
0.1	0.8535	0.9767	0.5834	1.0744	0.09580	0.5897	0.18004	0.40334	0.30862
	0.8570	**0.9745**	**0.5751**	**1.0744**	**0.09534**	**0.5902**	**0.17872**	**0.39639**	**0.31076**
0.2	0.8773	0.9289	0.5702	0.9007	0.10221	0.5141	0.16531	0.40490	0.28991
	0.8773	**0.9288**	**0.5702**	**0.9007**	**0.10221**	**0.5141**	**0.16531**	**0.4049**	**0.28991**
0.3	0.8989	0.8614	0.5261	0.7469	0.10109	0.4433	0.14662	0.37949	0.27869
	0.8952	**0.8634**	**0.5323**	**0.7467**	**0.10157**	**0.4429**	**0.14761**	**0.38468**	**0.27732**
0.4	0.9183	0.7772	0.4655	0.6096	0.09463	0.3757	0.12616	0.33929	0.27105
	0.9130	**0.7796**	**0.4724**	**0.6093**	**0.09522**	**0.3752**	**0.12728**	**0.34513**	**0.26942**
0.5	0.9357	0.6782	0.3955	0.4856	0.08429	0.3104	0.10497	0.29051	0.26543
	0.9318	**0.6797**	**0.3995**	**0.4854**	**0.08464**	**0.3101**	**0.10561**	**0.29386**	**0.26437**
0.6	0.9513	0.5657	0.3202	0.3728	0.07105	0.2467	0.08359	0.23662	0.26106
	0.9514	**0.5657**	**0.3202**	**0.3728**	**0.07105**	**0.2467**	**0.08359**	**0.23662**	**0.26105**
0.7	0.9654	0.4409	0.2419	0.2691	0.05560	0.1841	0.06230	0.17961	0.25753
	0.9705	**0.4398**	**0.2393**	**0.2693**	**0.05533**	**0.1843**	**0.06187**	**0.17738**	**0.25860**
0.8	0.9780	0.3046	0.1619	0.1731	0.03840	0.1223	0.04123	0.12071	0.25461
	0.9869	**0.3033**	**0.1591**	**0.1734**	**0.03810**	**0.1225**	**0.04077**	**0.11829**	**0.25630**
0.9	0.9895	0.1574	0.0811	0.08374	0.01978	0.0610	0.02046	0.06068	0.25212
	0.9975	**0.1569**	**0.0799**	**0.08385**	**0.01965**	**0.0611**	**0.02026**	**0.05966**	**0.25350**
0.99	0.9990	0.0162	0.0081	0.00814	0.00203	0.00609	0.00203	0.00609	0.25016
	0.9990	**0.0162**	**0.0081**	**0.00814**	**0.00203**	**0.00609**	**0.00203**	**0.00609**	**0.25015**
0.999	0.9999	0.0016	8.1×10^{-4}	8.1×10^{-4}	2.0×10^{-3}	6.1×10^{-3}	2.03×10^{-3}	6.1×10^{-4}	0.25062
	1.0011	**0.0016**	$\mathbf{8.1\times10^{-4}}$	$\mathbf{8.1\times10^{-4}}$	$\mathbf{2.0\times10^{-3}}$	$\mathbf{6.1\times10^{-3}}$	$\mathbf{2.03\times10^{-3}}$	$\mathbf{6.1\times10^{-4}}$	**0.25062**

$$a_1 = \sum_0^4 a_{1,i}\beta^i,\ a_3 = \sum_0^4 a_{3,i}\beta^i,\ b_1 = \sum_0^4 b_{1,i}\beta^i,\ c_1 = \sum_0^4 c_{1,i}\beta^i,\ d_1 = \sum_0^4 d_{1,i}\beta^i. \tag{5.33}$$

Values of the polynomial coefficients are $a_{1,i} = 0.5338, 0.6936, -3.0692, 2.7371, -0.8952$; $a_{3,i} = 1.2715, -2.1025, 1.3781, -0.7233, 0.1761$; $b_{1,i} = 0.07887, 0.2204, -0.5981, 0.4228, -0.1239$; $c_{1,i} = 0.6726, -0.8612, 0.3994, -0.2969, 0.08607$; $d_{1,i} = 0.1859, -0.03277, -0.4271, 0.4124, -0.1384$ for i=0, 1, 2, 3, 4, respectively.

Approximate values of Eqs. (5.33) were substituted into Eq. (5.29) for α and Eq. (5.32) for $\tau_{w\varphi}/\tau_{w\varphi0}$. Approximate values of α and $\tau_{w\varphi}/\tau_{w\varphi0}$ computed in such a way in [190] are given in Table 5.8 in boldface. In this Table, the second line of values for each parameter for a given β (bold font) corresponds to the approximate solution, Eqs. (5.29), (5.30), (5.31), (5.32) and (5.33). It is obvious that inaccuracies of the approximate solution are very small and do not exceed 1%.

For the comparison purpose, Fig. 5.7 represents data of an approximate zeroth-order solution for α and $\tau_{w\varphi}/\tau_{w\varphi0}$, Eqs. (5.29) and (5.32), based on an assumption that the profiles of g and f are independent of β. From the mathematical point of view, this means that the coefficients a_1, a_3, b_1, c_1 and d_1 were computed by Eqs. (5.33) with allowance for just the first terms of the polynomial power series (at i=0). It is obvious that inaccuracies of such a solution are very high and reach 34% for α and 42% for $\tau_{w\varphi}/\tau_{w\varphi0}$ for $\beta \leq 0.5$; for $\beta \rightarrow 1$ these inaccuracies are equal to 54.3% for α and 3200% for $\tau_{w\varphi}/\tau_{w\varphi0}$.

Integral method: heat transfer. The integral equation (2.75) of the thermal boundary layer is solved together with a transformed equation (5.16) [190]

$$(4+m)(1-\beta)\alpha D_1\delta/r + 2\beta\alpha A_1\delta/r = \frac{c_f}{2}(1-\beta)(1+\alpha^2)^{1/2}, \tag{5.34}$$

$$\frac{d}{dr}\left[r^2\delta\bar{\delta}_T^{***}\left(T_w - T_\infty\right)\right] + \frac{dT_\infty}{dr}\delta r^2\int_0^1 \frac{v_r}{\omega r}d\xi = \chi\frac{c_f}{2}r^2(1-\beta)(1+\alpha^2)^{1/2}\left(T_w - T_\infty\right). \tag{5.35}$$

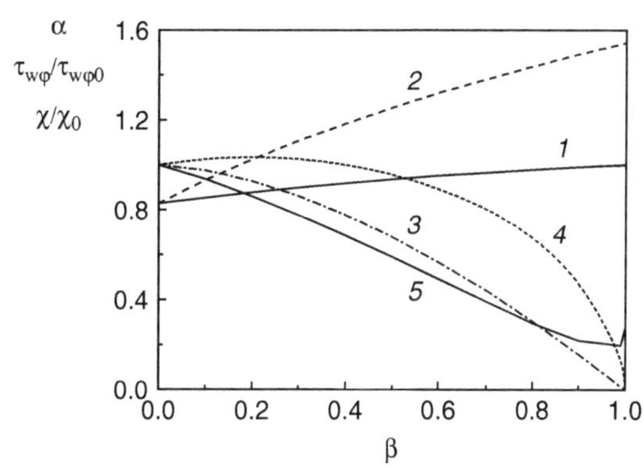

Fig. 5.7 Effect of the swirl parameter β on functions α (curves *1* and *2*), $\tau_{w\varphi}/\tau_{w\varphi0}$ (curves *3* and *4*) and χ/χ_0 (curve *5*) computed based on the exact solution (*1*, *3* and *5*) and approximate zeroth-order approximation (*2* and *4*)

For laminar flow, a form of the approximate relation for the enthalpy thickness $\overline{\delta_T^{**}}$ is found from generalization of the "turbulent" solution, Eq. (5.5), at the expense of implementation of coefficients b_1 and b_2 [190]:

$$\overline{\delta_T^{**}} = -(1-\beta)\alpha(-A_1/b_2 + \chi Pr^{n_p} D_2 b_1/b_2). \tag{5.36}$$

Constants b_1 (do not confuse with Eq. (5.21)) and b_2, as well as n_p, depending for laminar flow on the Prandtl number are accepted to be the same as those found for a free rotating disk in Sect. 3.4. For turbulent flow, b_1=1, b_2=1.

Then a joint solution of Eqs. (5.34) and (5.35) with allowance for Eq. (5.36) takes the following form for the boundary condition (2.36) [190]

$$\chi = \frac{(2+m+n_*) + b_2 n_* \beta/(1-\beta)_*}{(2+m+n_*)Pr^{n_p}(1-K_V)b_1 + (4+m)K_V b_2 + b_2 2\beta/(1-\beta)}, \tag{5.37}$$

where, according to Eq. (3.47), $K_V = 1 - D_2/A_1 = 1 - d_2/a_1$. Solution (5.37) generalizes Eq. (5.9) valid for turbulent flow. For the boundary condition (2.35), Eq. (5.37) reduces to

$$\chi = \frac{2+m+n_*}{(2+m+n_*)Pr^{n_p}(1-K_V)b_1 + (4+m)K_V b_2 + b_2 2\beta/(1-\beta)}. \tag{5.38}$$

For laminar flow m=0. Accepting values b_1=1, b_2=1 for turbulent flow, Eq. (5.38) reduces to Eq. (5.10).

Parameter K_V, together with the constants a_1 and d_2 (see Eq. (3.47)), for laminar flow depends on the value of β. One should also determine the dependence of coefficients b_1 and b_2 on β. In doing so, it is assumed that dependence of the parameters b_2 and n_p on the Prandtl number (b_1 being independent of Pr) is the same as that for a free disk (Sect. 3.4, Eqs. (3.65), (3.66) and (3.67)).

Based on Eqs. (2.52), (3.5) and (5.24), the Reynolds analogy parameter χ for the self-similar solution can be calculated from the following relation:

$$\chi = -K_1(1-\beta)\left[Pr\,(dG/d\zeta)_{\zeta=0}\right]^{-1}. \tag{5.39}$$

Equating values of χ obtained by Eq. (5.38) to those computed by the exact solution (5.39) for any β at Pr=1, n_*=0, n_*=2, one can obtain values of the coefficients b_1 and b_2, for the sake of convenience in use approximated with the fourth-order polynomials [190]

$$b_1 = \sum_0^4 b_{1,i}\beta^i, \quad b_2 = b_{2,0}\left(1 + \sum_1^4 b_{2,i}\beta^i\right), \tag{5.40}$$

where $b_{1,0}$=0.6827, $b_{1,1}$=0.1694, $b_{1,2}$=0.1115, $b_{1,3}$=0.7931 and $b_{1,4}$=–0.7569. The value of $b_{2,0}$ for β=0 is calculated from Eq. (3.65), while $b_{2,1}$=0.2866,

$b_{2,2}$=–2.5420, $b_{2,3}$=2.7414 and $b_{2,4}$=–0.8586. Used as asymptotic values for β=0.999 were b_1=1 and b_2=1.

Given a value of χ by Eq. (5.38) based on the integral method, the constant K_1 can be found from the following relation:

$$K_1 = 0.6159 \frac{\tau_{w\varphi}}{\tau_{w\varphi 0}} \chi \frac{Pr}{1-\beta}, \tag{5.41}$$

where the ratio $\tau_{w\varphi}/\tau_{w\varphi 0}$ is specified by Eq. (5.32).

Values of K_1 and χ calculated for Pr=1 and 0.71 in accordance with the exact solution and the integral method, Eqs. (5.38) and (5.41), are given in Tables 5.9, 5.10, 5.11 and 5.12 [190].

Dependence of the constant K_1 on the swirl parameter β has a point of maximum at β=0.1, though the value K_1 at this point for Pr=0.71–1.0 just by 1–2% exceeds the value of K_1 at β=0. With the further increase in β, the constant K_1, at first slowly (for $\beta \leq 0.4$), and then rapidly decreases. Authors [155], who studied experimentally heat transfer at a stagnation point of a swirling jet impinging on a stationary surface, also noticed a point of maximum of the dependence of the Nusselt number on the swirl parameter for small values of β. Remarkable decrease in convective

Table 5.9 Constant K_1 for Pr=1.0 based on the exact solution and the integral method (bold font)

β	n_*=–2	n_*=–1.5	n_*=–1	n_*=–0.5	n_*=0	n_*=0.5	n_*=1	n_*=2	n_*=4
0.0	0.0	0.1305	0.2352	0.3221	0.3962	0.4608	0.5180	0.6159	0.7693
	0.0	**0.1261**	**0.2311**	**0.3199**	**0.3961**	**0.4620**	**0.5197**	**0.6159**	**0.7557**
0.1	0.0	0.1333	0.2396	0.3275	0.4022	0.4672	0.5246	0.6228	0.7764
	0.0	**0.1287**	**0.2354**	**0.3254**	**0.4022**	**0.4686**	**0.5266**	**0.6228**	**0.7622**
0.2	0.0	0.1305	0.2351	0.3221	0.3962	0.4607	0.5178	0.6155	0.7685
	0.0	**0.1259**	**0.2308**	**0.3196**	**0.3958**	**0.4618**	**0.5195**	**0.6158**	**0.7559**
0.3	0.0	0.1123	0.2243	0.3086	0.3809	0.4442	0.5003	0.5967	0.7479
	0.0	**0.1198**	**0.2205**	**0.3064**	**0.3804**	**0.445**	**0.5017**	**0.5969**	**0.7367**
0.4	0.0	0.1138	0.208	0.288	0.3574	0.4184	0.4729	0.5669	0.7151
	0.0	**0.1111**	**0.2055**	**0.2867**	**0.3574**	**0.4194**	**0.4742**	**0.5669**	**0.7045**
0.5	0.0	0.1013	0.1868	0.2607	0.3255	0.3832	0.4351	0.5253	0.6688
	0.0	**0.0998**	**0.1858**	**0.2607**	**0.3264**	**0.3847**	**0.4366**	**0.5253**	**0.6591**
0.6	0.0	0.0873	0.1613	0.2268	0.2852	0.3379	0.3858	0.4702	0.6064
	0.0	**0.0856**	**0.1608**	**0.2272**	**0.2865**	**0.3395**	**0.3874**	**0.4703**	**0.5983**
0.7	0.0	0.0721	0.1319	0.1861	0.2356	0.2811	0.323	0.3983	0.5229
	0.0	**0.0683**	**0.1297**	**0.1852**	**0.2356**	**0.2816**	**0.3237**	**0.3983**	**0.5174**
0.8	0.0	0.0610	0.1032	0.1426	0.1796	0.2144	0.2472	0.3075	0.4116
	0.0	**0.0478**	**0.0921**	**0.1333**	**0.1717**	**0.2077**	**0.2413**	**0.3027**	**0.4058**
0.9	0.0	0.0643	0.0314	0.1065	0.1266	0.1461	0.1649	0.2009	0.2670
	0.0	**0.0247**	**0.0484**	**0.0712**	**0.0933**	**0.1145**	**0.1350**	**0.1739**	**0.2443**
0.99	0.0	0.0987	0.1008	0.1028	0.1048	0.1068	0.1088	0.1128	0.121
	0.0	**0.0025**	**0.0050**	**0.0075**	**0.0099**	**0.0124**	**0.0149**	**0.0197**	**0.0294**
0.999	0.0	0.1417	0.1418	0.1421	0.1423	0.1424	0.1426	0.1429	0.1437
	0.0	**0.0003**	**0.0005**	**0.0008**	**0.0010**	**0.0012**	**0.0015**	**0.0020**	**0.0030**

Table 5.10 Value χ for Pr=1.0 based on the exact solution and the integral method (bold font)

β	n_*=-2	n_*=-1.5	n_*=-1	n_*=-0.5	n_*=0	n_*=0.5	n_*=1	n_*=2	n_*=4
0.0	0.0	0.2119	0.3818	0.5230	0.6433	0.7482	0.8410	1.0	1.1335
	0.0	**0.2047**	**0.3752**	**0.5195**	**0.6431**	**0.7502**	**0.8439**	**1.0**	**1.1249**
0.1	0.0	0.1995	0.3585	0.49	**0.6018**	0.699	**0.7849**	**0.9318**	1.055
	0.0	**0.193**	**0.353**	**0.4879**	**0.6031**	**0.7027**	**0.7895**	**0.9339**	**1.049**
0.2	0.0	0.1824	0.3288	0.4504	**0.554**	0.6442	**0.724**	**0.8608**	**0.9755**
	0.0	**0.1760**	**0.3228**	**0.4470**	**0.5535**	**0.6458**	**0.7265**	**0.8612**	**0.9689**
0.3	0.0	0.1632	0.296	0.4072	0.5026	0.586	0.6601	0.7873	0.8943
	0.0	**0.1577**	**0.2903**	**0.4033**	**0.5008**	**0.5858**	**0.6605**	**0.7858**	**0.8867**
0.4	0.0	0.1426	0.2608	0.3611	0.4479	0.5245	0.5928	0.7106	0.8101
	0.0	**0.1388**	**0.2568**	**0.3583**	**0.4466**	**0.5240**	**0.5926**	**0.7083**	**0.8024**
0.5	0.0	0.1213	0.2236	0.312	0.3897	0.4587	0.5208	0.6288	0.7206
	0.0	**0.1192**	**0.2219**	**0.3113**	**0.3899**	**0.4595**	**0.5215**	**0.6274**	**0.7144**
0.6	0.0	0.1002	0.1852	0.2603	0.3274	0.3879	0.4429	0.5397	0.623
	0.0	**0.0983**	**0.1846**	**0.2609**	**0.3288**	**0.3898**	**0.4448**	**0.5399**	**0.6194**
0.7	0.0	0.0796	0.1457	0.2056	0.2603	0.3105	0.3569	0.4400	0.5129
	0.0	**0.0756**	**0.1436**	**0.2051**	**0.2609**	**0.3119**	**0.3586**	**0.4411**	**0.5118**
0.8	0.0	0.0651	0.1100	0.1521	0.1915	0.2286	0.2635	0.3278	0.3859
	0.0	**0.0511**	**0.0986**	**0.1427**	**0.1839**	**0.2223**	**0.2584**	**0.3240**	**0.3823**
0.9	0.0	0.0664	0.0885	0.1099	0.1306	0.1507	0.1701	0.2073	0.2423
	0.0	**0.0255**	**0.0501**	**0.0737**	**0.0965**	**0.1185**	**0.1397**	**0.1800**	**0.2176**
0.99	0.0	0.099	0.1011	0.1031	0.1051	0.1071	0.1091	0.1131	0.1171
	0.0	**0.0025**	**0.0050**	**0.0075**	**0.0100**	**0.0124**	**0.0149**	**0.0198**	**0.0247**
0.999	0.0	0.1417	0.1419	0.1421	0.1423	0.1424	0.1426	0.143	0.1434
	0.0	**0.00025**	**0.0005**	**0.00075**	**0.0010**	**0.0012**	**0.0015**	**0.0020**	**0.0025**

heat transfer along the axis z (as well as the axis r) is caused by the decrease in absolute shear of the tangential velocity v_z for $\beta \to 1$. For $\beta \geq 0.8$, the exact solution for the dimensionless temperature θ and consequently for the coefficients b_1 and b_2 becomes to be affected by prevailing influence of heat conduction over convective heat transfer along the axis z. In accordance with the heat conduction equation, one can obtain for $\beta \to 1$ that

$$K_1 = -\theta'_{\zeta=0} = 1/\zeta_\infty, \tag{5.42}$$

where ζ_∞ is the coordinate at which θ=0. It is worth noting that this solution is independent of the values Pr and n_*. Asymptotic values b_1=1 and b_2=1 for β=0.999 correspond to $\zeta_\infty=\infty$. Data of the exact solution were calculated for β=0.999 at ζ_∞=7.05, and for β=0.99 at ζ_∞=9.93 (Tables 5.9, 5.10, 5.11 and 5.12) [190]. These data agree well with the exact solution and the integral method.

An analysis of the data in Tables 5.9, 5.10, 5.11 and 5.12 show that the exact and approximate solutions agree well for $\beta \leq 0.7$. For β=0.7, their maximal disagreements at the value of n_*=–1.5 (rarely occurred in practice) reach 5.6% for Pr=1 and 13.2% for Pr=0.71. These inaccuracies of the approximate solution reduce with an increasing n_*. Over the range β=0–0.7 and $n_*=-1$ to 4, they do not exceed 1.6% for Pr=1 and 2.5% for Pr=0.71. Inaccuracies of the approximate solution somewhat increase following deviation of the Prandtl number from unity, the value for

Table 5.11 Constant K_1 for Pr=0.71 based on the exact solution and the integral method (bold font)

β	n_*=–2	n_*=–1.5	n_*=–1	n_*=–0.5	n_*=0	n_*=0.5	n_*=1	n_*=2	n_*=4
0.0	0.0	0.1035	0.1893	0.2624	0.3259	0.3818	0.4319	0.5185	0.6555
	0.0	**0.1009**	**0.187**	**0.2612**	**0.3259**	**0.3828**	**0.4332**	**0.5185**	**0.6457**
0.1	0.0	0.1060	0.1934	0.2675	0.3316	0.3881	0.4384	0.5254	0.6627
	0.0	**0.1031**	**0.1906**	**0.2658**	**0.3312**	**0.3885**	**0.4392**	**0.5248**	**0.6518**
0.2	0.0	0.1033	0.189	0.2621	0.3255	0.3815	0.4315	0.518	0.6549
	0.0	**0.1008**	**0.1867**	**0.2609**	**0.3256**	**0.3825**	**0.4329**	**0.5183**	**0.6458**
0.3	0.0	0.0971	0.1789	0.2493	0.3108	0.3654	0.4145	0.4997	0.6350
	0.0	**0.0958**	**0.1782**	**0.2498**	**0.3125**	**0.3680**	**0.4174**	**0.5015**	**0.6281**
0.4	0.0	0.0886	0.1643	0.2303	0.2888	0.341	0.3883	0.4710	0.6034
	0.0	**0.0888**	**0.1659**	**0.2334**	**0.2930**	**0.3461**	**0.3936**	**0.4751**	**0.5992**
0.5	0.0	0.0795	0.1469	0.2067	0.2605	0.3091	0.3535	0.4320	0.5594
	0.0	**0.0797**	**0.1497**	**0.2117**	**0.2670**	**0.3165**	**0.3613**	**0.4387**	**0.5585**
0.6	0.0	0.0709	0.1279	0.1795	0.2267	0.2701	0.3102	0.3821	0.5013
	0.0	**0.0683**	**0.1293**	**0.1840**	**0.2334**	**0.2783**	**0.3192**	**0.3910**	**0.5045**
0.7	0.0	0.0616	0.1065	0.1481	0.1869	0.2231	0.2572	0.3195	0.4257
	0.0	**0.0544**	**0.1039**	**0.1493**	**0.1910**	**0.2295**	**0.2651**	**0.3288**	**0.4328**
0.8	0.0	0.0584	0.0891	0.1183	0.1462	0.1728	0.1982	0.246	0.3308
	0.0	**0.038**	**0.0735**	**0.1069**	**0.1383**	**0.1679**	**0.1959**	**0.2473**	**0.3355**
0.9	0.0	0.0668	0.0821	0.0970	0.1116	0.1258	0.1397	0.1667	0.2172
	0.0	**0.0195**	**0.0385**	**0.0568**	**0.0745**	**0.0916**	**0.1083**	**0.1401**	**0.1983**
0.99	0.0	0.0993	0.1007	0.1022	0.1036	0.1050	0.1065	0.1093	0.1149
	0.0	**0.0020**	**0.0039**	**0.0059**	**0.0079**	**0.0098**	**0.0118**	**0.0156**	**0.0233**
0.999	0.0	0.1417	0.1418	0.1420	0.1421	0.1422	0.1424	0.1426	0.1432
	0.0	**0.0002**	**0.0004**	**0.0006**	**0.00079**	**0.00099**	**0.0012**	**0.0016**	**0.0024**

which the dependence of coefficients b_1 and b_2 on β was obtained. However, this increase in the inaccuracies may be regarded insignificant, which means that Eqs. (3.65), (3.66) and (3.67) describing dependence of b_2 and n_p on the Pr number obtained earlier for a free disk (Sect. 3.4) hold also for the solid-body rotation of fluid. For β=0.8, inaccuracies of the approximate solution reach 10.6% at n_*=–0.5 and decrease to 1.5% at n_*=4, which is an acceptable result. For β=0.9, the present integral method can be recommended only at n_*=1–4, while for β>0.9 heat conduction effects in fluid dominate over convective heat transfer, and the integral method becomes unacceptable [190].

Thus, in the present sub-section an exact self-similar solution of the problem of laminar heat transfer of a disk rotating in a co-rotating fluid for boundary conditions (2.32), (2.33), (2.34), (2.35) and (2.36) and β<1 was found. Based on the exact solution, an integral method was developed and an approximate analytical solution of this problem was found. Inaccuracies of the approximate solution for β=0–0.7, $n_*=-1.0$–4.0 and Pr=0.71–1 do not exceed 2.5%. Following an increase in the parameter β to values β=0.8–0.9, the range of the values of n_*, which can provide low inaccuracies of the approximate solution, becomes narrower and shifts to the region of the positive values of n_*=1–4. For β>0.9, heat conduction in fluid dominates over convective heat transfer, and the integral method becomes unacceptable. Dependence of the constant K_1 (i.e. the Nusselt number) on the parameter β exhibits

Table 5.12 Value χ for Pr=0.71 based on the exact solution and the integral method (bold font)

β	n_*=–2	n_*=–1.5	n_*=–1	n_*=–0.5	n_*=0	n_*=0.5	n_*=1	n_*=2	n_*=4
0.0	0.0	0.2366	0.4330	0.6001	0.7452	0.8732	0.9876	1.1856	1.3533
	0.0	**0.2308**	**0.4276**	**0.5973**	**0.7453**	**0.8753**	**0.9906**	**1.1857**	**1.3446**
0.1	0.0	0.2233	0.4076	0.5637	0.6988	0.8177	0.9238	1.1071	1.2619
	0.0	**0.2177**	**0.4025**	**0.5614**	**0.6995**	**0.8206**	**0.9276**	**1.1084**	**1.2551**
0.2	0.0	0.2034	0.3723	0.5162	0.6411	0.7513	0.8499	1.0203	1.1646
	0.0	**0.1985**	**0.3678**	**0.5139**	**0.6413**	**0.7534**	**0.8527**	**1.0209**	**1.1580**
0.3	0.0	0.1805	0.3325	0.4632	0.5776	0.6791	0.7702	0.9286	1.0630
	0.0	**0.1777**	**0.3304**	**0.4631**	**0.5794**	**0.6823**	**0.7738**	**0.9298**	**1.0577**
0.4	0.0	0.1564	0.2900	0.4066	0.5098	0.6021	0.6855	0.8315	0.9563
	0.0	**0.1563**	**0.2919**	**0.4107**	**0.5157**	**0.6090**	**0.6926**	**0.8361**	**0.9547**
0.5	0.0	0.1340	0.2476	0.3486	0.4391	0.5212	0.596	0.7284	0.8426
	0.0	**0.1341**	**0.2519**	**0.3561**	**0.4491**	**0.5325**	**0.6077**	**0.7381**	**0.8471**
0.6	0.0	0.1146	0.2067	0.2903	0.3666	0.4367	0.5015	0.6179	0.7199
	0.0	**0.1104**	**0.2090**	**0.2975**	**0.3774**	**0.4500**	**0.5161**	**0.6322**	**0.7308**
0.7	0.0	0.0959	0.1656	0.2304	0.2908	0.3472	0.4002	0.4971	0.5839
	0.0	**0.0848**	**0.1621**	**0.2329**	**0.2980**	**0.3580**	**0.4135**	**0.5129**	**0.5993**
0.8	0.0	0.0877	0.1338	0.1777	0.2195	0.2595	0.2977	0.3694	0.4355
	0.0	**0.0572**	**0.1109**	**0.1612**	**0.2086**	**0.2532**	**0.2954**	**0.3729**	**0.4427**
0.9	0.0	0.0970	0.1192	0.1409	0.1621	0.1828	0.2030	0.2421	0.2796
	0.0	**0.0285**	**0.0561**	**0.0827**	**0.1086**	**0.1336**	**0.1579**	**0.2042**	**0.2479**
0.99	0.0	0.1403	0.1423	0.1443	0.1464	0.1484	0.1504	0.1544	0.1584
	0.0	**0.0028**	**0.0056**	**0.0084**	**0.0111**	**0.0139**	**0.0166**	**0.0221**	**0.0275**
0.999	0.0	0.1997	0.1998	0.2000	0.2002	0.2004	0.2006	0.2009	0.2013
	0.0	**0.00029**	**0.00056**	**0.00084**	**0.0011**	**0.0014**	**0.0017**	**0.0022**	**0.0028**

a point of maximum at β=0.1. Afterwards, the constant K_1 decreases with increasing β [190].

5.2 Accelerating Radial Flow Without Imposed External Rotation

5.2.1 Flow Impingement onto an Orthogonal Rotating Disk: Experimental and Computational Data of Different Authors

Flow impingement onto an orthogonal surface, whose particular case are impinging jets, is known as a high-performance technique for cooling or heating a surface [44, 99, 114, 149, 150, 157, 196]. In particular, impinging jet flows of various types are used in rotating-disk systems for cooling end-wall surfaces of gas turbine rotors, etc.

Fundamentals of modelling axisymmetric round jets impinging orthogonally onto a stationary flat disk are documented, e.g. in works [22, 41, 93, 158]. These authors described, in particular, a known theoretical solution for the case where an axisymmetric flow with a uniform axial velocity profile impinges orthogonally

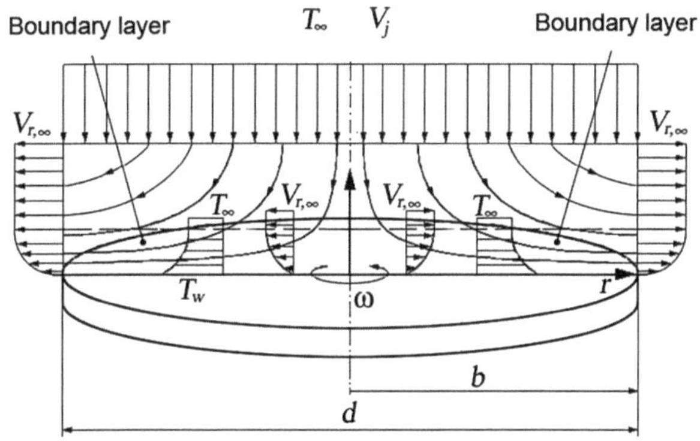

Fig. 5.8 Hydrodynamics and heat transfer between a rotating disk and a uniform flow perpendicular to it [158, 213]

onto a flat surface (Fig. 5.8). For real axisymmetric single jets impinging onto a stationary disk of a diameter d [14, 44, 57, 158, 213], fluid flow in the region of impingement and radial turn of a jet on a disk with sufficiently high accuracy can be considered laminar, and its peculiarities around the stagnation point are close to those of uniform axisymmetric impingement flow over a disk.

Fluid flow in axisymmetric air jets impinging onto rotating disks has been studied quite in detail [11, 15, 24, 26, 41, 59, 94, 123, 124, 147, 154–156, 198, 207–211]. As shown in these works, one can accept that a constant thickness boundary layer develops in the vicinity of the stagnation point, while the radial $v_{r,\infty}$ and axial $v_{z,\infty}$ velocity components at the outer edge of the boundary layer are described by Eqs. (2.32).

In the case where potential uniform flow with the velocity at infinity V_j impinges orthogonally onto a disk, constant a in Eqs. (2.32) may be calculated by the following relation [41]:

$$a = 2V_j/(\pi b), \quad A = 4/\pi. \tag{5.43}$$

The second of Eqs. (5.43) takes into consideration the fact that flow in this case impinges onto the whole surface of the disk; hence, in the definition of the parameter a, it was accepted that $d_j{=}d$.

The stagnation pressure p_s on a disk at a local radius r can be written as

$$p_s = p(r) + (1/2)\rho v_{r,\infty}^2, \tag{5.44}$$

where velocity $v_{r,\infty}$ is given by Eqs. (2.32) and (5.43). On the other hand,

$$p_s = p_\infty + (1/2)\rho V_j^2. \tag{5.45}$$

As a result, based on Eqs. (5.44) and (5.45), one can derive the following relation [116]:

$$\frac{p - p_\infty}{(1/2)\rho V_j^2} = 1 - \left(\frac{2}{\pi}\right)^2 \left(\frac{r}{b}\right)^2. \tag{5.46}$$

The theoretical equation (5.46) agrees well with experimental data [116] up to the radial location r/b=0.85 that confirms validity of the model assumptions, Eqs. (2.32) and (5.43), for the radial velocity $v_{r,\infty}$ over a disk subjected to uniform orthogonal impingement. Disagreement of Eq. (5.46) with experimental data [116] for pressure distribution at r/b>0.85 is caused by effects of flow separation at the outer rim of the disk.

For the case where a single jet impinges onto an orthogonal disk, velocity $v_{r,\infty}$ in the vicinity of the stagnation point follows the first of Eqs. (2.32), while the parameter a is initially almost constant. Further, the parameter a gradually decreases with an increasing r till the value of $v_{r,\infty}$ reaches its maximum. Having reached its maximum, the component $v_{r,\infty}$ begins to decrease. This means beginning of a principally different region of a turbulent wall jet, which is not studied in the present work. Empirical correlations, which allow predicting a radial location of the point of maximum of $v_{r,\infty}$ and its value at this point are given in [44].

In the case of real impinging jets, a significant role belongs to the jet-to-disk diameter ratio d_j/d and dimensionless jet-to-disk distance h_j/d_j. Value of A for real jets is different from $4/\pi$ and depends on the parameter h_j/d_j (as well as on the velocity profile and turbulence level at the nozzle outlet). Experimental data of different authors describing dependence of A on h_j/d_j differ quite noticeably from each other. We accept the physically justifiable point of view of the authors [14, 213] that the value A (together with the Nusselt number at the stagnation point) decreases with increasing h_j/d_j for high enough values of Re_j. For laminar flow with a uniform velocity profile at the nozzle outlet, authors of [14] proposed an empirical correlation

$$A = 1.5 \cdot (h_j/d_j)^{-0.22}, \tag{5.47}$$

which is valid over the range h_j/d_j=2–6.

Heat and mass transfer of impinging jets were experimentally investigated in works [5, 11, 24, 59, 147, 154–156]. Theoretical studies were undertaken in works [26, 41, 94, 198, 209, 210]. Authors of [94] obtained an exact self-similar solution for heat transfer for uniform axisymmetric flow impingement onto a disk for Pr=0.7, 1, 10 and T_w=*const*. A self-similar solution for mass transfer was obtained in the work [198] for Sc=0.2, 3.0 and C_w=*const*.

Self-similar form of the Navier–Stokes and energy equations for the problem under consideration reduces to Eqs. (2.37), (2.38), (2.39), (2.40) and (2.41) for β=0. These equations are solved for boundary conditions (2.32), (2.33), (2.34) and (2.35). Based on this solution, the Nusselt numbers for a stationary disk are given by the following equations [44]:

$$Nu_d = K_1 \cdot Re_a^{1/2}, \quad Nu_d = 0.763 \cdot Re_j^{1/2} \cdot Pr^{0.4} \cdot A^{1/2}. \tag{5.48}$$

The second of Eqs. (5.48) is valid for a uniform flow impingement onto a disk, with $Re_a = Re_j \cdot A$, $d_j = d$ and $T_w = const$.

Authors of the work [116] investigated both theoretically and experimentally the problem of laminar fluid flow and heat transfer at uniform orthogonal impingement of air stream ($Pr = 0.71$) onto a rotating disk with constant surface temperature $T_w = const$ over a wide range of variation of the ratio between the impingement velocity and angular velocity of rotation $a/\omega = 0\ldots\infty$. As shown in [116], the self-similar theoretical solution for the Nusselt number (considered in detail below in Sect. 3.2.3) agrees well with original experimental data.

Authors of the work [26] approximated results of their self-similar solution for mass transfer of a rotating disk by the following equation:

$$\left[\frac{Sh}{\left(Re_a^2 + Re_\omega^2\right)^{1/4}}\right]^m = \left[\beta^{1/4}\frac{Sh_a}{Re_a^{1/2}}\right]^m + \left[(1+\beta)^{1/4}\frac{Sh_\omega}{Re_\omega^{1/2}}\right]^m, \qquad (5.49)$$

valid for $C_w = const$ over a range of the Schmidt numbers $Sc = 0\ldots\infty$. Sherwood numbers for a free rotating disk Sh_ω are calculated by Eqs. (3.4) (for $n_R = 1/2$) and (3.7), and substituting in doing so the values of Nu with Sh, and Pr numbers with the Schmidt numbers Sc, respectively. Sherwood numbers Sh_a for uniform axisymmetric flow impingement onto a stationary disk are calculated by the following formula [26]:

$$Sh_a/Re_a^{1/2} = K_1 = 0.65693Sc^{1/2}/(0.3098 + 1.015Sc^{1/2} + Sc)^{1/6}, \qquad (5.50)$$

where the local radius r is the characteristic length. Power exponent m in Eq. (5.49) is specified by empirical values $m = 2.65$ for $Sc = 0.1$–1.0 or $m = 3.85$ for $Sc = 1$–$10{,}000$. Authors [26] asserted that inaccuracies in calculation of the Sherwood number by Eq. (5.49) comparing to the exact solution do not exceed 7.55% for $Sc = 0.1$–1.0 and 3.85% for $Sc = 1$–$10{,}000$. An analysis shows however that inaccuracies of Eq. (5.49) with respect to the exact solution reach 12% for $Sc = 1$, primarily, because of the inaccuracy of Eq. (5.50) and, probably, in the choice of the numerical values of exponents m.

Experimental data [154–156] for local heat transfer of impinging jets reveal that both rotation and flow impingement result in heat transfer augmentation (Fig. 5.9). A detailed analysis of data [154–156] in comparison with computations performed by the author of the present work is given below.

Experiments [11, 147] for heat transfer of impinging jets were summarized in a form of average Nusselt numbers for the *entire* disk and do not include data for local heat transfer at the stagnation point to be modelled below. Data [24, 59] also represent results for mass transfer in naphthalene sublimation in air for an *entire* disk.

For turbulent flow and boundary conditions (2.32) at $\beta = 0$ and $N = const$, the author of the work [212] using model (2.42), (2.43), (2.44), (2.45) and (2.46) and

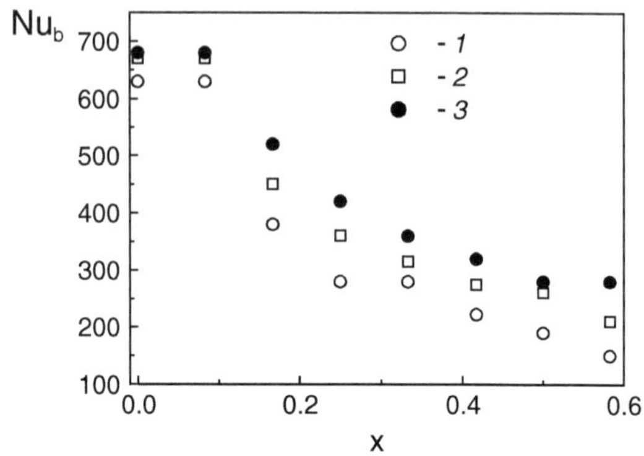

Fig. 5.9 Effect of disk rotation on the local Nu_b values in the case of axisymmetric single jet impingement. Experiments for Re_j=6.8×10^3, h_j/d_j=2 [155]: *1* – Re_φ=5.35×10^4, 2 – 2.0×10^5, *3* – 3.53×10^5

Eq. (2.48) for *tanφ* obtained a solution of Eqs. (2.22) and (2.23) for n=1/7. This solution was generalized in the work [138] for arbitrary values of n

$$\alpha^2 + R\alpha N = S + TN^2, \quad \gamma = K^{\frac{n+1}{3n+1}} \left[\frac{(3n+1)(1+\alpha^2)^{0.5(1-n)/(1+n)}}{(5+11n)(G\alpha + HN)} \right]^{(n+1)/(3n+1)}, \tag{5.51}$$

$$C_M = \gamma \, (G\alpha + HN) \, Re_\varphi^{-2n/(3n+1)}. \tag{5.52}$$

Functions R, S, T, G, H and K are tabulated in [138]. Numerical data for selected parameters of this solution computed in [138] and listed in Table 5.13 are analysed below in comparison with results obtained using the present integral method (described in Sect. 2.4).

Table 5.13 Constants in Eqs. (2.82), (2.83), (2.84), (2.85) and (2.86) for N=*const*, β=0

	Solution (2.88), (2.89), (2.90) and (2.91) [163], n=1/7				Solution (5.51), (5.52) [138, 212], n=1/7			
N	α	γ	ε_m	ε_M	α	γ	ε_m	ε_M
0	0.2087	0.5299	0.1806	0.1466	0.162	0.526	0.219	0.146
0.2	0.2818	0.2892	0.3566	0.1727	0.2587	0.2918	0.3648	0.172
0.4	0.4509	0.191	0.4509	0.191	0.496	0.1916	0.4389	0.198
0.6	0.6408	0.1484	0.4994	0.2255	0.6288	0.1486	0.5010	0.224
0.8	0.8373	0.1258	0.5611	0.252	0.8263	0.1259	0.5621	0.25
1	1.037	0.1122	0.6236	0.2794	1.026	0.1122	0.6242	0.278
2	2.046	0.0853	0.9446	0.422	2.033	0.0853	0.9448	0.42
3	3.062	0.076	1.262	0.563	3.044	0.076	1.262	0.56
4	4.073	0.0708	1.567	0.699	4.056	0.0708	1.568	0.696

5.2.2 Turbulent Flow

In uniform flow impinging onto a perpendicular surface, as well as in the stagnation region of an impinging jet, the radial velocity component at the outer edge of the boundary layer increases linearly according to Eq. (2.32), which means $N=const$. As mentioned above, flow in such cases is laminar or close to this. However, flow for $N=const$ can also be turbulent, e.g. in converging channels. Such cases were studied in works [41, 138, 212]. An investigation of turbulent flow under condition $N=const$ is also actual here in view of the ideology of the present integral method that assumes extending a solution obtained for turbulent flow onto the case of laminar regime.

Table 5.13 presents numerical values of constants α, γ, ε_m and ε_M computed by Eqs. (2.88), (2.89), (2.90) and (2.91) based on the present integral method for $n=1/7$, as well as analogous data obtained in [138, 212] using the model (2.42), (2.43), (2.44), (2.45), (2.46) and (2.48) resulting in solutions (5.51) and (5.52). To remind, these constants are substituted into Eqs. (2.82), (2.83), (2.84), (2.85) and (2.86) for respective fluid flow parameters. Differences between both solutions disappear very fast with an increasing N; it is worth noting also that the respective values α and ε_m for both methods practically coincide already for $N>0.5$.

The solution (2.93) of the thermal boundary layer equation for $\beta=0$ can be rewritten as follows:

$$\begin{aligned}\frac{\overline{\delta_T^{**}}}{\alpha A_1} &= -\left[\left(-1+\chi Pr^{n_p}(1-K_{V0})\right)+\tfrac{\kappa}{\alpha}\left(-\tfrac{A_2}{A_1}+\chi Pr^{n_p}\tfrac{D_4}{A_1}\right)\right]-\\ &\tfrac{\kappa}{\alpha}\tfrac{1}{A_1}\left(\tfrac{(\chi Pr^{n_p})^{-1/n}-\chi Pr^{n_p}}{n+1}-(\chi Pr^{n_p})^{-1/n}+1\right).\end{aligned} \quad (5.53)$$

Based on this, a transcendental equation for χ takes the following form:

$$\begin{aligned}&-1+\chi Pr^{n_p}(1-K_{V0})+\tfrac{\kappa}{\alpha}\left(-\tfrac{A_2}{A_1}+\chi Pr^{n_p}\tfrac{D_4}{A_1}\right)-\\ &\tfrac{\kappa}{\alpha}\tfrac{1}{A_1}\left(\tfrac{-n(\chi Pr^{n_p})^{-1/n}-\chi Pr^{n_p}}{n+1}+1\right)=\chi\tfrac{4+m}{2+m+n_*}K_{V0}\left[1+\tfrac{\kappa}{\alpha}\left(\tfrac{B_5}{D_1}-1\right)\right].\end{aligned} \quad (5.54)$$

Here $K_{V0}=1-D_2/A_1$, $B_5=C_2/2=D_1+D_3$, whereas values χ and Δ are interrelated by Eq. (5.8) for $\beta=0$.

Since algebraic equation (5.54) is transcendental, it seems actual to find its approximate analytical solutions. Assuming $\chi Pr^{n_p}\approx 1$ and $(\chi Pr^{n_p})^{-1/n}\neq 1$ for $Pr=0.5\ldots1$, one can obtain

$$\chi = Pr^{-n_p}\left[1-\frac{n+1}{n}\left(\frac{\alpha}{\kappa}D_1+D_3\right)\left(1-Pr^{-n_p}\frac{4+m}{2+m+n_*}\right)\right]^{-n}. \quad (5.55)$$

Another approximate solution of Eq. (5.54) can be obtained for small values of Pr. In this case, assuming $\chi Pr^{n_p}\to 0$ and $(\chi Pr^{n_p})^{-1/n}\to\infty$, one can obtain

$$\chi = Pr^{-n_p}\left[\frac{n+1}{n}Pr^{-n_p}\left(\frac{\alpha}{\kappa}D_1 + D_3\right)\frac{4+m}{2+m+n_*}\right]^{\frac{-n}{n+1}}. \qquad (5.56)$$

Computed values of the constants K_1 and χ for Pr=0.71 (air) are presented in Tables 5.14 and 5.15 depending on the parameters κ and n_*.

These results allow making the following important conclusions.

First, coefficient K_1 increases steadily with an increasing parameter κ, though the values of K_1 increase much slower than κ. Such a behaviour of the coefficient K_1 is caused by the increase in the coefficient $A_c(1 + \alpha^2)^{1/2}$ according to Eq. (5.8).

Second, the order of magnitude of the Reynolds analogy parameter χ remains the same regardless of variation of κ over the range κ=0–∞. The most noticeable variation of the parameter χ (33.5%) is observed for n_*=–1.5, while the least noticeable (1.5%) for n_*=4.

Third, results of computations are presented in Tables 5.14 and 5.15 only for κ=0–5. The reason for this is that the values of χ remain unchanged within the limits of four meaningful digits already for κ>5. Moreover, it is evident that with inaccuracy of 2% sufficient for practical computations the parameter χ becomes asymptotically constant already for κ=0.2. This means that the turbulent thermal boundary layer (strictly saying, in frames of the model used) becomes very rapidly adapted to the negative pressure gradient superimposed onto it.

Table 5.14 Values of the constant K_1 according to a numerical solution of Eq. (5.54) and its approximate solutions (5.55) and (5.56) for turbulent flow and Pr=0.71

$\kappa=a/\omega$	κ/α	n_*=–1.5	n_*=–1	n_*=–0.6	n_*=0	n_*=1	n_*=2	n_*=3	n_*=4
0.0 (5.54)		0.0125	0.0152	0.0167	0.0185	0.0205	0.0218	0.0227	0.0234
0.1 (5.54)	0.4491	0.0197	0.0207	0.0213	0.0221	0.0231	0.0239	0.0246	0.0251
0.1 (5.55)		0.0189	0.0200	0.0207	0.0216	0.0227	0.0237	0.0246	0.0254
0.1 (5.56)		0.0194	0.0203	0.0209	0.0216	0.0225	0.0232	0.0238	0.0243
0.2 (5.54)	0.7098	0.0219	0.0229	0.0236	0.0243	0.0253	0.0261	0.0267	0.0273
0.2 (5.55)		0.0213	0.0225	0.0232	0.0241	0.0252	0.0260	0.0267	0.0273
0.2 (5.56)		0.0220	0.0230	0.0237	0.0244	0.0255	0.0262	0.0269	0.0275
0.3 (5.54)	0.8295	0.0238	0.0249	0.0256	0.0264	0.0274	0.0282	0.0288	0.0294
0.3 (5.55)		0.0232	0.0245	0.0252	0.0261	0.0273	0.0281	0.0288	0.0294
0.3 (5.56)		0.0240	0.0251	0.0258	0.0267	0.0278	0.0287	0.0294	0.0300
0.5 (5.54)	0.9183	0.0273	0.0285	0.0292	0.0301	0.0313	0.0322	0.0329	0.0335
0.5 (5.55)		0.0267	0.0281	0.0289	0.0299	0.0312	0.0321	0.0329	0.0335
0.5 (5.56)		0.0275	0.0289	0.0297	0.0307	0.0313	0.0329	0.0337	0.0344
1.0 (5.54)	0.9649	0.0359	0.0375	0.0385	0.0397	0.0412	0.0423	0.0433	0.0441
1.0 (5.55)		0.0352	0.0370	0.0381	0.0394	0.0411	0.0423	0.0433	0.0441
1.0 (5.56)		0.0363	0.0381	0.0392	0.0405	0.0422	0.0435	0.0445	0.0455
5.0 (5.54)	0.9813	0.1068	0.1116	0.1145	0.1180	0.1224	0.1258	0.1286	0.1310
5.0 (5.55)		0.1048	0.1102	0.1134	0.1173	0.1221	0.1257	0.1286	0.1309
5.0 (5.56)		0.1082	0.1134	0.1166	0.1204	0.1255	0.1294	0.1326	0.1353

Table 5.15 Values of the Reynolds analogy parameter χ according to a numerical solution of Eq. (5.54) and its approximate solutions (5.55) and (5.56) for turbulent flow and Pr=0.71

$\kappa=a/\omega$	κ/α	n_*=–1.5	n_*=–1	n_*=–0.6	n_*=0	n_*=1	n_*=2	n_*=3	n_*=4
0.0 (5.54)	0	0.6580	0.7971	0.8788	0.9707	1.0747	1.144	1.193	1.231
0.1 (5.54)	0.4491	0.954	1.004	1.035	1.072	1.122	1.160	1.192	1.218
0.1 (5.55)		0.916	0.970	1.004	1.046	1.104	1.151	1.193	1.231
0.1 (5.56)		0.941	0.986	1.014	1.048	1.091	1.125	1.153	1.177
0.2 (5.54)	0.7098	0.976	1.022	1.051	1.085	1.129	1.163	1.191	1.215
0.2 (5.55)		0.950	1.002	1.033	1.072	1.122	1.160	1.192	1.218
0.2 (5.56)		0.978	1.025	1.054	1.089	1.134	1.170	1.197	1.224
0.3 (5.54)	0.8295	0.983	1.028	1.055	1.089	1.131	1.164	1.191	1.214
0.3 (5.55)		0.960	1.028	1.042	1.079	1.126	1.162	1.191	1.215
0.3 (5.56)		0.990	1.037	1.067	1.102	1.148	1.184	1.213	1.238
0.5 (5.54)	0.9183	0.987	1.031	1.058	1.091	1.133	1.165	1.191	1.213
0.5 (5.55)		0.966	1.017	1.047	1.084	1.129	1.164	1.191	1.213
0.5 (5.56)		0.997	1.045	1.074	1.110	1.156	1.192	1.222	1.247
1.0 (5.54)	0.9649	0.989	1.033	1.060	1.092	1.133	1.165	1.191	1.213
1.0 (5.55)		0.969	1.019	1.050	1.086	1.131	1.164	1.191	1.213
1.0 (5.56)		1.001	1.048	1.078	1.114	1.160	1.196	1.226	1.252
5.0 (5.54)	0.9813	0.989	1.033	1.060	1.093	1.133	1.165	1.191	1.213
5.0 (5.55)		0.970	1.020	1.050	1.086	1.131	1.164	1.191	1.212
5.0 (5.56)		1.002	1.050	1.079	1.115	1.162	1.198	1.228	1.253

Fourth, both approximate solutions (5.55) and (5.56) agree rather well with the exact numerical solution (5.54) for χ. Inaccuracies of both approximate solutions do not exceed 2%, with Eq. (5.55) being more accurate for positive values of n_* and Eq. (5.56) showing better performance for negative values of n_*.

5.2.3 Laminar Flow

Exact solution. It is possible to obtain an exact self-similar solution of the Navier–Stokes and energy equations in cylindrical polar coordinates in the case of co-axial orthogonal uniform flow impingement onto a rotating disk (Fig. 5.8) for the boundary conditions (2.32), (2.33), (2.34), (2.35) and (2.36) using self-similar functions and variables (2.31). Such solution was obtained by the authors in [180, 181, 183, 184]. As a result, the problem statement can be reduced to Eqs. (2.37), (2.38), (2.39), (2.40) and (2.41) (where β=0).

Rearranged boundary conditions can be expressed as

$$\begin{aligned} &\zeta = 0; \quad F = H = 0, \quad G = 1/(1+\kappa), \quad \theta = 1, \\ &\zeta \to \infty: \quad G = 0, \quad F = \kappa/(1+\kappa), \quad H \to const, \quad \theta = 0, \end{aligned} \tag{5.57}$$

where $\kappa = N = a/\omega$. Equation for the Nusselt number takes the following form:

$$Nu_d = K_1 \cdot (Re_{\omega d} + Re_a)^{1/2}, \quad K_1 = -\theta'_{\zeta=0}, \tag{5.58}$$

which is a generalization of Eqs. (3.4) and (5.48).

Setting $\omega \to 0$ (or $a/\omega \to \infty$), $\varphi = -H/2$ and making some transformations, one can reduce system (2.37), (2.38), (2.39), (2.40) and (2.41) to equations for axisymmetric impinging flow over a stationary orthogonal surface solved in works [22, 158] and others. For $a \to 0$ and finite value of ω, Eqs. (2.37), (2.38), (2.39), (2.40) and (2.41) are reduced to the case of a free rotating disk discussed in Chap. 3.

Numerical solution of (2.37), (2.38), (2.39), (2.40) and (2.41) was performed using the MathCAD software. Dimensionless profiles of the velocity components F, G, H and temperature θ are depicted in Figs. 5.10, 5.11, 5.12 and 5.13.

Dimensionless temperature profiles are qualitatively similar to each other (Fig. 5.13), while the radial velocity profiles vary with an increasing κ from a form typical for wall-jets characteristic in flow over a rotating disk to boundary-layer profiles already for $\kappa > 0.5$ (Fig. 5.10).

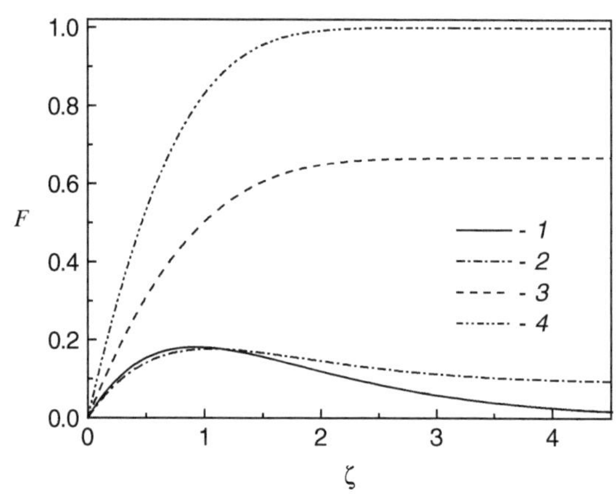

Fig. 5.10 Radial velocity profiles for laminar co-axial uniform orthogonal flow impingement onto a rotating disk [183]. *1* – $\kappa=0$; *2* – $\kappa=0.1$; *3* – $\kappa=2$; *4* – $\kappa=4000$

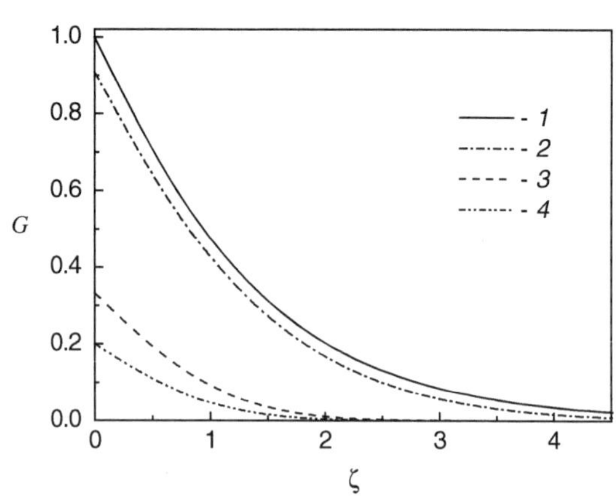

Fig. 5.11 Tangential velocity profiles for laminar co-axial uniform orthogonal flow impingement onto a rotating disk. *1* – $\kappa=0$; *2* – $\kappa=0.1$; *3* – $\kappa=2$; *4* – $\kappa=4$

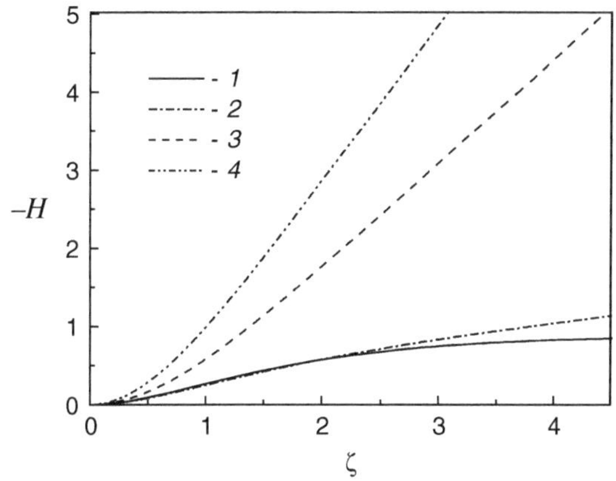

Fig. 5.12 Axial velocity profiles for laminar co-axial uniform orthogonal flow impingement onto a rotating disk. *1* – κ=0; *2* – κ=0.1; *3* – κ=2; *4* – κ=4000

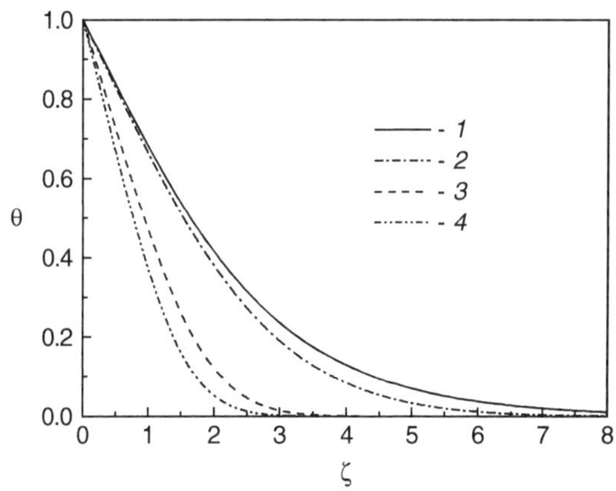

Fig. 5.13 Dimensionless temperature profiles for laminar co-axial uniform orthogonal flow impingement onto a rotating disk, Pr=0.71 (air) and n_*=1 [183]. *1* – κ=0; *2* – κ=0.1; *3* – κ=2; *4* – κ=4000

This qualitative reshaping of the radial velocity profiles causes drastic changes in the heat transfer regime over a rotating disk subjected to flow impingement described below. Profiles of the axial and tangential velocity components do not change their qualitative behaviour (Figs. 5.11 and 5.12). It is noteworthy that, with an increase in κ, noticeable decrease in the boundary thickness and simultaneous increase in the tangential shear stresses on the disk occur. Values of derivatives of the radial and tangential velocity components (with respect to the coordinate ζ) and relative values of the tangential shear stresses on the disk are presented in Table 5.16 (subscript "0" stands for a free disk).

Computed values of the proportionality coefficient K_1 in Eq. (5.58) for the Nusselt number are given in Tables 5.17 and 5.18 for various values of the parameters $\kappa = a/\omega$ and n_* and two values of the Prandtl numbers Pr =0.71 and 1.0. As $\kappa \to \infty$, the value of K_1 for n_*=0 (or $T_w = const$) agrees with Eq. (5.48), whereas for $\kappa = a/\omega = 0$ the data of Tables 5.17 and 5.18 coincide with the results for a free rotating disk (Chap. 3).

Table 5.16 Values of constants of the exact self-similar solution of the Navier–Stokes equations for uniform flow impingement onto a rotating disk [183]

$\kappa=a/\omega$	$\alpha=-F'_w/G'_w$	$\tau_{w\varphi}/\tau_{w\varphi 0}$	F'_w	$-G'_w$
0.0	0.8284	1.0	0.5102	0.6159
0.1	0.8127	1.0438	0.4529	0.5573
0.2	0.8172	1.1124	0.4259	0.5212
0.4	0.8939	1.2834	0.4266	0.4772
0.6	1.0365	1.4642	0.4619	0.4456
0.8	1.2167	1.6383	0.5084	0.4179
1.0	1.4180	1.8022	0.5565	0.3924
1.5	1.9664	2.1692	0.6646	0.3380
2.0	2.5440	2.4888	0.7505	0.2950
3.0	3.7312	3.0338	0.8715	0.2336
4.0	4.9349	3.4972	0.9508	0.1927
5.0	6.1454	3.9070	1.0062	0.1637
6.0	7.3593	4.2781	1.0470	0.1423
8.0	9.7922	4.9378	1.1030	0.1126
10.0	12.229	5.5195	1.1395	0.0932
50	61.043	12.338	1.2736	0.0209
100	122.08	17.448	1.2925	0.0106
4.10^3	4883.1	110.35	1.3114	0.00027

Table 5.17 Values of K_1 according to the exact self-similar solution for Pr=0.71 [183]

$\kappa=a/\omega$	$n_*=-2$	$n_*=-1.5$	$n_*=-1$	$n_*=-0.5$	$n_*=0$	$n_*=0.5$	$n_*=1$	$n_*=2$	$n_*=3$	$N_*=4$
0.0	0.0	0.1035	0.1893	0.2624	0.3259	0.3818	0.4319	0.5185	0.5918	0.6555
0.1	0.0	0.1174	0.2078	0.2811	0.3429	0.3964	0.4436	0.5246	0.5927	0.6519
0.2	0.0	0.1306	0.2258	0.3007	0.3626	0.4154	0.4617	0.5405	0.6065	0.6637
0.3	0.0	0.1424	0.2425	0.3197	0.3826	0.4360	0.4825	0.5612	0.6269	0.6836
0.4	0.0	0.1524	0.2573	0.3371	0.4018	0.4562	0.5035	0.5833	0.6496	0.7068
0.5	0.0	0.1612	0.2705	0.3530	0.4194	0.4752	0.5236	0.6049	0.6723	0.7305
0.6	0.0	0.1687	0.2821	0.3671	0.4354	0.4926	0.5421	0.6252	0.6940	0.7532
0.7	0.0	0.1752	0.2922	0.3797	0.4498	0.5084	0.5590	0.6439	0.7141	0.7745
0.8	0.0	0.1809	0.3013	0.3910	0.4627	0.5226	0.5743	0.6609	0.7325	0.7941
0.9	0.0	0.186	0.3093	0.4011	0.4743	0.5354	0.5882	0.6764	0.7493	0.8120
1.0	0.0	0.1905	0.3165	0.4101	0.4847	0.547	0.6007	0.6905	0.7646	0.8283
1.5	0.0	0.2071	0.3432	0.444	0.5241	0.5908	0.6483	0.7442	0.8234	0.8914
2.0	0.0	0.2177	0.3604	0.4659	0.5498	0.6196	0.6796	0.7798	0.8625	0.9334
3.0	0.0	0.2304	0.3812	0.4926	0.5811	0.6546	0.7179	0.8235	0.9106	0.9853
4.0	0.0	0.2378	0.3934	0.5082	0.5994	0.6752	0.7404	0.8492	0.9389	1.0158
5.0	0.0	0.2426	0.4013	0.5184	0.6114	0.6887	0.7551	0.8661	0.9575	1.0359
6.0	0.0	0.2461	0.4069	0.5256	0.6198	0.6982	0.7656	0.878	0.9707	1.0502
8.0	0.0	0.2505	0.4143	0.5351	0.631	0.7108	0.7793	0.8938	0.9881	1.069
10	0.0	0.2533	0.4189	0.5411	0.6381	0.7187	0.788	0.9037	0.9991	1.0808
50	0.0	0.2630	0.4349	0.5618	0.6624	0.7462	0.8181	0.9382	1.0372	1.122
1000	0.0	0.2654	0.4390	0.5671	0.6687	0.7532	0.8258	0.9471	1.0469	1.1331

Table 5.18 Values of K_1 according to the exact self-similar solution for $Pr=1$ [183]

$\kappa=a/\omega$	$n_*=-2$	$n_*=-1.5$	$n_*=-1$	$n_*=-0.5$	$n_*=0$	$n_*=0.5$	$n_*=1$	$n_*=2$	$n_*=3$	$n_*=4$
0.0	0.0	0.1305	0.2352	0.3221	0.3963	0.4608	0.5180	0.6159	0.6982	0.7693
0.1	0.0	0.1414	0.2483	0.3339	0.4055	0.4669	0.5209	0.6130	0.6901	0.7568
0.2	0.0	0.1542	0.2652	0.3517	0.4227	0.4832	0.5360	0.6254	0.7001	0.7647
0.3	0.0	0.1663	0.2821	0.5744	0.4426	0.5034	0.5562	0.6453	0.7195	0.7836
0.4	0.0	0.1771	0.2978	0.3891	0.4626	0.5244	0.5779	0.6681	0.7428	0.8073
0.5	0.0	0.1867	0.3121	0.4061	0.4815	0.5447	0.5993	0.6910	0.7670	0.8324
0.6	0.0	0.1950	0.3248	0.4216	0.4989	0.5636	0.6194	0.7129	0.7903	0.8569
0.7	0.0	0.2023	0.3361	0.4355	0.5147	0.5809	0.6379	0.7334	0.8123	0.8801
0.8	0.0	0.2088	0.3462	0.4480	0.5290	0.5966	0.6548	0.7521	0.8325	0.9016
0.9	0.0	0.2145	0.3552	0.4592	0.5419	0.6108	0.6701	0.7693	0.8511	0.9214
1.0	0.0	0.2196	0.3633	0.4693	0.5536	0.6237	0.6840	0.7849	0.8681	0.9395
1.5	0.0	0.2385	0.3935	0.5075	0.5977	0.6728	0.7373	0.8450	0.9337	1.0099
2.0	0.0	0.2506	0.4130	0.5323	0.6268	0.7052	0.7726	0.8850	0.9776	1.0571
3.0	0.0	0.2652	0.4368	0.5626	0.6622	0.7449	0.8159	0.9343	1.0318	1.1154
4.0	0.0	0.2737	0.4506	0.5804	0.6830	0.7682	0.8413	0.9633	1.0637	1.1499
5.0	0.0	0.2792	0.4597	0.5920	0.6966	0.7835	0.8580	0.9824	1.0848	1.1726
6.0	0.0	0.2831	0.4661	0.6002	0.7063	0.7943	0.8699	0.9959	1.0997	1.1887
8.0	0.0	0.2883	0.4745	0.6110	0.7190	0.8086	0.8855	1.0138	1.1194	1.2099
10	0.0	0.2915	0.4798	0.6178	0.7270	0.8176	0.8953	1.0250	1.1318	1.2233
50	0.0	0.3026	0.4981	0.6414	0.7547	0.8488	0.9295	1.0641	1.1749	1.2699
1000	0.0	0.3055	0.5028	0.6475	0.7619	0.8568	0.9383	1.0741	1.1860	1.2819

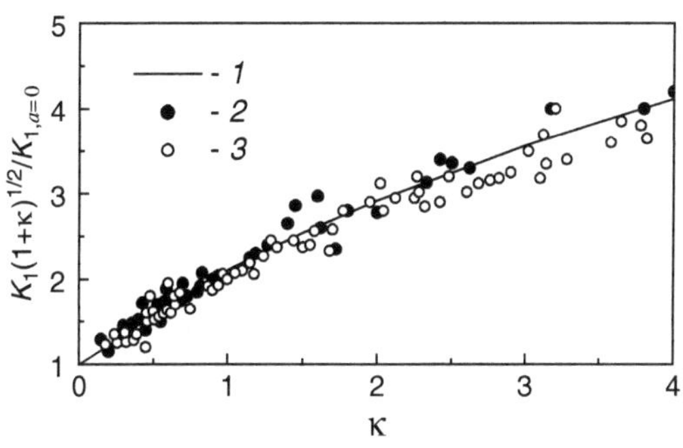

Fig. 5.14 Effect of the parameter $\kappa=a/\omega$ on heat transfer rate at orthogonal uniform air flow impingement onto an isothermal rotating disk ($Pr=0.71$, $T_w=const$). *1* – self-similar solution, Table 5.17. Experiments [116]: *2* – b=75.3 mm; *3* – b=25 mm

The self-similar solution for an isothermal disk ($T_w=const$) rotating in air was validated in the work [116] via comparisons with original experimental data over a wide range of variation of the parameter $a/\omega=0\ldots\infty$. As can be seen from Figs. 5.14 and 5.15, the self-similar solution agrees fairly well with the experimental data taken within the area of a disk $r/b\leq 0.8$ in view of the necessity to fulfil the requirements of the boundary conditions, Eqs. (2.32) and (5.43), for the radial velocity $v_{r,\infty}$ (see comments to Eqs. (5.44), (5.45) and (5.46)).

The constant K_1 increases significantly with κ. The expression for the Nusselt number (5.58) can be transformed as follows:

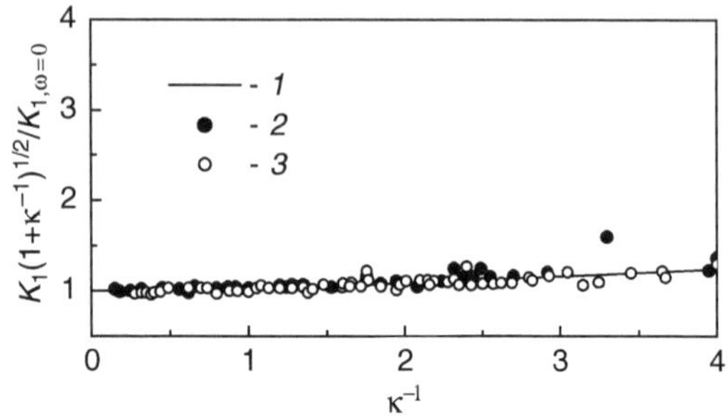

Fig. 5.15 Effect of the parameter $\kappa^{-1}=\omega/a$ on heat transfer rate at orthogonal uniform air flow impingement onto an isothermal rotating disk ($Pr=0.71$, $T_w=const$). *1* – self-similar solution, Table 5.17. Experiments [116]: *2* – $b=75.3$ mm; *3* – $b=25$ mm

$$Nu_d = K_1 \cdot (1 + \omega/a)^{1/2}\, Re_a^{1/2}. \tag{5.59}$$

The value of the product $K_{1*} = K_1 \cdot (1 + \omega/a)^{1/2}$ is given in Tables 5.19 and 5.20. It is obvious that the parameter K_{1*} is conservative and almost constant for $\kappa=1.5$–∞, all values of the Prandtl number and the parameter n_*. This indicates that, when $\kappa=a/\omega$ is larger than a certain threshold value, heat transfer of a rotating disk subjected to flow impingement is determined only by the impinging-flow parameters and is independent of the rotation velocity.

The data in Tables 5.19 and 5.20 may be interpreted in such a way that, in the case of the flow impingement on a disk, surface rotation leads to increase in heat transfer for $Re_a = idem$ only for values of κ less than the aforementioned threshold values.

Tables 5.21 and 5.22 represent data for the Reynolds analogy parameter χ (defined by Eq. (5.8)) for $Pr=1$ and 0.71 (and $\beta=0$). It is obvious that the values of χ vary depending on κ much weaker than the coefficient K_1. Like in turbulent flow, in case where $Pr=0.71$ the most noticeable (32%) variation of χ is observed for $n_*=-1.5$, whilst the weakest (0.9%) takes place for $n_*=4$. Apparently, in laminar flow relative variation of χ depending on κ is somewhat weaker than in turbulent flow, however, this variation takes place over much larger range of the parameter κ. Tables 5.21 and 5.22 contain data of calculations for $\kappa=0$–50, however, it is evident that χ is practically constant already for $\kappa=1.5$–∞ for any values of n_* (and for any Prandtl number).

Expression (5.59) is useful from a practical point of view in order to compare computations with experimental data. It is convenient that the values of K_{1*} become equal to K_1 for $\kappa\to\infty$. However, the constant K_{1*} becomes infinite for $\kappa=0$, which makes parameter K_{1*} less valuable than χ in view of the necessity to further develop the present integral method for laminar flow.

Comparisons with experiments for a case of co-axial orthogonal single jet impingement onto a rotating disk. As mentioned in the literature [22, 41, 93, 158], flow in a stagnation region of the jet depicted schematically in Fig. 5.16 is analogous to the case of uniform flow impingement onto a disk studied above. Authors [15] found that the regime where the impinging jet dominates over fluid flow on a rotating disk emerges under condition

Table 5.19 Values of $K_{1*} = K_1 \cdot (1 + \omega/a)^{1/2}$ according to the exact self-similar solution for Pr=1 [181, 183]

$\kappa=a/\omega$	n_*=–2	n_*=–1.5	n_*=–1	n_*=–0.5	n_*=0	n_*=0.5	n_*=1	n_*=2	n_*=3	n_*=4
0.0	0.0	∞	∞	∞	∞	∞	∞	∞	∞	∞
0.1	0.0	0.4689	0.8234	1.1076	1.3447	1.5486	1.7278	2.0330	2.2887	2.5100
0.2	0.0	0.3777	0.6495	0.8615	1.0355	1.1835	1.3128	1.532	1.7149	1.8731
0.3	0.0	0.3462	0.5872	0.7716	0.9213	1.0478	1.1578	1.3434	1.4978	1.6311
0.4	0.0	0.3314	0.5572	0.7278	0.8654	0.9810	1.0812	1.2498	1.3897	1.5103
0.5	0.0	0.3233	0.5405	0.7034	0.8340	0.9434	1.0380	1.1968	1.3284	1.4417
0.6	0.0	0.3184	0.5304	0.6884	0.8147	0.9203	1.0114	1.1642	1.2906	1.3993
0.7	0.0	0.3153	0.5238	0.6787	0.8021	0.9052	0.9941	1.1429	1.2658	1.3715
0.8	0.0	0.3132	0.5193	0.6720	0.7935	0.8949	0.9821	1.1282	1.2488	1.3524
0.9	0.0	0.3117	0.5161	0.6672	0.7874	0.8875	0.9736	1.1177	1.2366	1.3388
1.0	0.0	0.3106	0.5137	0.6637	0.7829	0.8820	0.9674	1.1100	1.2277	1.3287
1.5	0.0	0.3079	0.5080	0.6551	0.7717	0.8686	0.9519	1.0909	1.2055	1.3038
2.0	0.0	0.3069	0.5059	0.6520	0.7676	0.8637	0.9462	1.0839	1.1973	1.2946
3.0	0.0	0.3062	0.5043	0.6497	0.7646	0.8601	0.9421	1.0788	1.1914	1.2880
4.0	0.0	0.3060	0.5038	0.6489	0.7636	0.8588	0.9406	1.0770	1.1893	1.2856
5.0	0.0	0.3059	0.5035	0.6485	0.7631	0.8583	0.9399	1.0762	1.1883	1.2845
6.0	0.0	0.3058	0.5034	0.6483	0.7628	0.8579	0.9396	1.0757	1.1878	1.2839
8.0	0.0	0.3057	0.5033	0.6481	0.7626	0.8576	0.9392	1.0753	1.1873	1.2833
10	0.0	0.3057	0.5032	0.6480	0.7624	0.8575	0.9390	1.0750	1.1870	1.2830
50	0.0	0.3057	0.5031	0.6478	0.7622	0.8572	0.9387	1.0747	1.1866	1.2825
100.	0.0	0.3057	0.5031	0.6478	0.7622	0.8572	0.9387	1.0747	1.1866	1.2825

$$Re_j/Re_{rj} = V_j d_j/\omega r_i^2 > 0.125, \tag{5.60}$$

$$a/\omega > 0.125A(r_i/d_j)^2, \tag{5.61}$$

where $Re_{ri} = \omega r_i^2/\nu$ is the rotational Reynolds number based on an impingement radius r_i.

Strictly saying, this parameter becomes senseless for r_i=0. However, substituting b instead of r_i (because authors [15] dealt with jets impinging much closer to the rim of the disk than to the axis of rotation), one can rewrite Eq. (5.60) as

$$a/\omega > 0.03125A(d/d_j)^2. \tag{5.62}$$

According to Eq. (5.47), the value of A varies in experiments [155, 156] over the range of A=1.0–1.29. Hence, applicably to experiments [155, 156] (where d_j/d=0.09), the regime of impinging jet domination over heat transfer emerges, in accordance with Eq. (5.62), for a/ω>3.86–5. Allowing for the fact that Eq. (5.60) was obtained for jets located far enough from the axis of rotation, its agreement with our estimate at the beginning of the impingement domination regime (a/ω>1.5) is quite fair. However, it is worth noting that our results are more accurate for the co-axial impingement case.

For co-axial impingement of jets, whose diameter is significantly less than the disk diameter, the above-obtained self-similar solution is valid only inside the

Table 5.20 Values of $K_{1*} = K_1 \cdot (1 + \omega/a)^{1/2}$ according to the exact self-similar solution for Pr=0.71 [180, 181, 183]

$\kappa=a/\omega$	$n_*=-2$	$n_*=-1.5$	$n_*=-1$	$n_*=-0.5$	$n_*=0$	$n_*=0.5$	$n_*=1$	$n_*=2$	$n_*=3$	$n_*=4$
0.0	0.0	∞	∞	∞	∞	∞	∞	∞	∞	∞
0.1	0.0	0.3896	0.6891	0.9323	1.1373	1.3147	1.4714	1.7398	1.9657	2.1620
0.2	0.0	0.3199	0.5531	0.7366	0.8881	1.0176	1.1310	1.3240	1.4856	1.6257
0.3	0.0	0.2965	0.5048	0.6654	0.7965	0.9076	1.0044	1.1683	1.3049	1.4231
0.4	0.0	0.2850	0.4813	0.6307	0.7516	0.8535	0.9420	1.0912	1.2153	1.3223
0.5	0.0	0.2793	0.4685	0.6114	0.7265	0.8231	0.9069	1.0477	1.1645	1.2652
0.6	0.0	0.2755	0.4606	0.5995	0.7110	0.8045	0.8853	1.0209	1.1333	1.2300
0.7	0.0	0.2730	0.4554	0.5918	0.7010	0.7923	0.8712	1.0034	1.1128	1.2069
0.8	0.0	0.2713	0.4519	0.5865	0.6941	0.7839	0.8615	0.9914	1.0988	1.1911
0.9	0.0	0.2702	0.4494	0.5827	0.6891	0.7780	0.8546	0.9828	1.0887	1.1798
1.0	0.0	0.2694	0.4476	0.5800	0.6855	0.7736	0.8495	0.9765	1.0813	1.1714
1.5	0.0	0.2674	0.4431	0.5732	0.6766	0.7628	0.8369	0.9608	1.063	1.1508
2.0	0.0	0.2666	0.4414	0.5707	0.6733	0.7588	0.8323	0.9551	1.0563	1.1432
3.0	0.0	0.2661	0.4402	0.5689	0.6709	0.7559	0.829	0.9509	1.0514	1.1377
4.0	0.0	0.2659	0.4398	0.5682	0.6701	0.7549	0.8278	0.9495	1.0497	1.1357
5.0	0.0	0.2658	0.4396	0.5679	0.6697	0.7544	0.8272	0.9488	1.0489	1.1348
6.0	0.0	0.2658	0.4395	0.5677	0.6695	0.7542	0.8269	0.9484	1.0485	1.1343
8.0	0.0	0.2657	0.4394	0.5676	0.6693	0.7539	0.8266	0.948	1.048	1.1338
10	0.0	0.2657	0.4393	0.5675	0.6692	0.7538	0.8265	0.9478	1.0478	1.1336
50	0.0	0.2656	0.4393	0.5674	0.669	0.7536	0.8262	0.9475	1.0475	1.1332
100.	0.0	0.2656	0.4392	0.5674	0.669	0.7536	0.8262	0.9475	1.0475	1.1332

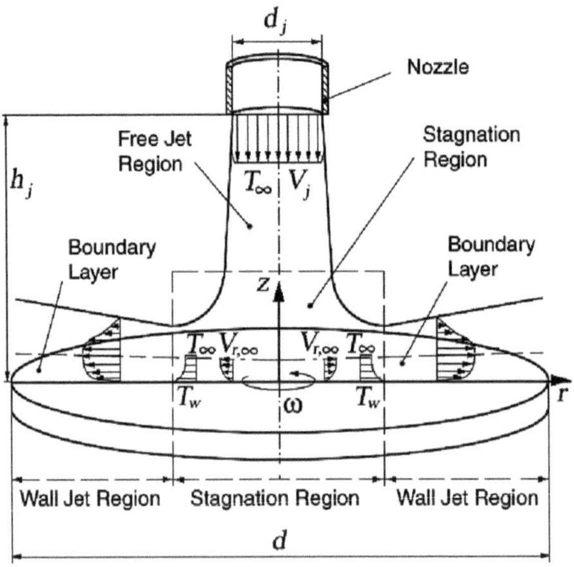

Fig. 5.16 Geometric arrangement, fluid flow and heat transfer of a single impinging round jet perpendicular to and co-axial with a rotating disk [181]

Table 5.21 Values of χ according to the exact self-similar solution and the present integral method for $Pr=1$ (bold font)

$\kappa=a/\omega$	$n_*=-2$	$n_*=-1.5$	$n_*=-1$	$n_*=-0.5$	$n_*=0$	$n_*=0.5$	$n_*=1$	$n_*=2$	$n_*=3$	$n_*=4$
0.0	0.0	0.2119	0.3818	0.5230	0.6433	0.7482	0.8410	1.0000	1.1335	1.2490
0.0	**0.0**	**0.2047**	**0.3752**	**0.5195**	**0.6431**	**0.7502**	**0.8439**	**1.0000**	**1.1249**	**1.2270**
0.1	0.0	0.2306	0.4050	0.5448	0.6614	0.7617	0.8499	1.0000	1.1257	1.2346
0.1	**0.0**	**0.2310**	**0.4026**	**0.5446**	**0.6643**	**0.7665**	**0.8549**	**1.0000**	**1.1142**	**1.2063**
0.2	0.0	0.2465	0.4240	0.5623	0.6759	0.7726	0.8570	1.0000	1.1194	1.2227
0.2	**0.0**	**0.2478**	**0.4206**	**0.5613**	**0.6783**	**0.7773**	**0.8621**	**1.0000**	**1.1073**	**1.1931**
0.4	0.0	0.2651	0.4458	0.5824	0.6924	0.7850	0.8651	1.0000	1.1120	1.2085
0.4	**0.0**	**0.2657**	**0.4402**	**0.5794**	**0.6935**	**0.7889**	**0.8699**	**1.0000**	**1.1000**	**1.1792**
0.5	0.0	0.2701	0.4516	0.5877	0.6968	0.7882	0.8673	1.0000	1.1099	1.2046
0.5	**0.0**	**0.2704**	**0.4454**	**0.5842**	**0.6975**	**0.7920**	**0.8719**	**1.0000**	**1.0981**	**1.1756**
0.6	0.0	0.2735	0.4556	0.5913	0.6998	0.7905	0.8688	1.0000	1.1086	1.2020
0.6	**0.0**	**0.2737**	**0.4490**	**0.5875**	**0.7003**	**0.7941**	**0.8733**	**1.0000**	**1.0968**	**1.1731**
0.8	0.0	0.2776	0.4603	0.5956	0.7034	0.7932	0.8706	1.0000	1.1069	1.1988
0.8	**0.0**	**0.2777**	**0.4534**	**0.5915**	**0.7036**	**0.7966**	**0.8750**	**1.0000**	**1.0952**	**1.1702**
1.0	0.0	0.2798	0.4628	0.5980	0.7053	0.7946	0.8715	1.0000	1.1060	1.1970
1.0	**0.0**	**0.2798**	**0.4557**	**0.5937**	**0.7055**	**0.7980**	**0.8759**	**1.0000**	**1.0944**	**1.1686**
1.5	0.0	0.2823	0.4656	0.6005	0.7074	0.7962	0.8726	1.0000	1.1050	1.1951
1.5	**0.0**	**0.2822**	**0.4584**	**0.5961**	**0.7075**	**0.7995**	**0.8770**	**1.0000**	**1.0934**	**1.1668**
5.0	0.0	0.2842	0.4679	0.6026	0.7091	0.7975	0.8734	1.0000	1.1042	1.1936
5.0	**0.0**	**0.2842**	**0.4605**	**0.5981**	**0.7091**	**0.8008**	**0.8778**	**1.0000**	**1.0927**	**1.1654**
10.0	0.0	0.2844	0.4681	0.6028	0.7092	0.7976	0.8735	1.0000	1.1042	1.1935
10.0	**0.0**	**0.2843**	**0.4607**	**0.5983**	**0.7093**	**0.8009**	**0.8778**	**1.0000**	**1.0926**	**1.1653**
50.0	0.0	0.2844	0.4681	0.6028	0.7093	0.7976	0.8735	1.0000	1.1041	1.1934
50.0	**0.0**	**0.2844**	**0.4608**	**0.5983**	**0.7093**	**0.8009**	**0.8779**	**1.0000**	**1.0926**	**1.1653**

stagnation region for $r \leq d_j/2$. For the conditions of co-axial jet impingement onto a rotating disk, Eq. (5.59) should be rewritten so that

$$Nu_{dj} = K_{1*} \cdot Re_j^{1/2} \cdot A^{1/2}. \tag{5.63}$$

Here d_j instead of d is used as a characteristic length in both the Nusselt number and the Reynolds number.

Experimental data [155, 156] are presented in a form of the Nusselt numbers $Nu_b = Nu_{dj} \cdot (b/d_j)$. Therefore, to simplify comparisons with experiments [155, 156], Eq. (5.63) should be rearranged in the final form

$$Nu_b = K_{1*} \cdot Re_j^{1/2} \cdot A^{1/2} \cdot (b/d_j). \tag{5.64}$$

It should also be pointed out that the disk surface in experiments [155, 156] in the stagnation region was practically isothermal at $n_*=0$.

Authors [155, 156] kept the ratio $d_j/d = 0.09$ constant, while intensity of rotation was characterized by the Re_φ number. Using the definition of parameter A, one can obtain

Table 5.22 Values of χ according to the exact self-similar solution and the present integral method for Pr=0.71 [184] (bold font)

$\kappa=a/\omega$	$n_*=-2$	$n_*=-1.5$	$n_*=-1$	$n_*=-0.5$	$n_*=0$	$n_*=0.5$	$n_*=1$	$n_*=2$	$n_*=3$	$n_*=4$
0.0	0.0	0.2366	0.4330	0.6001	0.7452	0.8732	0.9876	1.1856	1.3533	1.4990
0.0	**0.0**	**0.2308**	**0.4276**	**0.5973**	**0.7453**	**0.8753**	**0.9906**	**1.1857**	**1.3446**	**1.4766**
0.1	0.0	0.2699	0.4774	0.6459	0.7879	0.9108	1.0193	1.2053	1.3618	1.4978
0.1	**0.0**	**0.2698**	**0.4727**	**0.6430**	**0.7883**	**0.9137**	**1.0231**	**1.2048**	**1.3495**	**1.4674**
0.2	0.0	0.2941	0.5085	0.6772	0.8165	0.9355	1.0398	1.2172	1.3658	1.4946
0.2	**0.0**	**0.2944**	**0.5020**	**0.6728**	**0.8162**	**0.9385**	**1.0440**	**1.2168**	**1.3525**	**1.4618**
0.4	0.0	0.3212	0.5424	0.7108	0.8470	0.9619	1.0616	1.2297	1.3695	1.4902
0.4	**0.0**	**0.3207**	**0.5337**	**0.7049**	**0.8461**	**0.9648**	**1.0660**	**1.2293**	**1.3556**	**1.4560**
0.5	0.0	0.3286	0.5513	0.7195	0.8549	0.9687	1.0672	1.2329	1.3704	1.4889
0.5	**0.0**	**0.3276**	**0.5421**	**0.7133**	**0.8540**	**0.9717**	**1.0717**	**1.2326**	**1.3563**	**1.4546**
0.6	0.0	0.3334	0.5572	0.7253	0.8602	0.9733	1.0710	1.2351	1.3710	1.4880
0.6	**0.0**	**0.3324**	**0.5479**	**0.7191**	**0.8593**	**0.9764**	**1.0756**	**1.2347**	**1.3569**	**1.4536**
0.8	0.0	0.3388	0.5641	0.7322	0.8665	0.9787	1.0755	1.2377	1.3717	1.4870
0.8	**0.0**	**0.3382**	**0.5549**	**0.7262**	**0.8658**	**0.9820**	**1.0803**	**1.2374**	**1.3575**	**1.4524**
1.0	0.0	0.3418	0.5679	0.7359	0.8698	0.9816	1.0779	1.2390	1.3721	1.4864
1.0	**0.0**	**0.3414**	**0.5587**	**0.7300**	**0.8694**	**0.9851**	**1.0828**	**1.2388**	**1.3579**	**1.4517**
1.5	0.0	0.3453	0.5721	0.7400	0.8736	0.9848	1.0805	1.2405	1.3725	1.4857
1.5	**0.0**	**0.3449**	**0.5630**	**0.7343**	**0.8733**	**0.9885**	**1.0856**	**1.2404**	**1.3583**	**1.4510**
5.0	0.0	0.3479	0.5754	0.7433	0.8765	0.9874	1.0826	1.2417	1.3728	1.4852
5.0	**0.0**	**0.3478**	**0.5664**	**0.7377**	**0.8765**	**0.9913**	**1.0879**	**1.2417**	**1.3586**	**1.4505**
10.0	0.0	0.3480	0.5756	0.7435	0.8767	0.9876	1.0828	1.2418	1.3728	1.4852
10.0	**0.0**	**0.3480**	**0.5666**	**0.7380**	**0.8767**	**0.9914**	**1.0881**	**1.2418**	**1.3586**	**1.4504**
50.0	0.0	0.3481	0.5757	0.7436	0.8768	0.9876	1.0829	1.2418	1.3728	1.4852
50.0	**0.0**	**0.3481**	**0.5668**	**0.7381**	**0.8768**	**0.9916**	**1.0882**	**1.2418**	**1.3586**	**1.4504**

$$\kappa = \frac{A \cdot V_j}{\omega \cdot d_j} = A \cdot \frac{Re_j}{Re_{\omega j}}, \tag{5.65}$$

where $Re_{\omega j} = \omega d_j^2/\nu$ is the rotational Reynolds number based on the nozzle diameter. It is obvious that $Re_{\omega j} = Re_\varphi \cdot b^2/d_j^2 = 0.25 \cdot Re_\varphi \cdot (d/d_j)^2$.

In accordance with Eq. (5.47), parameter A takes values A =1.29, 1.14 and 1.01 for h_j/d_j=2, 4 and 6, respectively. As shown beneath, these values allow reaching good agreement of predictions with experiments [155, 156] for the Nusselt number for Re_j=24,700. For Re_j=6800, experimental results do not confirm the tendency of decrease in Nu_b with an increasing h_j/d_j. The best agreement with experiments for Re_j=6800 yields the value A=1.12 independent of h_j/d_j.

Comparisons of the present predictions and experiments are shown in Fig. 5.17 in a form of dependence of $K_{1*} = Nu_b/[Re_j^{1/2}A^{1/2}(b/d_j)]$ (transformed equation (5.64)) on the parameter $\kappa=a/\omega$, which for experimental conditions [155, 156] was calculated by Eq. (5.65) using the aforementioned values of the parameter A. Predictions and experiments agree well both quantitatively (with inaccuracy of 5%) and qualitatively clearly exhibiting tendency of increase with decreasing values of κ.

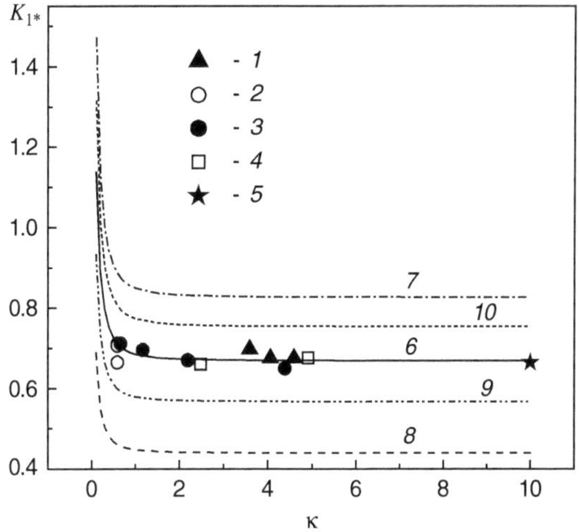

Fig. 5.17 Effect of the parameter $\kappa = a/\omega$ on the constant K_{1*} for $Pr=0.71$ [183]. Experiments [155, 156]: *1* – $Re_j=24{,}700$, $Re_b=214{,}000$, $h_j/d_j=2$, 4 and 6; *2* – $Re_j=6800$, $Re_b=396{,}000$, $h_j/d_j=2$, 4 and 6; *3* – $Re_j=6800$, $Re_b=53{,}500$, 107,000, 200,000 and 353,000, $h_j/d_j=2$; *4* – $Re_j=24{,}700$, $Re_b=2000$ and 396,000, $h_j/d_j=2$. Experiments [44, 200]: *5* – $n_*=0$, various Re_j and h_j/d_j. Predictions: *6* – $n_*=0$; *7* – $n_*=1$; *8* – $n_*=-1$; *9* – $n_*=-0.5$; *10* – $n_*=0.5$

Three experimental points fall out of this generally good agreement. This fact can be explained in the following way. Too high experimental value of K_{1*} for $Re_j=24{,}700$, $Re_b=214{,}000$ and $h_j/d_j=6$ ($\kappa=3.6$) is probably caused by the too low value $A=1.01$ used in the recalculation of experimental data for Nu_b using Eq. (5.64). Perhaps, real dependence of A on h_j/d_j is weaker than that predicted by Eq. (5.47). Too low experimental value of K_{1*} for $Re_j=6800$, $Re_b=53{,}000$ and $h_j/d_j=2$ ($\kappa=4.4$) is probably caused by the fact that we used the constant value $A=1.12$ in the recalculation of Nu for $Re_j=6800$ and varying Re_φ. Rotation may affect A for low values of Re_j, hence one has to find this dependence in an empirical way in the future. At last, too low value of K_{1*} for $Re_j=6800$, $Re_b=396{,}000$ and $h_j/d_j=4$ ($\kappa=0.6$) is probably explained by experimental inaccuracy. Otherwise, it is impossible to account for by any physical reason why the value of Nu_b for $h_j/d_j=4$ decreases noticeably in comparison with its value for $h_j/d_j=2$ and then increases again for $h_j/d_j=6$ under conditions of $Re_j=idem$, $Re_b=idem$. In order to clarify the dependence of A on h_j/d_j and, possibly, on the Reynolds numbers Re_j and Re_b, an additional experimental research is needed.

Integral method: fluid flow. An integral method for this problem was developed by the authors in works [180, 181, 184]. Integral equations of the boundary layer are used in the form of Eqs. (2.26), (2.27), (2.28), (2.29) and (2.30) for $v_{\varphi,\infty}=0$ (or $\beta=0$). Model assumptions for the velocity profiles are

$$v_\varphi/(\omega\, r) = g, \quad v_r/(\omega\, r) = (1 - g_*) \cdot tan\varphi, \tag{5.66}$$

$$(tan\varphi - \kappa)\, /\, (\alpha - \kappa) = T. \tag{5.67}$$

For $\kappa=0$, one can write down

$$\left[\mathrm{v}_r/(\omega\, r)\right]_{\kappa=0} = \alpha_0\,(1-g_*)\,T = f, \tag{5.68}$$

then

$$\mathrm{v}_r/\mathrm{v}_{r,\infty} = (1-g_*)\,\kappa + (\alpha-\kappa)f/\alpha_0. \tag{5.69}$$

Here g, g_*, f, $tan\varphi$ and T are universal functions of the only variable ξ; $g = G_0 = G_{a=0}$ and $f = F_0 = F_{a=0}$ are determined from the exact solution for a free disk at $\kappa = 0$; g_*, $tan\varphi$ and T are unknown; $\alpha = tan\varphi_w$ for $z=0$; $\alpha_0=0.8284$; subscript "0" indicates the free rotating disk ($\kappa=0$).

As a result, one can obtain the following relations for integral parameters:

$$\overline{\delta}_r^* = A_4 - A_1 X, \quad \overline{\delta}_r^{**} = B_7 + B_8 X - B_1 X^2, \tag{5.70}$$

$$\overline{\delta}_{\varphi r}^{**} = \kappa\,(B_6 + D_1 X) = \alpha D_1 + \kappa\,(B_6 - D_1)\,, \quad \int_0^1 \mathrm{v}_\varphi^2 d\xi = (\omega r)^2\, C_1, \tag{5.71}$$

$$A_4 = \int_0^1 g_*\, d\xi, \quad B_6 = \int_0^1 (1-g_*)\, g\, d\xi, \quad B_7 = \int_0^1 (1-g_*)\, g_*\, d\xi, \tag{5.72}$$

$$B_8 = -A_1 + 2D_5, \quad D_5 = \frac{1}{\alpha_0}\int_0^1 f\, g_*\, d\xi, \quad X = \alpha/\kappa - 1, \tag{5.73}$$

while constants A_1, B_1, C_1, D_1 are specified in Eqs. (5.17), (5.18) and (5.19). Equations (5.66), (5.67), (5.68), (5.69), (5.70), (5.71), (5.72) and (5.73) are valid both for laminar and turbulent flow. In the latter case, Eqs. (5.20) and $g = g_*$ should be used. Constants A_2, $A_3 = 1 - A_1 - A_2$, B_2, B_3, $B_4 = -A_1 + 2D_1$, $B_5 = \int_0^1 (1-g)\, g\, d\xi$, D_2, D_3 and D_4 are used in Sects. 2.4 and 2.5 to obtain the solution for turbulent flow. Whereas $g = g_*$, the value of B_8 equals to B_4, while the values of B_6 and B_7 are equal to B_5.

It is evident that constants (5.72) and (5.73) are universal for a given flow regime, whilst integral parameters $\overline{\delta}_r^*$, $\overline{\delta}_r^{**}$ and $\overline{\delta}_{\varphi r}^{**}$ are functions of the only variable X and do not depend on the boundary layer thickness δ.

Finding out the value of δ in numerical solution of Eqs. (2.37), (2.38), (2.39) and (2.40) depends on the desired accuracy of satisfying boundary conditions (5.57) for the values of F and G at $z\to\infty$. Therefore, numerical integration of the functions f and g (and g_*) here, like in Sect. 5.1.2, was conducted using the variable ζ and then reduced to the conditions for a free disk ($\kappa=N=0$ and $\beta=0$) using the variable $\zeta_0 = z\sqrt{\omega/\nu} = \zeta_{a=0}$. The upper limit of integration $\zeta_\delta = \delta\sqrt{\omega/\nu}$ was chosen high enough to find four significant digits of the definite integrals. As a result, for laminar flow Eqs. (5.21), (5.22) and (5.23) for the respective constants at $\beta=0$ remain in force, while the rest of the constants are defined by the following relations:

$$A_4 = a_4/\zeta_{\delta 0},\quad B_6 = b_6/\zeta_{\delta 0},\quad B_7 = b_7/\zeta_{\delta 0},\quad B_8 = b_8/\zeta_{\delta 0},\quad D_5 = d_5/\zeta_{\delta 0}, \tag{5.74}$$

$$a_4 = \int\limits_0^{\zeta_{\delta 0}} g_*\, d\zeta_0,\quad b_6 = \int\limits_0^{\zeta_{\delta 0}} (1-g_*)\, g\, d\zeta_0,\quad b_7 = \int\limits_0^{\zeta_{\delta 0}} (1-g_*)\, g_*\, d\zeta_0, \tag{5.75}$$

$$b_8 = -a_1 + 2d_5,\quad d_5 = \frac{1}{\alpha_0}\int\limits_0^{\zeta_{\delta 0}} f g_*\, d\zeta_0. \tag{5.76}$$

Values of all the above constants are universal and independent of κ. Exact values of a_1, b_1, c_1 and d_1 according to data of Table 5.8 for β=0 are as follows:

$$a_1 = 0.5338,\quad b_1 = 0.07887,\quad c_1 = 0.6726,\quad d_1 = 0.18587. \tag{5.77}$$

Constants (5.77) coincide with respective values obtained in [139]. The rest of the constants (5.75) and (5.76) are yet unknown and will be found below.

Shear stresses are determined by Eqs. (5.24), (5.25) and (5.26) and the following relations:

$$\tau_{wr} = \frac{\mu v_{r,\infty}}{\delta}\left(\frac{d\bar{v}_r}{d\xi}\right)_{\xi=0},\quad \left(\frac{d\bar{v}_r}{d\xi}\right)_{\xi=0} = -\left(\frac{dg_*}{d\xi}\right)_{\xi=0} + \frac{X}{\alpha}\left(\frac{df}{d\xi}\right)_{\xi=0},$$

$$\left(\frac{df}{d\xi}\right)_{\xi=0} = \left(\frac{df}{d\zeta_0}\right)_{\zeta=0}\cdot \zeta_{\delta 0},\quad \left(\frac{dg_*}{d\xi}\right)_{\xi=0} = \left(\frac{dg_*}{d\zeta_0}\right)_{\zeta=0}\cdot \zeta_{\delta 0}.$$

Values $(dg/d\zeta_0)_{\zeta=0}$ and $(df/d\zeta_0)_{\zeta=0}$ are given in Eqs. (3.1) based on the solution for a free disk.

Given α=*const*, δ=*const* and X=*const* [41], one can transform Eqs. (2.26) and (2.27) as follows:

$$a^2\delta\left[3\left(B_7 + B_8 X - B_1 X^2\right) + A_4 - A_1 X\right] + \delta\omega^2 C_1$$
$$= \frac{\nu a}{\delta}\left[-\left(\frac{dg_*}{d\xi}\right)_{\xi=0} + \frac{X}{\alpha_0}\left(\frac{df}{d\xi}\right)_{\xi=0}\right],\ 4a\delta\left[B_6 + D_1 X\right] = -\frac{\nu}{\delta}\left(\frac{dg}{d\xi}\right)_{\xi=0}. \tag{5.78}$$

Analysis of the exact solution shows that, as applied to Eqs. (5.78) the following is true:

$$(dg_*/d\xi)_{\xi=0} = (dg/d\xi)_{\xi=0}\,. \tag{5.79}$$

Solving Eqs. (5.78) jointly, one can obtain the following quadratic equation:

$$X^2 + \frac{h_2}{h_3}X + \frac{h_4 + c_1/\kappa^2}{h_3} = 0, \tag{5.80}$$

$$h_3 = -3b_1 - 4d_1,\quad h_2 = 3b_8 - a_1 - 4b_6 - 4d_1, \tag{5.81}$$

$$h_1 = h_4 + c_1/\kappa^2,\quad h_4 = 3b_7 + a_4 - 4b_6. \tag{5.82}$$

Its solution looks as follows:

$$X = X_{s1} + \left(X_{s2} + X_{s3}/\kappa^2\right)^{1/2}, \tag{5.83}$$

$$\alpha = (X_{s1} + 1)\,\kappa + \left(X_{s2}\kappa^2 + X_{s3}\right)^{1/2}, \tag{5.84}$$

$$X_{s1} = -h_2/\,(2h_3)\,, \quad X_{s2} = X_{s1}^2 - h_4/h_3, \quad X_{s3} = -c_1/h_3. \tag{5.85}$$

It is noteworthy that namely constants (5.75), (5.76) and (5.77) rather than (5.72) and (5.73) are used in Eqs. (5.80), (5.81), (5.82), (5.83), (5.84) and (5.85), which means that the value $\zeta_{\delta 0}$ was eliminated as a result of the transformations.

Constants of Eqs. (5.83), (5.84) and (5.85) can be found via agreeing these formulas with the exact solution. Equation (5.84) satisfies conditions for κ=0, since constants h_3 and X_{s3} are based on the data for a free disk. Imposing a requirement that the approximate solution (5.83), (5.84) and (5.85) must coincide with the exact solution for κ=4000 (a stationary disk) and κ=1.65 (optimal value), one can obtain

$$h_2 = -2.79818, \quad h_3 = -0.98009, \quad h_4 = 0.66556, \tag{5.86}$$

$$X_{s1} = -1.42751, \quad X_{s2} = 2.71687, \quad X_{s3} = 0.68625. \tag{5.87}$$

Based on the second of Eqs. (5.78), one can write

$$\frac{\tau_{w\varphi}}{\tau_{w\varphi 0}} = \frac{\delta_0}{\delta} = \left[\frac{\alpha + \kappa\,(b_6/d_1 - 1)}{\alpha_0}\right]^{1/2}. \tag{5.88}$$

Agreeing Eq. (5.88) with the exact solution, one can obtain

$$b_6 = 0.42776, \quad b_6/d_1 = 2.30137. \tag{5.89}$$

The constant a_4 can be found using Eq. (5.70) for $\overline{\delta}_r^*$ at κ=4000

$$a_4 = \overline{\delta}_{r\zeta}^*\,(\delta_0/\delta)\,(1+\kappa)^{-1/2}, \quad \text{where } \overline{\delta}_{r\zeta}^* = \int\limits_0^\infty \left(1 - F\frac{1+\kappa}{\kappa}\right) d\zeta. \tag{5.90}$$

Parameter $\overline{\delta}_{r\zeta}^*$ can also be calculated from the exact solution. The remaining constants b_7, b_8 and d_5 can be determined from equations (5.75), (5.76) and (5.82). As a result, one can obtain

$$a_4 = 1.11052, \quad b_7 = 0.42202, \quad b_8 = 0.063373, \quad d_5 = 0.29859. \tag{5.91}$$

For κ=4000, the value of b_8 determined by means of Eq. (5.70)

$$b_8 = \left[\bar{\delta}_{r\zeta}^{**}(\delta_0/\delta)(1+\kappa)^{-1/2} - b_7 + X^2 b_1\right] X^{-1}, \tag{5.92}$$

Table 5.23 Hydrodynamic parameters of the problem by the data of the exact solution and the integral method [180]

$\kappa=a/\omega$	$\overline{\delta}^{*}_{r\zeta}$, exact	$\overline{\delta}^{*}_{r\zeta}$	$\overline{\delta}^{**}_{r\zeta}$, exact	$\overline{\delta}^{**}_{r\zeta}$	α, exact	α	$\tau_{w\varphi}/\tau_{w\varphi0}$, exact	$\tau_{w\varphi}/\tau_{w\varphi0}$
0.1	–1.760	–2.607	–3.057	–2.985	0.8127	0.8019	1.0438	1.0607
0.2	0.0290	–0.4896	–0.1604	–0.1064	0.8172	0.8061	1.1124	1.1346
0.3	0.4849	0.1461	0.2281	0.2652	0.8452	0.8365	1.1947	1.2170
0.4	0.6491	0.4171	0.3207	0.3467	0.8939	0.8877	1.2834	1.3038
0.5	0.7150	0.5499	0.3459	0.3646	0.9591	0.9548	1.3741	1.3921
0.7	0.7497	0.6571	0.3487	0.3593	1.1232	1.1211	1.5525	1.5662
0.9	0.7457	0.6882	0.3391	0.3457	1.3153	1.3143	1.7216	1.7322
1.0	0.7399	0.6931	0.3340	0.3394	1.4180	1.4172	1.8022	1.8116
1.5	0.7072	0.6867	0.3132	0.3156	1.9664	1.9663	2.1692	2.1749
2.0	0.6821	0.6708	0.2999	0.3012	2.5440	2.5441	2.4888	2.4926
4.0	0.6326	0.6300	0.2761	0.2764	4.9349	4.9350	3.4972	3.4987
6.0	0.6130	0.6119	0.2672	0.2673	7.3593	7.3593	4.2781	4.2790
8.0	0.6026	0.6020	0.2625	0.2626	9.7922	9.7923	4.9380	4.9384
10	0.5962	0.5958	0.2597	0.2597	12.229	12.229	5.5197	5.5201
50	0.5746	0.5745	0.2501	0.2501	61.043	61.033	12.338	12.338
100.	0.5717	0.5717	0.2489	0.2489	122.08	122.06	17.449	17.449
1000	0.5692	0.5692	0.2478	0.2478	1220.8	1220.8	55.178	55.178
4000	0.5690	0.5690	0.2477	0.2477	4883.1	4883.1	110.36	110.36

$$\overline{\delta}^{**}_{r\zeta} = \int\limits_0^\infty \left(1 - F\frac{1+\kappa}{\kappa}\right) F\frac{1+\kappa}{\kappa} d\zeta, \tag{5.93}$$

coincides with the value b_8 computed by Eq. (5.91). Here $\overline{\delta}^{**}_{r\zeta}$ is calculated based on the exact solution.

Table 5.23 presents data for some parameters of fluid flow calculated both by the exact solution and the integral method. Maximal discrepancies of the data for α and $\tau_{w\varphi}/\tau_{w\varphi0}$ by the integral method with respect to the exact solution do not exceed 1.7% and decrease with increasing κ. Integral parameters $\overline{\delta}^{**}_{r\zeta}$ for $\kappa>0.5$ and $\overline{\delta}^{*}_{r\zeta}$ for $\kappa>1$ calculated using the integral method also agree well with the exact solution.

Integral method: heat transfer. The integral equation (2.28) of the thermal boundary layer jointly with transformed equation (2.27) can be reduced to a form analogous to Eqs. (3.52) and (3.53)

$$\frac{d}{dr}\left[r^4\delta\overline{\delta}^{**}_{\varphi r}\right] = \frac{c_f}{2} r^4 (1+\alpha^2)^{1/2}, \tag{5.94}$$

$$\frac{d}{dr}\left[r^2\delta\overline{\delta}^{**}_T\,(T_w - T_\infty)\right] = \chi\frac{c_f}{2} r^2 (1+\alpha^2)^{1/2}\,(T_w - T_\infty)\,. \tag{5.95}$$

The solution (2.82), (2.83), (2.84), (2.85), (2.86) and (2.87) for the parameters of the laminar and turbulent boundary layers remains in force and will be used further below.

Then Eqs. (5.94) and (5.95), in view of the boundary condition (2.35), can be transformed in the following way:

$$(2 + m + n_*)\,\frac{\overline{\delta_T^{**}}}{\alpha A_1} = \chi\,(4 + m)\,\frac{\overline{\delta_{\varphi r}^{**}}}{\alpha A_1}, \tag{5.96}$$

$$(2 + m + n_*)\,\frac{\overline{\delta_T^{**}}}{\alpha A_1} = \chi\,(4 + m)\,K_{V0}\left[1 + \frac{\kappa}{\alpha}\left(\frac{b_6}{d_1} - 1\right)\right]. \tag{5.97}$$

In Eqs. (5.96) and (5.97), relations (5.70), (5.71), (5.72), (5.73), (5.74), (5.75), (5.76), (5.77) and (5.78) are taken into account. For laminar flow $K_{V0} = 0.3482$ (Eq. (3.55)). For turbulent flow $K_{V0} = 1 - D_2/A_1$ (Eq. (3.47)), and B_5/D_1, where $B_5 = C_2/2 = D_1 + D_3$, is used instead of b_6/d_1.

To remind, this problem for turbulent flow was already solved in Sect. 5.2.2. The form of approximating expression for $\overline{\delta_T^{**}}$ for laminar flow is found by generalization of Eq. (5.53) obtained for turbulent flow in Sect. 5.2.2. For laminar flow n=1, and an approximate solution for $\overline{\delta_T^{**}}$ at κ=0 must be agreed with the solution for a free disk (Chap. 3). As a result of these considerations, one can transform Eq. (5.53) to the following relation:

$$\frac{\overline{\delta_T^{**}}}{\alpha A_1} = \frac{1}{b_2} - \chi Pr^{n_p}\,(1 - K_{V0})\,\frac{b_1}{b_2} + \frac{\kappa}{\alpha}\left(e_1\chi^{-1} + e_2\chi + e_3\right), \tag{5.98}$$

where constants b_1 (again not to confuse with Eq. (5.77)) and b_2, as well as n_p, depending on the Prandtl number are given by Eqs. (3.65), (3.66) and (3.67) for a free disk. Then Eq. (5.97) for laminar flow takes the following form:

$$\frac{1}{b_2} - \chi Pr^{n_p}\,(1 - K_{V0})\,\frac{b_1}{b_2} + \frac{\kappa}{\alpha}\left(e_1\chi^{-1} + e_2\chi + e_3\right) = \chi\,\frac{4}{2 + n_*}K_{V0} \left[1 + \frac{\kappa}{\alpha}\left(\frac{b_6}{d_1} - 1\right)\right]. \tag{5.99}$$

The coefficients e_1, e_2 and e_3 are assumed to be dependent on the Prandtl number alone. Equating values of χ in Eq. (5.99) to the results of the exact solution for n_*=–1.5, 0 and 2, one can obtain values e_1, e_2 and e_3 given in Table 5.24. The value of χ for the exact solution can be calculated by the formula

$$\chi = -K_1\left[(1 + \kappa)\,Pr\,(dG/d\zeta)_{\zeta=0}\right]^{-1}. \tag{5.100}$$

Given e_1, e_2 an e_3, Eq. (5.99) can be transformed to the quadratic equation

$$\chi^2 f_1 + \chi f_2 + f_3 = 0, \tag{5.101}$$

Table 5.24 Values of the coefficients e_1, e_2 and e_3 [180, 181, 184]

Pr	e_1	e_2	e_3
1	0.1426	–0.7227	1.0333
0.9	0.1591	–0.7313	1.1280
0.8	0.1800	–0.7404	1.2427
0.72	0.2007	–0.7495	1.3556
0.71	0.2042	–0.7497	1.3696
0.6	0.2425	–0.7631	1.5708
0.5	0.2957	–0.7733	1.8104
0.4	0.3618	–0.7915	2.1676
0.3	0.5152	–0.7756	2.6166
0.2	0.7072	–0.8166	3.6207
0.1	1.4528	–0.8052	5.7291

where

$$f_1 = \tfrac{\kappa}{\alpha} e_2 - \tfrac{4}{2+n_*} K_{V0} \left[1 + \tfrac{\kappa}{\alpha}\left(\tfrac{b_6}{d_1} - 1\right)\right] - Pr^{n_p}\left(1 - K_{V0}\right)\tfrac{b_1}{b_2},$$
$$f_2 = \tfrac{1}{b_2} + \tfrac{\kappa}{\alpha} e_3, \quad f_3 = \tfrac{\kappa}{\alpha} e_1.$$

The solution of Eq. (5.101) has the following form:

$$\chi = \left[-f_2 - D^{1/2}\right]/(2f_1), \quad D = f_2^2 - 4f_1 f_3. \tag{5.102}$$

Constants K_1 and K_{1*} at known χ can be calculated from the relations

$$K_1 = -\chi \frac{\tau_{w\varphi}}{\tau_{w\varphi 0}} \left(\frac{dg}{d\zeta_0}\right)_{\zeta=0} (1+\kappa)^{-1/2}\, Pr, \tag{5.103}$$

$$K_{1*} = -\chi \frac{\tau_{w\varphi}}{\tau_{w\varphi 0}} \left(\frac{dg}{d\zeta_0}\right)_{\zeta=0} \kappa^{-1/2} Pr, \tag{5.104}$$

with allowance for Eqs. (3.1) and (5.88).

Numerical values of the parameter χ computed using Eq. (5.102) for Pr=0.71 and 1 are presented in Tables 5.21 and 5.22 in comparison with the exact solution. Maximal deviations of the values χ by Eq. (5.102) from the exact solution do not exceed 2.4% (for n_*=4). As data in Table 5.25 indicate, the same conclusion is valid also for K_{1*} (as well as for the constant K_1). Computations for other values of the Prandtl number in the range Pr=0.1–1 also confirmed this conclusion.

Table 5.26 represents the values of K_{1*} calculated based on the results of the exact solution and the integral method for a stationary disk ($\kappa \geq 50$) for different values of Pr. In particular, for a stationary disk at Pr=0.71, it follows from Eqs. (5.102), (5.103) and (5.104) that

Table 5.25 Values of K_{1*} computed by the integral method for Pr=0.71 (the first line represents K_1) [180]

$\kappa=a/\omega$	n_*=–2	n_*=–1.5	n_*=–1	n_*=–0.5	n_*=0	n_*=1	n_*=2	n_*=3	n_*=4
0.0	0.0	0.1009	0.187	0.2612	0.3259	0.4332	0.5185	0.5880	0.6457
0.1	0.0	0.3957	0.6933	0.9431	1.1562	1.5007	1.7671	1.9793	2.1524
0.2	0.0	0.3266	0.5569	0.7464	0.9055	1.1582	1.3499	1.5004	1.6217
0.3	0.0	0.3015	0.5063	0.6725	0.8106	1.0273	1.1897	1.3159	1.4169
0.4	0.0	0.2891	0.4811	0.6354	0.7628	0.9610	1.1082	1.2220	1.3126
0.5	0.0	0.2821	0.4667	0.6141	0.7352	0.9226	1.0611	1.1677	1.2523
0.7	0.0	0.2749	0.4518	0.5920	0.7065	0.8827	1.0120	1.1110	1.1893
0.9	0.0	0.2715	0.4448	0.5815	0.6929	0.8637	0.9886	1.0841	1.1593
1.0	0.0	0.2704	0.4426	0.5783	0.6887	0.8578	0.9814	1.0757	1.1500
1.5	0.0	0.2678	0.4371	0.5702	0.6781	0.8430	0.9632	1.0547	1.1268
2.0	0.0	0.2669	0.4351	0.5672	0.6742	0.8376	0.9565	1.0470	1.1181
4.0	0.0	0.2659	0.4331	0.5642	0.6703	0.8321	0.9498	1.0393	1.1096
6.0	0.0	0.2657	0.4328	0.5636	0.6696	0.8311	0.9485	1.0378	1.1080
8.0	0.0	0.2657	0.4326	0.5634	0.6693	0.8307	0.9481	1.0373	1.1074
10.	0.0	0.2657	0.4326	0.5634	0.6692	0.8306	0.9479	1.0371	1.1071
50.	0.0	0.2656	0.4325	0.5632	0.6690	0.8303	0.9475	1.0366	1.1067

$$K_{1*} = K_1 = 0.763\chi, \quad \chi = \frac{-1.8205 - \left[3.8863 + \frac{1.9253}{n_*+2}\right]^{1/2}}{-1.7102 + \frac{5.7551}{n_*+2}}. \qquad (5.105)$$

Obviously, inaccuracies of the integral method do not exceed 2.4% for any value of Pr.

Thus, in the present sub-section an exact solution of the problem of fluid flow and heat transfer under conditions of uniform laminar forced air cooling of a rotating disk under boundary conditions (2.32), (2.33), (2.34) and (2.35) for β=0 was obtained. Nusselt numbers were computed for the Prandtl number values equal to 1 and 0.71, and a series of discrete values of the exponent n_* (in the range n_*=–2 to 4) in the power law for the radial variation of the disk temperature T_w and ratios of the impingement-to-rotation velocity $\kappa=a/\omega$=0–4000. Boundaries of the flow regime were determined, where heat transfer depends only on the impinging flow velocity.

Such a regime emerges for $\kappa=a/\omega$=1.5–∞ for any values of Pr and n_*. These results can be interpreted in such a way that, if a disk is subjected to flow impingement, disk rotation leads to increase in heat transfer for $Re_a = idem$ only for values of κ less than the aforementioned threshold values.

Results of the computations performed in this sub-section agree overall well with the experiments [155, 156] in the stagnation region. An improvement of the agreement of the experimental points that fall out of this generally good trend can be attained via finding a more accurate dependence of the parameter A on h_j/d_j and the Reynolds numbers Re_j and Re_φ.

An integral method was developed and an approximate analytical solution of this problem in the range n_*=–2 to 4 was obtained, whose constants were determined

Table 5.26 Values of K_{1*} the data of the exact solution (boldface) and the integral method for different values of Pr and $\kappa \geq 50$ for a stationary disk [180]

Pr	$n_*=-2$	$n_*=-1.5$	$n_*=-1$	$n_*=-0.5$	$n_*=0$	$n_*=1$	$n_*=2$	$n_*=3$	$n_*=4$
1.0	**0.0**	**0.3057**	**0.5031**	**0.6478**	**0.7622**	**0.9387**	**1.0747**	**1.1866**	**1.2825**
	0.0	0.3056	0.4952	0.6430	0.7623	0.9434	1.0747	1.1742	1.2523
0.9	**0.0**	**0.2928**	**0.4827**	**0.6221**	**0.7324**	**0.9028**	**1.034**	**1.1421**	**1.2348**
	0.0	0.2928	0.4751	0.6175	0.7325	0.9073	1.0340	1.1302	1.2057
0.8	**0.0**	**0.279**	**0.4607**	**0.5944**	**0.7003**	**0.864**	**0.9902**	**1.0942**	**1.1833**
	0.0	0.2790	0.4535	0.5900	0.7003	0.8683	0.9902	1.0829	1.1556
0.72	**0.0**	**0.2671**	**0.4417**	**0.5705**	**0.6726**	**0.8306**	**0.9525**	**1.0529**	**1.139**
	0.0	0.2671	0.4349	0.5663	0.6726	0.8347	0.9524	1.0419	1.1123
0.6	**0.0**	**0.2476**	**0.4105**	**0.5311**	**0.6269**	**0.7754**	**0.8901**	**0.9846**	**1.0657**
	0.0	0.2476	0.4042	0.5272	0.6269	0.7792	0.8901	0.9745	1.0409
0.5	**0.0**	**0.2295**	**0.3812**	**0.494**	**0.5839**	**0.7235**	**0.8314**	**0.9203**	**0.9967**
	0.0	0.2295	0.3755	0.4905	0.5839	0.7269	0.8314	0.9110	0.9737
0.4	**0.0**	**0.2083**	**0.3477**	**0.4517**	**0.5348**	**0.6641**	**0.7642**	**0.8468**	**0.9178**
	0.0	0.2083	0.3424	0.4484	0.5348	0.6672	0.7641	0.8381	0.8965
0.3	**0.0**	**0.1838**	**0.3082**	**0.4017**	**0.4766**	**0.5936**	**0.6845**	**0.7595**	**0.8240**
	0.0	0.1838	0.3027	0.3975	0.4753	0.5956	0.6844	0.7527	0.8069
0.2	**0.0**	**0.1536**	**0.2592**	**0.3393**	**0.4039**	**0.5053**	**0.5843**	**0.6498**	**0.7061**
	0.0	0.1535	0.2554	0.3369	0.4038	0.5076	0.5842	0.6432	0.6901
0.1	**0.0**	**0.1124**	**0.1911**	**0.2519**	**0.3015**	**0.3802**	**0.4421**	**0.4936**	**0.5381**
	0.0	0.1124	0.1886	0.2503	0.3015	0.3819	0.4420	0.4888	0.5262

using the data of the exact solution. Maximal errors of the integral method relative to the exact solution do not exceed 2.4% (for $n_*=4$) for any values of Pr.

5.3 Non-symmetric Flow over a Parallel Rotating Disk

In this section, theoretical and experimental data of different authors are considered for the case of forced external flow parallel to flat surfaces of a disk and orthogonal to its cylindrical surface (Fig. 5.18). The main difference of this type of flow from the case considered in Sect. 5.2 is that the flow pattern in Fig. 5.18 is non-symmetric, as fluid flows parallel to the disk only from one side (from left to right in Fig. 5.18).

For the first time, this problem was studied experimentally by the authors of [36]. They studied only average heat transfer of an *entire* isothermal ($T_w=const$) disk rotating in external flow of air.

According to [36], average heat transfer of a free rotating disk and a stationary disk subjected to non-symmetric external parallel flow in turbulent regime is described by the following empirical relations, respectively,

$$Nu_{av} = 0.015 Re_{\varphi}^{0.8}, \tag{5.106}$$

$$Nu_{av} = 0.036 Re_{\infty}^{0.8}, \tag{5.107}$$

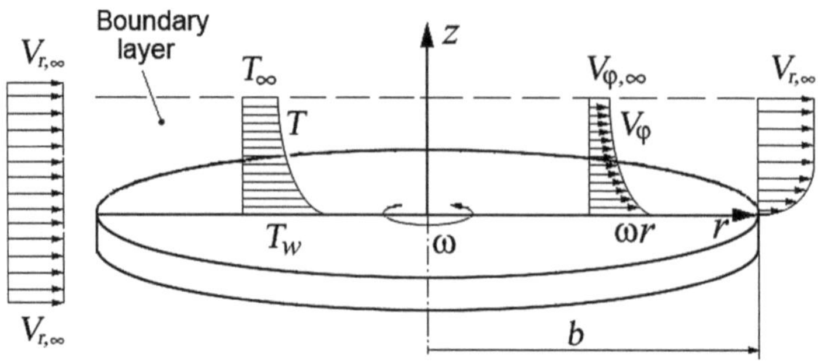

Fig. 5.18 Fluid flow and heat transfer in non-symmetric flow over a parallel rotating disk

where $Re_\infty = v_{r,\infty}b/\nu$. In laminar flow regime ($Re_\omega \leq 2.4 . 10^5$), the authors of [36] obtained for a free rotating disk

$$Nu_{av} = 0.4Re_\varphi^{0.5}. \tag{5.108}$$

Data by Eq. (5.107) exceed values by 33%, which can be calculated using the relation

$$Nu_{av} = 0.027Re_\infty^{0.8} \tag{5.109}$$

that was obtained in [36] based on a model given in monograph [158] as applied to a circular flat plate with a heated starting length. According to the authors of [36], the deviation of Eq. (5.107) from Eq. (5.109) is caused by separation and subsequent reattachment of flow impinging on the cylindrical rim of the disk with sharp non-round edges between the rim and the flat surfaces of the disk. Another consequence of such disk geometry was that external flow over the disk was always turbulent.

Based on the experimental results of [36], total average heat transfer for simultaneous existence of external flow and disk rotation may be calculated as a root-mean-square of Eqs. (5.106), (5.107) and (5.108), i.e. by the following relations:

$$Nu_{av} = \sqrt{(0.4Re_\varphi^{0.5})^2 + (0.036Re_\infty^{0.8})^2}, \quad Re_\varphi \leq 4 \times 10^5, \tag{5.110}$$

$$Nu_{av} = \sqrt{(0.015Re_\varphi^{0.8})^2 + (0.036Re_\infty^{0.8})^2}, \quad Re_\varphi > 4 \times 10^5. \tag{5.111}$$

Equations (5.110) and (5.111) agree with experiments [36] with an error not exceeding 2.3%. From Eqs. (5.110) and (5.111), one can also determine boundaries of regimes with dominant effects of rotation or external flow, as well as of the regime where both effects are in force.

The author of [220, 221] considered a problem where the disk depicted in Fig. 5.18 was very thin (actually, it had zero thickness). Because of this, flow separation and subsequent reattachment causing significant flow turbulization and heat transfer enhancement did not occur in the investigation [220, 221]. Therefore, results of [220, 221] have rather academic than applied value. The problem was modelled numerically using large eddy simulation (LES) approach also for the boundary condition T_w=*const* and air flow, Pr=0.7. Distributions of the local Nusselt numbers on the disk and temperature fields in the wake behind the disk presented in these works reveal strong non-symmetry of the problem.

According to generalizations of the computational data [220, 221], average Nusselt numbers for the disk can be described by the following relations:

(a) for $10^3 \leq Re_\infty \leq 5.0\times 10^4$

$$Nu_{\rm av} = \begin{cases} 0.417Re_\infty^{1/2}, & Re_\varphi/Re_\infty \leq 1.4, \\ 0.33Re_\varphi^{1/2}, & Re_\varphi/Re_\infty > 1.4, \end{cases} \tag{5.112}$$

(b) for $Re_\infty > 5.0\times 10^4$

$$Nu_{\rm av} = \begin{cases} \sqrt{(0.015Re_\varphi^{0.8})^2 + (0.0127Re_\infty^{0.8})^2}, & Re_\varphi > 2\times 10^5, \\ \sqrt{(0.33Re_\varphi^{0.5})^2 + (0.0127Re_\infty^{0.8})^2}, & Re_\varphi < 2\times 10. \end{cases} \tag{5.113}$$

Equations (5.112) and (5.113) agree with the results of numerical simulations with inaccuracy not exceeding 10% over the entire range of the Reynolds numbers considered in the study.

It follows from Eqs. (5.112) and (5.113) that case (a) and first of Eq. (5.112) for $Nu_{\rm av}$ are valid for a stationary disk in laminar flow. For turbulent flow, case (b) is in force, and the following equation should be used to calculate $Nu_{\rm av}$ over a stationary disk:

$$Nu_{\rm av} = 0.0127Re_\infty^{0.8}. \tag{5.114}$$

The Nusselt numbers computed by Eq. (5.114) are significantly lower than experimental data [36] described by both Eqs. (5.107) and (5.109), which again means that Eqs. (5.112), (5.113) and (5.114) can be hardly used in engineering practice.

Mass transfer in naphthalene sublimation in air from a surface of a rotating disk of a finite thickness for the problem depicted in Fig. 5.18 was studied experimentally in [59]. Authors of [59] obtained the following empirical relation for the Sherwood number and corresponding relation for the Nusselt number at Pr=0.72 and T_w=*const*

$$Sh_{\rm av} = 0.00952Re_\Sigma^{0.925}, \tag{5.115}$$

$$Nu_{\rm av} = 0.00588Re_\Sigma^{0.925}, \tag{5.116}$$

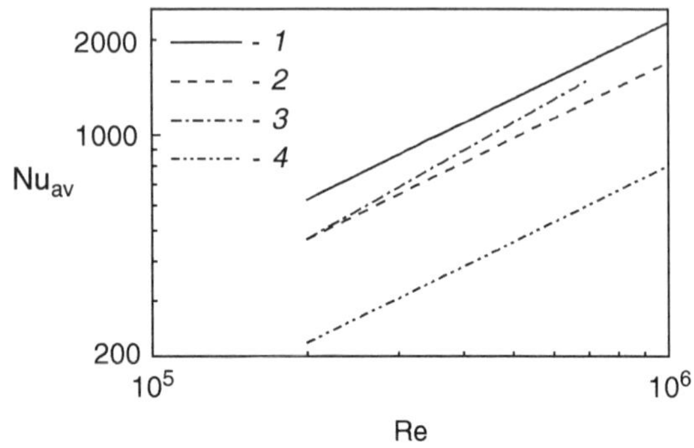

Fig. 5.19 Heat transfer in non-symmetric flow over a parallel stationary disk (see Fig. 5.18) for Pr=0.72, T_w=$const$. *1* – Eq. (5.107) [36]; *2* – Eq. (5.109) [36]; *3* – Eq. (5.116) [59]; *4* – Eq. (5.114) [220, 221]

where the Reynolds number Re_Σ is based on a characteristic velocity taking into account both the rotation velocity ω and the velocity of external flow $v_{r,\infty}$. Equations (5.115) and (5.116) are valid for Re_Σ=(2.03–6.86)×10^5. Unfortunately, the authors of [59] did not provide the reader with an exact definition of the characteristic velocity. It is however clear that $Re_\Sigma \rightarrow Re_\infty$ in the case where effects of rotation become negligible, which justifies use of the equation $Nu = Sh(Pr/Sc)^{0.4}$ for the recalculation of Eq. (5.115) into Eq. (5.116). This enables comparisons made in Fig. 5.19, which show that the curve based on (5.116) is in good agreement with standard equation (5.109). Curve based on experimental equation (5.107) for the reasons mentioned above lies higher than these data, while data from Eq. (5.114) [220, 221] as expected are too much lower than the experimental results.

Thus, it may be concluded that the average Nusselt number for the case of simultaneous disk rotation and non-symmetrical parallel external flow over its surface can be calculated as a root-mean-square value of the Nusselt numbers for a free rotating disk and a stationary disk in non-symmetrical parallel external flow over its surface, taking into account also the actual ratio of the characteristic Reynolds numbers of rotation and external flow.

Chapter 6
Outward Underswirled and Overswirled Radial Flow Between Parallel Co-rotating Disks

6.1 Flow in the Ekman Layers

This section overviews in brief the model and results for the Ekman layers, which are important for understanding and simulations of the flow regimes that develop in rotating-disk cavities.

If a disk rotates in a fluid under conditions $\beta \to 1$ and $v_{r,\infty}=0$, the so-called Ekman layers develop over the disk [138, 139], in which velocity components v_r, v_z and $(v_\varphi - \omega r)$ are negligibly small in comparison with ωr. Given this, the boundary layer equations in the differential and integral form can be reduced to the following [139]:

$$-2\rho\omega(v_\varphi - v_{\varphi,\infty}) = \frac{\partial \tau_r}{\partial z},\quad 2\rho\omega v_r = \frac{\partial \tau_\varphi}{\partial z}, \tag{6.1}$$

$$2\omega\,\rho\,\delta\int_0^1 \left(v_\varphi - v_{\varphi,\infty}\right) d\xi = \tau_{wr},\quad 2\omega\,\rho\,\delta\int_0^1 v_r d\xi = -\tau_{w\varphi}. \tag{6.2}$$

For laminar flows, the following solutions of Eqs. (6.1) and (6.2) were obtained from the work [139]:

$$v_r = \omega\, r(1-\beta)\exp(-\zeta)\sin(\zeta),\quad v_\varphi - v_{\varphi,\infty} = \omega\, r(1-\beta)\exp(-\zeta)\cos(\zeta), \tag{6.3}$$

$$\dot{m}_d/(\mu r) = \pi(1-\beta)Re_\omega^{1/2},\quad \delta/r = \pi Re_\omega^{1/2}, \tag{6.4}$$

$$\tau_{wr}/(\rho\omega^2 r^2) = (1-\beta)Re_\omega^{-1/2},\quad \alpha_E = 0, \tag{6.5}$$

where $\zeta = z\sqrt{\omega/\nu}$.

For turbulent flow, a solution of the boundary layer equations was obtained for Ekman layers in [139] using an integral method and the power-law approximation of the velocity profiles (2.45), (2.46) (and n=1/7). This solution has the form of Eqs. (2.82), (2.83), (2.84), (2.85) and (2.86), (2.90) and (5.2) provided that α_E=0.553, γ_{*E}=0.0983 and ε^*_{mE}=0.1395. These values α_E, γ_{*E} and ε^*_{mE} coincide with the respective data that can be obtained from Eqs. (5.1) for N=0 and β=1.

I.V. Shevchuk, *Convective Heat and Mass Transfer in Rotating Disk Systems*, Lecture Notes in Applied and Computational Mechanics 45, DOI 10.1007/978-3-642-00718-7_6,
© Springer-Verlag Berlin Heidelberg 2009

Table 6.1 Constants α_E, γ_E and ε^*_{mE} for Ekman layers [163]

Coefficient	n=1/7	n=1/8	n=1/9	n=1/10
α_E	0.6936	0.6442	0.604	0.5705
γ_E	0.1224	0.0978	0.0807	0.0677
ε^*_{mE}	0.1386	0.106	0.084	0.0678

Based on the model for power-law profiles (2.59), (2.60), (2.61), (2.62) and (2.63), the authors of [163] using the present integral method (see Chap. 2) obtained a similar solution for the Ekman layers for the case of turbulent flow:

$$\alpha_E = \left[A_1^{-1} n/(n+1)\right]^{1/2}, \tag{6.6}$$

$$\tau_{wr}/\rho = \left[2n/(n+1)\right](1-\beta)\omega^2 r\delta, \tag{6.7}$$

$$\tau_{w\varphi}/\rho = 2A_1\alpha_E(1-\beta)\omega^2 r\delta. \tag{6.8}$$

In the solution in [163], parameters δ/r, $\dot{m}_d/(\rho\omega r^3)$ and C_M for the Ekman layers are also given by Eqs. (2.82), (2.83), (2.84), (2.85) and (2.86), (2.89), (2.90) and (2.91) at N=0 and substitution of H_9 with $2A_1\alpha_E$.

Numerical values of α_E, γ_E and ε^*_{mE} for the Ekman layers for different n are documented in Table 6.1. It is obvious that these values coincide with respective constants for N=0 and β=1 for the case of solid-body rotation (see Sect. 5.1).

In spite of the rather modest differences in the numerical values of α_E, solution for ε^*_{mE} obtained in the present work at n=1/7 agrees with that given in [138].

Solutions for the Ekman layers are significantly simpler than the solutions of the full equations of the boundary layer for N=0 and β=*const*. Nevertheless, the formulas for the Ekman layers allow, with the accuracy sufficient for practical purposes, to calculate parameters of the fluid flow for $0.5\le\beta\le 2$; in doing so, no restrictions are imposed on the dependence of β on the coordinate r.

6.2 Radial Outflow Between Parallel Co-rotating Disks

6.2.1 Flow Structure, Experiments and Computations of Different Authors

Secondary air is widely used for cooling end surfaces of gas turbine rotors. Cooling air from a compressor is fed into a cavity between two rotating disks of a turbine (or between a disk and a cover plate) near the axis of rotation or at a certain radial location, then moves radially outwards and finally leaves the cavity (Figs. 6.1 and 6.2). Fluid flow and heat transfer in radially outward flow in the rotating-disk cavities

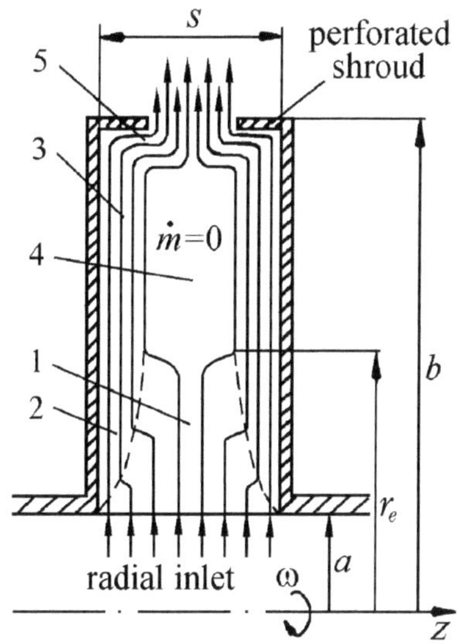

Fig. 6.1 Physical structure of radially outward flow between two parallel co-rotating disks for $\beta_i \leq 1$ and a uniform radial inlet [139]. *1* – source region; *2* – boundary layer entraining air from the source region; *3* – Ekman-type layer; *4* – internal core (no radial mass-flow); *5* – region of the outlet from the cavity

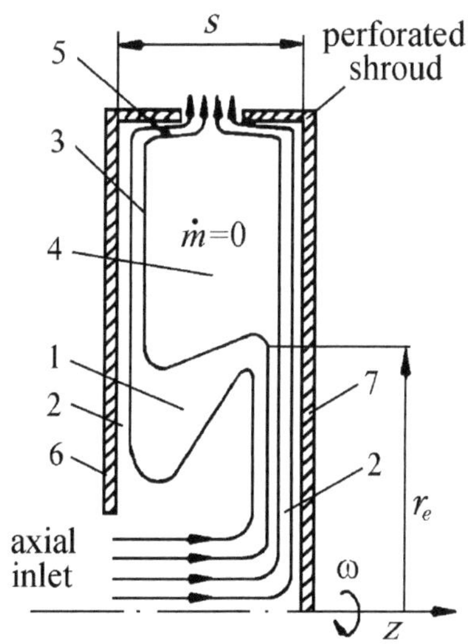

Fig. 6.2 Schematic of streamlines in a rotating cavity for $\beta_i \leq 1$ and axial inlet [139]. *1* – source region; *2* – boundary layer entraining air from the source region; *3* – Ekman-type layer; *4* – internal core (no radial mass-flow); *5* – region of the outlet from the cavity; *6* – upstream disk; *7* – downstream disk. Subscript "i" means parameters at the inlet into the cavity

were studied experimentally and numerically in many works [2, 25, 43, 72, 75–79, 86, 90, 114, 128, 129, 134, 135, 138, 139, 145, 148, 196, 225].

In the present work, cases of radial/axial inlet into the cavity are considered. Boundary layers on the disk are assumed to be separately developing (non-merging), and distributions of the disk surface temperature T_w were arbitrary functions (specified according to experimentally measured data).

For cases of the so-called underswirl of fluid flow, with values of the swirl coefficient at the inlet being less than unity ($\beta_i \leq 1$), experiments [139] showed that flow structure for the purely radial inlet into the cavity looks as depicted in Fig. 6.1. Starting from the inlet up to a certain radial coordinate r_e, air flows radially outwards over the entire cross-section of the cavity, while on the disks boundary layers develop, which entrain air from the potential flow (the so-called source region).

In the source region, tangential component of the potential flow velocity decreases according to a free-vortex law:

$$\mathrm{v}_{\varphi,\infty} = \left(\mathrm{v}_{\varphi,\infty} \cdot r\right)_i / r. \tag{6.9}$$

For the case of the axial inlet into the cavity (Fig. 6.2), incoming flow (totally or partially) in a form of a jet impinges over the downstream disk and moves further radially outwards in a form of an annular wall jet. This jet contains usually more than a half or even the entire amount of air fed into the cavity. However, this situation does not persist up to the end of the cavity. At the radial coordinate $r=r_e$, a part of air flowing over the downstream disk is released from the wall jet and is finally entrained by the boundary layer on the upstream disk.

As mentioned above, in the regions with $r<r_e$ on both disks, boundary layers develop that entrain air from the source region. For $r>r_e$, the entire mass flow of air is entrained into the so-called Ekman-type layers [139] with a constant mass flowrate $\dot{m}_d$ in them. For the case of the radial inlet into the cavity, in each Ekman-type layer the following relation holds: $\dot{m}_d = 0.5\dot{m} = const$ (the rate of entrainment into the layers is zero). For the case of the axial inlet, the fraction of air over the downstream disk in the impinging jet region is larger than that over the upstream disk. However, as indicated above, air is redistributed between flows over the disks even before the onset of the development of the Ekman-type layers, in which the condition $\dot{m}_d = 0.5\dot{m} = const$ remains in force for both the disks [139]. The tangential velocity in the internal core of flow begins to increase due to the spinning effect of the disks.

Over a long period of time, modelling of fluid flow and heat transfer in the rotating-disk cavities has been performed using integral methods.

Assuming that $\mathrm{v}_\varphi = \mathrm{v}_{\varphi,\infty}$ for any value of the axial coordinate z, integrating Eq. (2.23) from 0 to s and using, for β =0–0.9, the following relation for $\tau_{w\varphi}$

$$\left|\tau_{w\varphi}\right| = 0.0274\rho\,(\omega r)^2\,(1-\beta)^{1.2}\,Re_\omega{}^{-0.2}, \tag{6.10}$$

the authors of [196] derived the following differential equation for the swirl parameter β

$$\frac{d\beta}{dx} = -\frac{2\beta}{x} + 2Ax^{1.6}\,(1-\beta)^{1.2}, \tag{6.11}$$

where $A = 0.0274 \times 2\pi \times Re_\varphi^{0.8} / C_w$. As shown below, Eq. (6.11) agrees well with experimental data only in the beginning of the source region and at the end of the region of the Ekman-type layers.

Authors of [43], having further developed the approach of [196], obtained the following equation:

$$\frac{d\beta}{dx} = -\frac{2\beta}{x} + \frac{0.6c_1\,(1-\beta)^{c_2}\,[c_3\,(1-\beta)+\beta]\,Re_{\varphi}^{(n+1)/(3n+1)}x^{2(n+1)/(3n+1)}}{0.5C_w b/r_i - c_1\,(1-\beta)^{c_2}\,Re_{\varphi}^{(n+1)/(3n+1)}x^{(5n+3)/(3n+1)}}, \quad (6.12)$$

where $c_1 = 0.2054\pi e^{-0.175/n}$, $c_2 = 0.753n^{-0.424}$, $c_3 = 1.543n^{0.625}$, n=1/5. In their derivations, authors of [43] made a mistake, which resulted in that in the second term of the right-hand side of Eq. (6.12), the first term $0.5C_w b/r$ in the denominator was multiplied by ρ. As shown below, the "erroneous" solution of the authors of [43] agrees well enough with their own experiments. The corrected Eq. (6.12) agrees with the experiments of [43] already worse, and both variants of Eq. (6.12) disagree noticeably with experimental data [139] for the well-developed Ekman-type layers.

The most adequate model for the swirl parameter β was developed in the work [139] based on the physical understanding of the fluid flow structure discussed above while describing schematics in Figs. 6.1 and 6.2. Details of the model of [139] are outlined below in Sects. 6.2.2, 6.2.3 and 6.2.4.

Experiments of [77] for heat transfer in the cavity were performed for β_i =1 and separately developing boundary layers. In case of the axial inlet into the cavity, only the source region was formed, with the Nusselt numbers (Figs. 6.1 and 6.2) being described by the following correlation [77]:

$$Nu = 0.024Re_{\omega}^{0.8}Pr^{0.6}V_M, \quad (6.13)$$

$$V_M = 2.461+0.128\bar{r}^2+0.122\bar{r}\big/N_i + 0.02N_i^{-2} - 0.858\bar{r} - 0.392\big/N_i,\ \bar{r} = 1.2\ldots 2, \quad (6.14)$$

$$V_M = 1.08N_i^{0.07}, \quad \bar{r} = 2\ldots 3. \quad (6.15)$$

Here $N_i = \mathrm{v}_{r,i}/(\omega r_i)$=0.1667–0.5, $\bar{r} = r/r_i$, Re_{ω}=6×10^4–10^6, $T_w = const.$

It was obtained in work [196] that, given the boundary condition (2.35) for the disk temperature T_w, the Nusselt number can be described by the following correlation:

$$Nu = 0.0145\,(1-\beta)^{0.2}\,Re_{\omega}^{0.8}\,(n_* + 2.6)^{0.2}\left[1 - 0.45\,(1-\beta)^{0.6}\,Re_{\omega}^{-0.1}\right]^{-1}. \quad (6.16)$$

Experimental data [70] for flow between a disk and a co-rotating cover plate were generalized by the empirical correlation:

$$Nu = 0.046Re_{\omega}^{0.8}N^{0.328}, \quad (6.17)$$

where $N = \dot{m}/(2\pi \rho s r^2)$. The value of β at the inlet to the cavity and at the outlet from it was equal to unity; the disk was heated from the side of its cylindrical surface.

For the condition β_i=0, the author of [76] developed the following empirical equation:

$$Nu_{av} = 0.0306 Re_{\varphi}^{0.8} N_i^{0.1} \left(b/r_i\right)^{0.3}, \tag{6.18}$$

where Pr=0.72; the distribution of $T_w(r)$ was close to quadratic; $v_{r,\infty}=(v_{r,\infty} \cdot r)_i/r$; N_i=0.143–1.67; s/b=0.13; b/r_i=2.15–2.7; $Re_{\omega}=5\times10^5$ to 4×10^6.

Authors of [18] generalized their experiments for the case of β_i=0 with the correlation

$$Nu = 0.0339\, Re_{\omega}\, (1 + N^2)^{0.4} \exp(-0.148/N). \tag{6.19}$$

Here $r_i/b \rightarrow 0$; $N = \dot{m}/(2\pi \rho s \omega r^2)$; N_i =0.4–4.0; $Re_{\omega}(1 + N^2)^{0.4}$ exp $(-0.185/N) = 10^4$ to 2×10^5; $Re_{\omega} = \omega r r_h/\text{v}$; $Nu = q_w r_h/[k(\Delta T)]$; r_h is the coordinate of the beginning of a heated section of the disk; $\Delta T = T_w - T_{\infty}$; the distribution of $T_w(r)$ not specified (most probably, T_w =*const*).

While developing their integral methods, the authors of [73, 76, 77] recognized existence of only the source region and accepted a series of model simplifications, which significantly devalued their models. Theoretical dependencies given in these works are less reliable than empirical Eqs. (6.13), (6.14), (6.15), (6.16), (6.17), (6.18) and (6.19). Relation (6.16) takes into account effect of β, but disregards the effect of $v_{r,\infty}$ (i.e. the effect of superimposed radial flow) and that of the Prandtl number. Unfortunately, Eqs. (6.13), (6.14), (6.15), (6.17), (6.18) and (6.19) do not take into consideration the form of the distribution of $T_w(r)$ and differences in the flow structure in the source region and the Ekman-type layers.

As shown in the work [139], in the source region the flow temperature T_{∞} is constant and equals to T_i. In the region of the Ekman-type layers, entrainment of cold air from the core into the boundary layers ceases, and the latter, conversely, begin to release heat into the core. This leads to an increase in the temperature T_{∞}, which becomes itself one of the unknowns. The Nusselt number defined as $Nu = q_w r/[\lambda(T_w - T_i)]$, for the cases of $dT_w/dr \approx 0$ and $dT_w/dr < 0$ becomes negative in the region of the Ekman-type boundary layers [139]. This phenomenon is analysed below.

6.2.2 Computation of the Radial Variation of the Swirl Parameter Using the Integral Method

The present integral method described in Sect. 2.4 is used further in this section to simulate fluid flow in cavities between co-rotating disks. Since in the source region the swirl parameter β develops according to the free-vortex law (6.9), while in the Ekman-type layers the value of β becomes one of the unknowns, fluid flow in the entire cavity was simulated using the numerical version of the integral method based on Eqs. (2.77) and (2.79), which were solved using the Runge-Kutta method.

We estimated efficiency of the integral method by comparing its results with the experimental data available in the literature [43, 139]. The data depicted in Figs. 6.3, 6.4, 6.5 and 6.6 are based in part on the results of the work [169], while Fig. 6.7 was plotted partially using the data of [167]. The swirl parameter β was chosen as the one to be tested. Adequacy of modelling behaviour of the parameter β in the source region and in the Ekman-type layers and pinpointing the coordinate r_e of the boundary between these regions (minimum of the β function) is an evidence of adequacy of modelling fluid flow as a whole.

Experimental data [43] for a cavity open at its periphery (outlet into a large volume) indicate that air ingress from atmosphere into the cavity takes place near the centreplane of the gap between the disks ($z=s/2$). Because of this ingress, the mass flowrate in the boundary layers over the disks ceases to be constant at a certain radial location in the neighbourhood of the periphery and further increases with an increasing coordinate r. This results in decrease in the β parameter, like in the source region (Fig. 6.3). In this study, we did not intend to model the region of air ingress into the cavity; therefore, we compare our computations and the experiments [43] up to the point of maximum of β depending on r/r_i.

As seen from Fig. 6.3, curve *4* calculated for the dimensionless mass flowrates through the cavity $C_w = \dot{m}/(\mu b)$=2963 and 6173 agrees fairly well with the experimental data [43] up to the point of maximum of the parameter β. According to the data [43], fluid flow near the radial coordinate r/r_i =2.1 is transitional (the shape-factor $H = \delta^*/\delta^{**}$ calculated for the tangential velocity component is equal to 1.85), which induces certain inaccuracy due to the use of the model of the turbulent boundary layer in this region. Rather high mass flowrates C_w=2963–6173 are likely to favour turbulization of the boundary layer and reduction of the computational inaccuracy due to increase in C_w. However, differences between the calculated (curve *4*) and the experimental data for C_w = 1111 are more pronounced.

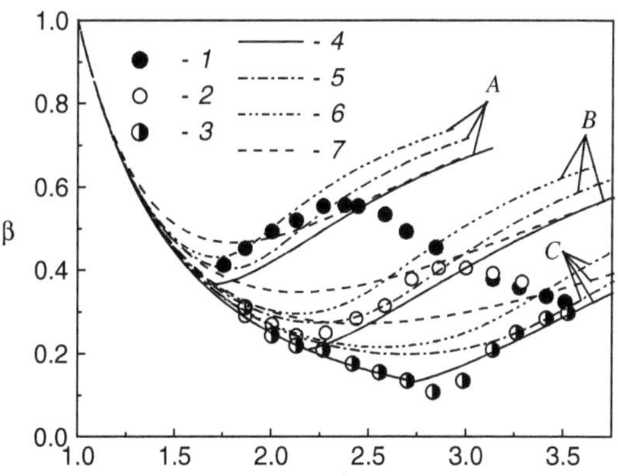

Fig. 6.3 Distribution of the swirl parameter β in the cavity open to atmosphere for $Re_\varphi=4.97\times10^5$, $\beta_i=1$, $s/b=0.068$, $r_i/b=0.285$ and $b=0.1$ m. *1–3* – experiments [43]; *4* – computation of the author [169], Eqs. (2.77) and (2.79), $n=1/7$; *5* – "erroneous" equation (6.12) [43]; *6* – corrected equation (6.12); *7* – Eq. (6.11) [196]. *Lines A* and experiments *1* – C_w =1111; *B* and *2* – 2963; *C* and *3* – 6173

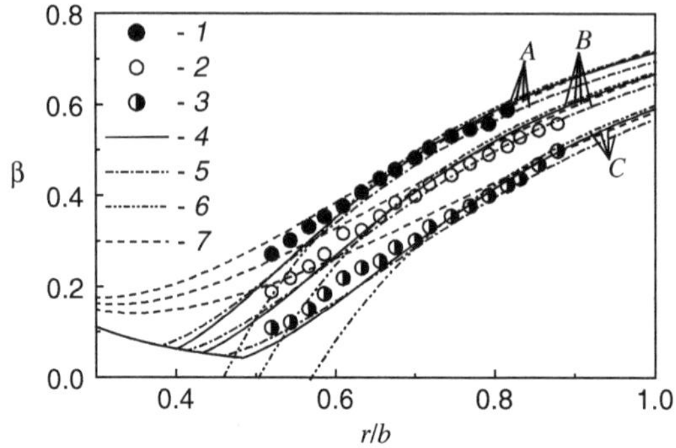

Fig. 6.4 Variation of the swirl parameter β in a cavity with a perforated shroud at the periphery for C_w=2500, β_i=1, s/b=0.1 and r_i/b=0.1. *1–3* – experiments [139]. Computations, present integral method, Eqs. (2.77) and (2.79): *4* – n=1/7 [169]; *5* – 1/9. *Line 6* – model of the Ekman-type layers (6.21), n=1/7; *line 7* – Eq. (6.11) [196]. *Lines A* and experiments *1* – $Re_\varphi=1.1\times10^6$; *B* and *2* – 6.177×10^5; *C* and *3* – 5.47×10^5

In [139], a cavity is closed at the periphery with a perforated shroud, which abruptly decreases the effect of air ingress from atmosphere in comparison with the cavity open at its periphery [43] considered above. Curves *4* and *5* (Fig. 6.4) calculated by the present integral method are in good agreement with the experiments [139]. The agreement becomes better with an increasing r/b, i.e. with increasing local Reynolds numbers Re_{V_*} indicating that flow acquires characteristics of developed turbulent regime. The data computed for n=1/9 lie somewhat lower than the curves for n=1/7 in the region of the developed Ekman-type layers, though transition to these layers begins earlier (which is explained by higher computational values of the mass flowrate through the boundary layer obtained at n=1/9, see Fig. 3.6).

Flow at the location of r/b=0.767 (Fig. 6.5) becomes turbulent even for rather small values of $Re_\varphi=2\times10^5$ [139]. For C_w=940, agreement between curve *4* calculated by the present integral method for n=1/9 and the experimental data [139] is excellent. The calculations for n=1/7 agree better with the experiments for C_w=1414 and 2500 (curve *5*).

The data calculated for C_w=2500 and the experiments of [139] differ rather significantly near the point of maximum of β. Similar differences in modelling this experimental case by an integral method were obtained in work [139], whose authors assumed that the differences were due to a mistake crept into records in the experimental log-book because, for this experiment, the ratio r_i/b should be 0.24 rather than 0.1.

The integral method of [139] is similar to the present integral method (described in Chap. 2). The differences between these two methods consist in the use of the von Karman's approximation in [139] for the v_r profile and the assumption $v_{r,\infty}=0$ at any point of the cavity. Moreover, authors of [139] always used the value n=1/7, since in the von Karman's method other values of n lead to substantial errors in

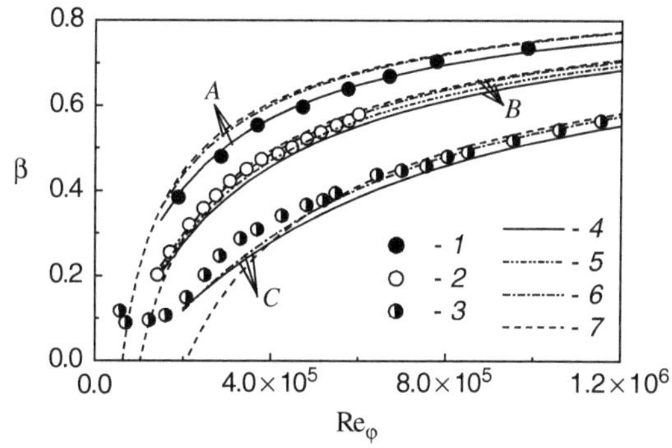

Fig. 6.5 Effects of Re_φ and C_w on the swirl coefficient β for $\beta_i=1$, $x=0.767$, $r_i/b=0.1$, $s/b=0.1$ (perforated shroud at the periphery). *1–3* – experiments [139]. Computations, present integral method, Eqs. (2.77) and (2.79) [169]: *4* – $n=1/9$; *5* – 1/8; *6* – 1/7; *7* – model of the Ekman-type layers (6.21), $n=1/7$. *Lines A* and experiments *1* – $C_w=940$; *B* and *2* – 1414; *C* and *3* – 2500

modelling the mass flowrate through the boundary layer (see Chap. 3). The data for β calculated in [139] are close to the data calculated using the present integral method for $n=1/7$; however, the advantages of the present method consist in the possibility of varying values of the exponent n from 1/7 to 1/9 to achieve the best agreement with the experimental data (see Figs. 6.4 and 6.5).

Based on the model for the Ekman layers, mass flowrate through the boundary layer can be calculated from the following equation:

$$\dot{m}_d/(\mu r) = \varepsilon_m^* \operatorname{sgn}(1-\beta)\,|1-\beta|^{2(n+1)/(3n+1)}\, Re_\omega^{(n+1)/(3n+1)}. \tag{6.20}$$

Transforming Eq. (6.20) and taking into account that $\dot{m}_d/\dot{m}=0.5$, one can obtain for the Ekman-type layers

$$\beta = 1 - \varepsilon_\beta Y^{-0.5(5n+1)/(n+1)}, \tag{6.21}$$

where $Y = x\lambda_T^{-(3n+1)/(5n+3)}$, $\lambda_T = C_w Re_\varphi^{-(n+1)/(3n+1)}$. The parameter $\varepsilon_\beta = (0.5/\varepsilon_m^*)^{0.5(2n+1)/(n+1)}$ for $n=1/7$; 1/8; 1/9; 1/10 takes the values $\varepsilon_\beta=2.23$; 2.581; 2.916; 3.256, respectively; in [139], $\varepsilon_\beta=2.21$ for $n=1/7$. The dependences of β calculated based on Eq. (6.21) for $n=1/7$ are given in Figs. 6.4 and 6.5 (curves *6* and *7*, respectively). Being noticeably different from the experimental data at the boundary between the source region and the Ekman-type layers, curves based on Eq. (6.21) merge very rapidly, following an increase in x, with the curves calculated by the integral method. To account for this rather unexpected agreement, the authors of work [139] calculated values of the individual non-linear terms in the integral equations; having relatively high absolute values, these terms compensate each other at $\beta>0.3$.

Relations (6.11) [196] and (6.12) [43] are analysed in Figs. 6.3, 6.4 and 6.6.

Figure 6.6 is a repetition of the conditions in Fig. 6.4; the comparisons are made with the curves plotted based on the present integral method [169] (as Fig. 6.4 shows, these curves agree well with the experiments of [139] and, hence, may be considered as the baseline data for comparisons).

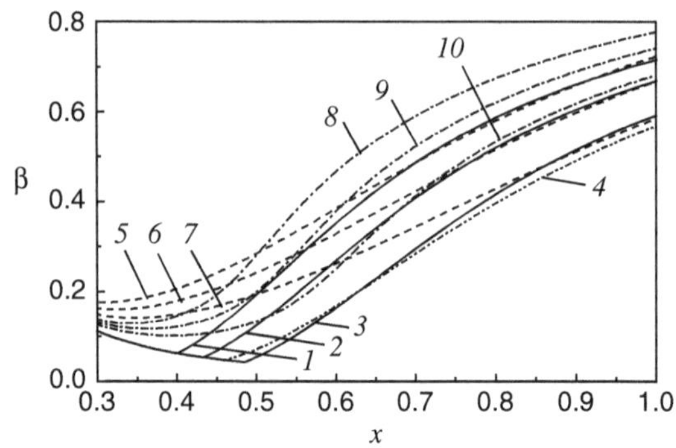

Fig. 6.6 Variation of the swirl parameter β in a cavity with a perforated shroud at the periphery for C_w=2500, β_i=1, s/b=0.1 and r_i/b=0.1 (experimental conditions of [139]): comparisons of the present integral method, Eqs. (2.77) and (2.79), with the computations based on the models (6.11) and (6.12). *1–4* – present integral method [169] (*1–3* – n=1/7; *4* – n=1/9); *5–7* – Eq. (6.11); *8–10* – Eq. (6.12). *1, 5, 8* – Re_φ=1.1×10^6; *2, 6, 9* – 6.177×10^5; *3, 4, 7, 10* – 5.47×10^5

Equation (6.12) uses a number of empirical assumptions adapted to the experiment conditions in the work [43]. A peculiarity of these conditions is a small span of the zone of the Ekman-type layers (Fig. 6.3). Therefore, being in good agreement with the experimental data of [43] (curves *5*, *6* in Fig. 6.3), this model disagrees substantially with the data of [139] for the conditions where well-developed Ekman-type layers exist (curves *8–10* in Fig. 6.6).

On the contrary, assumptions of model (6.11) turned out to perform well for the developed Ekman-type layers (curve *7* in Fig. 6.4, and curves *5–7* in Fig. 6.6) closer to the end of this region (the agreement becomes better with increasing Re_φ), as well as at the beginning of the source region (curve *7* in Fig. 6.3). In the vicinity of the boundary between these two regions (over rather wide radial span), inaccuracies of the calculated values of β are significant (Figs. 6.3, 6.4 and 6.6).

Strictly speaking, velocity $\mathrm{v}_{r,\infty}$ in the source region is described by relation $\mathrm{v}_{r,\infty} = (\dot{m} - 2\dot{m}_d)\big/(2\pi rs)$, where s is the distance between the disks. In doing so, the velocity $\mathrm{v}_{r,\infty}$ itself becomes unknown. This leads to the necessity to solve a third differential equation and thus complicates the integral method. Therefore, it is interesting to investigate the cases

$$\mathrm{v}_{r,\infty} = (\mathrm{v}_{r,\infty} \cdot r)_i / r, \tag{6.22}$$

$$\mathrm{v}_{r,\infty} = 0, \tag{6.23}$$

which serve as the upper and the lower limit of $\mathrm{v}_{r,\infty}$ distributions in the cavity, respectively.

As Fig. 6.7 indicates, dependencies for the swirl parameter β obtained using Eq. (6.22) are in better agreement with experiments [139] in the neighbourhood of the boundary between the source region and the Ekman-type layers [167]. However,

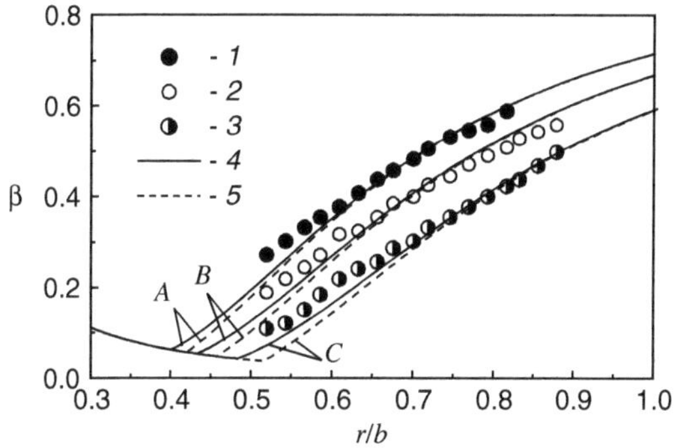

Fig. 6.7 Variation of the swirl parameter β in a cavity with a perforated shroud at the periphery for C_w=2500, β_i=1, s/b=0.1 and r_i/b=0.1 (experimental conditions of [139]). *1–3* – experiments [139]. Present integral method, Eqs. (2.77) and (2.79), n=1/7: *4* – Eq. (6.22), *5* – Eq. (6.23). *Lines* A and experiments *1* – $Re_\varphi=1.1\times10^6$; B and *2* – 6.177×10^5; C and *3* – 5.47×10^5

Eq. (6.23) also helps to obtain data that are in much better agreement with the experiments than the simplified approaches by other researchers analysed above.

6.2.3 Local Nusselt Numbers

A validation of the present integral method, Eqs. (2.77), (2.78), (2.79) and (2.80), has been performed via its comparisons against experimental data presented in works [77, 134, 135]. Data shown in Figs. 6.8, 6.9, 6.10, 6.11 and 6.12 are partially based on the results published by the author of the present work in papers [164] and [167], respectively.

In experiments [77] performed for β_i=1, T_w=*const* and axial inlet into a cavity, only a source region existed. The measured Nusselt numbers (data *1* and empirical curve *2* in Figs. 6.8 and 6.9) were approximated with Eqs. (6.13), (6.14) and (6.15).

Fig. 6.8 Variation of the local values of Nu versus Re_ω for β_i=1, N_i=0.3333, r_i/b=0.245, s/r_i=0.18, T_w=*const* and $Re_{\omega i}=6\times10^4$. *1*, *2* – experiments and their approximation [77]; *3* – Eq. (6.13) [77]; *4* – present integral method, Eqs. (2.77) and (2.78), *upper curve* for α_i=6.5, *lower curve* for α_i=4.4

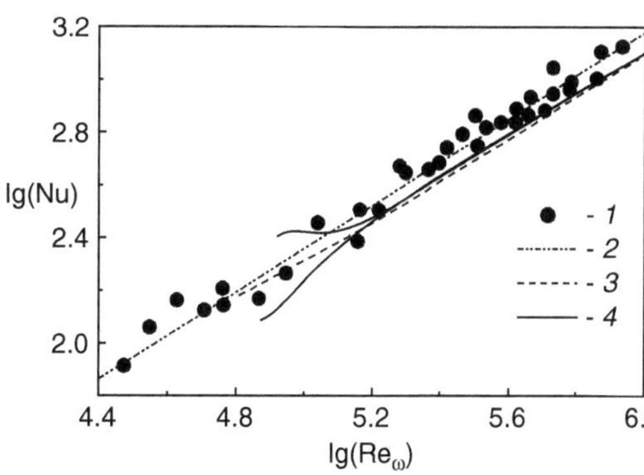

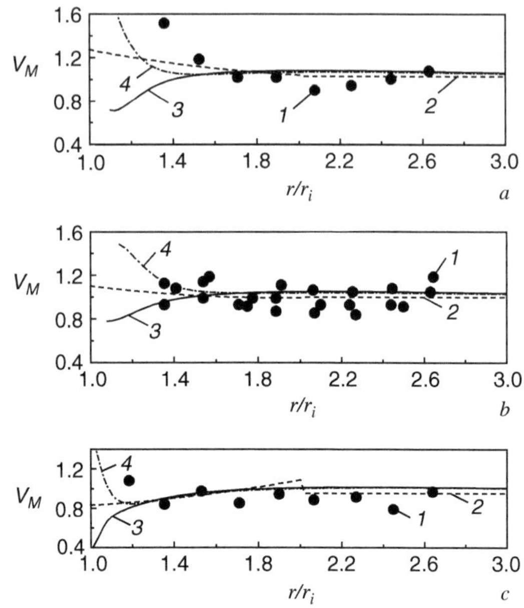

Fig. 6.9 Function V_M in Eq. (6.13): *a* – N_i=0.5; *b* – 0.333; *c* – 0.1667. *1* – experiments of [77]; *2* – Eqs. (6.14) and (6.15). *3, 4* – present integral method, Eqs. (2.77) and (2.78) for α_i=5.5 and 10.5 (**a**); 4.4 and 6.5 (**b**); 0.25 and 5.5 (**c**)

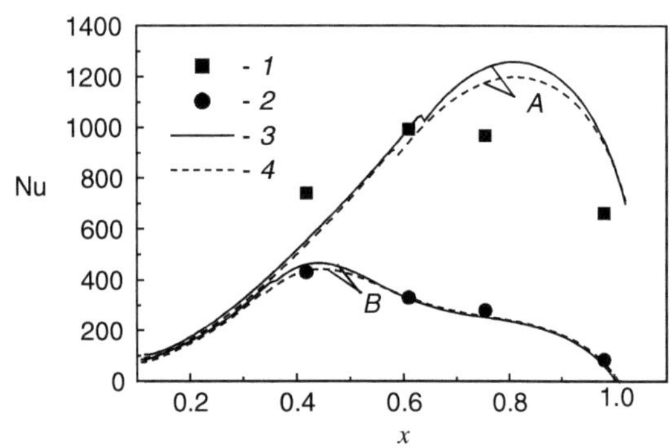

Fig. 6.10 Radial variation of the Nusselt numbers for β_i=0, $Re_\varphi=1.9\times10^6$, r_i/b=0.103, s/b=0.138, (s/r_i=0.776), b=0.428 m, dT_w/dr>0. *1, 2*– experiments [134, 135]. The present integral method, Eqs. (2.77), (2.78), (2.79) and (2.80): *3* – n=1/7; *4* – n=1/9. *Lines A* and experiments *1* – C_w=13,000; *B* and *2* – 2800

Experimental data for the Nusselt number obtained in [77] exceed those computed based on Eq. (6.13) (see curve *3* in Fig. 6.8) by 10–20%. This was caused by the turbulization effect of protrusions on the heaters glued on the disk, and thus creating an artificially roughened working surface [77].

Computational data for the Nusselt number obtained via a numerical solution of Eqs. (2.77), (2.78), (2.79) and (2.80) using the Runge-Kutta method for $n=n_T$=1/7 [164] are shown in Figs. 6.8 (curves *4*) and 6.9 (curves *3* and *4*).

For $Re_\omega\geq1.6\times10^5$, the computed curves practically coincide with curve *3* plotted in Fig. 6.8 based on Eq. (6.13) and curves *3* and *4* plotted in Fig. 6.9 based on Eqs. (6.14) and (6.15). For $Re_\omega=6\times10^4$–1.6×10^5, or in other words for r/r_i=1–1.5,

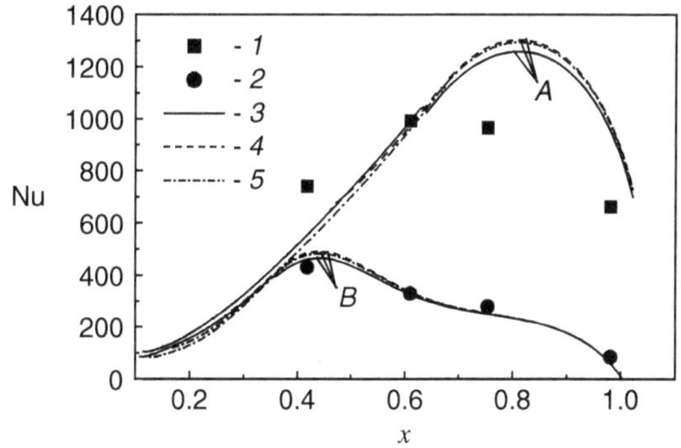

Fig. 6.11 Radial variation of the Nusselt numbers for Re_φ=1.9×10^6, β_i=0, r_i/b=0.103, s/b=0.138, b=0.428 m, dT_w/dr>0. *1*, *2*– experiments [134, 135]. Present integral method, Eqs. (2.77), (2.78), (2.79) and (2.80), n=1/7: *3* – k=0.5; *4* – 0.25; *5* – 0.03. *Lines A* and experiments *1* – C_w=13,000; *B* and *2* – 2800

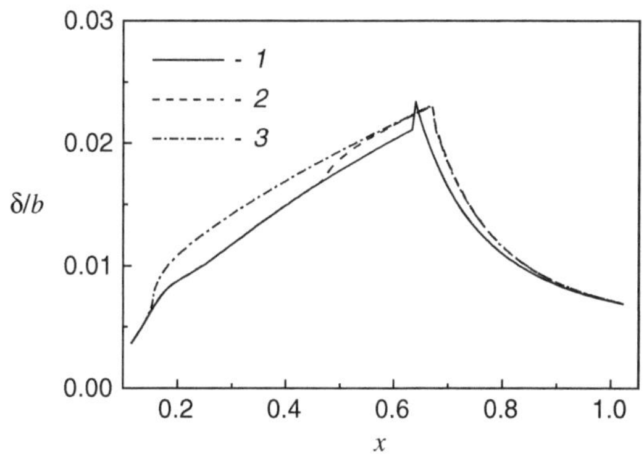

Fig. 6.12 Radial variation of the boundary layer thickness for C_w=13,000, Eqs. (2.77), (2.78), (2.79) and (2.80), n=1/7: *1* – k=0.5; *2* – 0.025; *3* – 0.03 (for the other conditions see Fig. 6.11)

computed values of *Nu* can lie either higher or lower than curves plotted based on Eqs. (6.13), (6.14) and (6.15) depending on an initial value of the tangent of the flow swirl angle on the wall α_i. Thus, an adjustment of the initial value α_i provides an opportunity in frames of the integral method to take into account the heat transfer augmentation in the area of the axial inlet into the cavity. As seen from Fig. 6.9, empirical equations (6.14) and (6.15) do not provide a solution for this problem, because they are valid just for the specific conditions of work [77]. It appeared also that Eqs. (6.14) and (6.15) fail to perform as expected for N_i=0.1667 (Fig. 6.9c).

Experimentally measured distributions $T_w(r)$ for dT_w/dr>0 for the downstream disk and an axial inlet into the cavity [134, 135] obtained for Re_φ=1.9×10^6, C_w=13,000 and 2800 were approximated by the seventh-order polynomial $T_w = \sum_0^7 C_i r^i$, where coefficients C_i are [164]:

(a) C_i = 293.305; 300.811; –242.70; 0; 0; 0; 0; 0 for i =0,. . .,7 in the region $0 \leq r \leq 0.138$ m;
(b) C_i = 660.529; –10406.3; 127631; –810784; 2957727; –6260010; 7207170; –3529663 for i =0,. . .,7 in the region 0.138 m $\leq r \leq$ 0.428 m.

In experiments [134, 135], non-swirling flow (β_i=0) was axially fed into the cavity. It is evident from Eq. (2.76) that, under condition β_i=0, the radial mass flowrate through the boundary layer in the source region significantly exceeds that in the above-analysed case β_i=1. Hence, in experiments [134, 135], authors observed appearance of the Ekman-type layers. In the region of entraining boundary layers, the temperature of the core T_∞ is constant and equal to T_i. In the region of the Ekman-type boundary layers, entrainment of cold air from the flow's core ceased, and these layers start to give the heat back to the flow's core; so, the temperature T_∞ increases, thus becoming one of the unknowns. Therefore, in the definition of the Nusselt number $Nu = q_w r/[\lambda(T_w - T_i)]$, authors [139] used the known temperature T_i instead of T_∞. This results in that local Nusselt numbers are to be calculated using the following relation [139]:

$$Nu = St \frac{V_* r}{\nu} Pr \frac{T_w - T_\infty}{T_w - T_i}. \tag{6.24}$$

Experimental data [134, 135] and predictions using the present integral method, Eqs. (2.77), (2.78), (2.79) and (2.80) (see Fig. 6.10), reveal qualitatively different behaviour of the Nusselt numbers calculated by Eq. (6.24) in the source region and in the Ekman-type layers. In the source region, $T_\infty = T_i$, the last factor in Eq. (6.24) is equal to unity, and the Nu values increase qualitatively in the very same way as data in Fig. 6.8. In the region of the Ekman-type layers, the temperature T_∞ increases (according to Eq. (2.80)), the last factor in Eq. (6.24) decreases with increasing T_∞, thus resulting in a corresponding decrease in the Nusselt number. Calculated data for the case C_w=2800 (or $C_w/Re_\varphi^{0.8} = 0.027$) agree fairly well with experiments [134, 135], while for the case C_w=13,000 (or $C_w/Re_\varphi^{0.8} = 0.222$) rather significant differences are observed.

In works [134, 139], experimental conditions of [134, 135] were simulated using an integral method based on a linear approximation of the tangent of the flow swirl angle in the boundary layer and model assumptions (2.56) for the enthalpy thickness modified in [134, 139] to the form of Eqs. (3.106), (3.107), (3.108) and (3.109). Simulations [134, 139] are in good consistency with computational curves in Fig. 6.10. This is an evidence of the fact that inexactitudes of the model assumptions [134, 139] produce weak effect on the Nusselt number for the same distribution of the disk temperature T_w. It is obvious that in the near-inlet region, which is dominated by an impinging jet making a 90 degree-turn, one can observe not only augmented Nusselt numbers (Fig. 6.10), but also increased radial mass flowrates through the boundary layer (in comparison with Eq. (2.76)), resulting in the earlier entrainment of air from the flow's core into the boundary layers and further formation of the Ekman-type layers. The latter phenomenon is evident from data in Fig. 6.10.

The effect of the exponent n taking values 1/7 or 1/9 is practically insignificant for the relatively low mass flowrate C_w=2800 and becomes more visible in the Ekman-type layers for the high mass flowrate C_w=13,000, with the value n=1/9 resulting in the lower Nusselt numbers.

A comparative investigation of effects of the boundary conditions (6.22) and (6.23) on the distributions of the Nusselt numbers in the cavity is elucidated in Fig. 6.11 [167]. The Nusselt numbers were calculated in such a way that over a certain initial radial length of the source region Eq. (6.22) was used, and further on the remaining radial length of the cavity Eq. (6.23) was applied. Transition between these lengths took place at a location where the condition $\dot{m}_d = k\dot{m}$ was satisfied. Here k=0.03, 0.25 and 0.5 (the Ekman-type layers started developing always for k=0.5). Curves for the Nusselt number in Fig. 6.11 are close to each other for C_w=*idem*. Note the deflection in the curve *A* following the transition from Eq. (6.22) to Eq. (6.23). At k=0.5, a jumpwise decrease in the *Nu* distribution is observed (the beginning of the Ekman-type layers).

An explanation of this phenomenon can be found in Fig. 6.12: variation of the boundary layer thickness δ/b exhibits a deflection or a jump at the radial location where the law of $v_{r,\infty}$ changes from Eqs. (6.22) to (6.23). The decreasing function δ/b of x corresponds to the Ekman-type layers. An upward jump at the beginning of this region for k=0.5 can be explained based on Eq. (2.76) for the mass flowrate in the boundary layer. At the end of the source region, the velocity $v_{r,\infty}$ decreases jumpwise from a non-zero value to zero; the smaller this non-zero value is, the smaller the jump of the *Nu* number occurs. These peculiarities are practically absent in curve *B* in Fig. 6.11, because for C_w=2800 (i.e. for Re_φ=*idem*) the value of $v_{r,\infty}$ at the location of the transition between the source region and the Ekman-type layers is close to zero. Graphical data for the dependence of δ/b on x at n=1/9 analogous to those depicted in Fig. 6.12 are documented in the work [167].

6.2.4 *Effect of the Radial Distribution of the Disk Surface Temperature*

It was shown above (see Chaps. 3, 4 and 5) that the law of radial variation of the disk surface temperature affects significantly heat transfer rate over the disk. It was therefore interesting to validate the present integral method via simulations of heat transfer for radially outward turbulent flow in a cavity between rotating disks for different laws of variation of the disk surface temperature T_w [170]. Distributions of T_w used as boundary conditions in the computations were those obtained experimentally in works [134, 139] and presented there in the form of smoothing curves. To employ these distributions of T_w in computations using the present integral method, they were approximated by the author [170] in a functional form of the seventh-order polynomial with respect to r. These distributions are shown in a transformed form in Fig. 6.13.

Variations of the centreplane temperature T_∞ and Nusselt numbers for positive, approximately constant and negative radial gradients of T_w are shown in Figs. 6.13 and 6.14.

Computations were compared with the experimental data both for T_w and the *Nu* numbers for the upstream disk presented in works [134, 139]. The experiments [134,

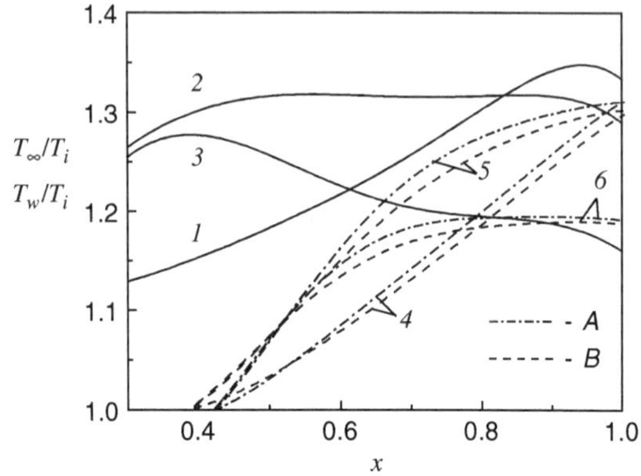

Fig. 6.13 Radial variation of the scaled wall temperature T_w/T_i (curves *1–3*, experiments [134, 139] transformed in [170]) and centreplane temperature T_∞/T_i (curves *4–6* computed based on the present integral method, Eqs. (2.77), (2.78), (2.79) and (2.80), n_T=1/7). *1, 4* – $dT_w/dr>0$; *2, 5* – $dT_w/dr\approx 0$; *3, 6* – $dT_w/dr<0$. *A* – n=1/7; *B* – n=1/10

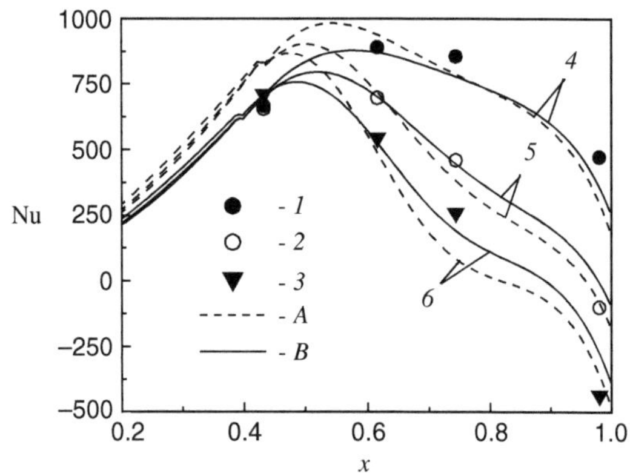

Fig. 6.14 Variation of the Nusselt numbers in a cavity ($Re_\varphi=3.2\times10^6$ for cases *1, 4*; and $Re_\varphi=3.3\times10^6$ for cases *2, 3, 5, 6*): *1–3* – experiments [134, 139]; *4–6* – present integral method, Eqs. (2.77), (2.78), (2.79) and (2.80), n_T=1/7. *1, 4* – $dT_w/dr>0$; *2, 5* – $dT_w/dr\approx 0$; *3, 6* – $dT_w/dr<0$. *A* – n=1/7; *B* – n=1/10

139] selected for the comparisons were carried out for β_i=0, Re_φ=(3.2–3.3)×10^6, r_i/b=0.103, s/b=0.138, b=0.428 m, C_w=7000 (or $C_w/Re_\varphi^{0.8}=0.044$). Majority of the computations were performed for n=1/10; this value corresponds to high Re_φ numbers observed in experiments. Use of a higher value n=1/7 leads to worse agreement of the computed Nusselt numbers with the experimental data (see Fig. 6.14). As mentioned in Chap. 3, temperature profiles are more conservative with respect to the Re_φ numbers; this caused a choice of the value n_T=1/7 for the parameters of the thermal boundary layer [170].

In the region of the entrainment boundary layers (source region), condition $T_\infty=T_i$ holds, while the Nusselt number increases like in the case of fluid flow over a free rotating disk. In the region of the Ekman-type layers, the rate of increase in the Nusselt number slows down, and finally the *Nu* values begin to decrease (Fig. 6.14) due to increase in the local values of T_∞ (Fig. 6.13). This decrease in the region of the Ekman-type layers is more distinct, and the smaller the difference between temperatures T_w and T_∞ is due to their actual radial distributions.

Both the experiments and the calculations using the present integral method show that for cases $dT_w/dr\approx 0$ and $dT_w/dr<0$, a region of negative Nusselt numbers emerges in the periphery. Physically this means that, because of the high rate of decrease in the wall temperature T_w and increase in the centreplane temperature T_∞, there emerges a region in which the temperature T_∞ becomes higher than the disk temperature T_w (Fig. 6.13). This leads to change in sign of heat flux on the disk surface: the air begins to heat up the disk rather than the reverse, as occurred practically in the entire cavity at lower values of the radial coordinate r.

Thus, based on the results of the application of the present integral method to simulations of fluid mechanics and heat transfer for the cases where underswirled ($\beta_i=1$) radially outward flow enters a cavity, one can make the following conclusions.

The integral method for the turbulent boundary layer based on the power-law approximation of the tangential velocity component v_φ and quadratic approximation of the tangent of the flow swirl angle $tan\varphi$ described in Sect. 2.4 allows obtaining a good agreement of simulations with known experimental data [77, 134, 135, 139] for a series of cases of radially outward flow in a cavity between parallel rotating disks, which cannot be attained with the help of integral methods developed earlier [43, 77, 196].

Advantages of the present integral method in comparison with the integral method developed in [139] consist in taking into account the radial flow outside of the boundary layer in the cavity and more accurate approximation of the radial velocity v_r component in the boundary layer (to remind, method [139] is based on the von Karman's linear approximation of v_r). This allows in a number of cases to increase accuracy of computations, in particular, at the expense of varying the value of the exponent n in the power-law approximation of the velocity profiles. Besides, the model of heat transfer based on the power-law approximation (2.58) of the temperature profiles in the thermal boundary layer also exceeds in its accuracy the approach based on the theory of local modelling, Eq. (2.56), used by the authors of [139] in the form of Eqs. (3.106), (3.107) and (3.108).

The present integral method, Eqs. (2.77), (2.78), (2.79) and (2.80), like the method of [138, 139], takes into account different conditions of fluid flow and heat transfer in the source region and the Ekman-type layers. Models (2.77), (2.78), (2.79) and (2.80) are valid for arbitrary radial distributions of the temperatures on the wall T_w and at the centreplane T_∞.

In case of high mass flowrates and axial inlet into the cavity (in particular, for $C_w=13{,}000$), the present integral method becomes inaccurate: apparently, in the region around the inlet into the cavity mass flowrates in the boundary layer exceed values computed from Eq. (2.76), and therefore the Ekman-type layers develop much earlier (at smaller local radii). Approaches to account for this phenomenon in the integral methods have not been developed so far.

Effect of the law of variation of the radial velocity outside of the boundary layer is rather weak. However, use of Eq. (6.22) leads to obtaining more accurate data, whereas possible jumpwise changes in the curve of the Nusselt number variation at the boundary between the source region and the Ekman-type layers can be rather easily eliminated using standard procedures of smoothing the computational data.

Effect of the law of radial variation of the disk surface temperature on the Nusselt number is weak in the source region and becomes however decisive in the Ekman-type layers. For negative or approximately constant radial gradients of the disk surface temperature, a region with negative Nusselt numbers (reversed direction of the wall heat flux) can emerge within the radial span of the Ekman-type layers.

6.3 Effect of the Flow Overswirl

Rotating cavities formed by parallel disks with radially outward flow therein function as radial rotating diffusers. They are used in gas turbines to compress air in the cooling system [86, 139, 166, 196]. If non-swirling flow enters a cavity, the increase in the pressure rate is limited by the geometric dimensions of the diffuser. Swirl of incoming flow ($\beta_i \neq 0$) enables, in particular, to attain an additional transformation of the velocity into pressure.

Overswirl of flow ($\beta_i > 1$) supplied radially outwards into a cavity changes its physical structure in the vicinity of the walls: a zone of reverse radially inward flow emerges on the disks in the region where $\beta_i \geq \beta \geq 1$, while near the centreplane of the cavity radially outward flow still persists. In other words, in the region where $\beta_i \geq \beta \geq 1$, recirculation flow emerges (Fig. 6.15). A radial coordinate, where $\beta = 1$, becomes a location where (a) ring-like impinging jets hit the surfaces of both disks and (b) because of this, two boundary layers originate on each disk developing further in the opposite directions. Flow structure in the region of fully radially outward flow for $\beta_i \leq 1$ is traditional: a part of the cavity is occupied by the source region in which entrainment boundary layers develop and afterwards the Ekman-type layers can originate (for $\dot{m} = 0.5\dot{m} = \text{const}$).

In works [79, 145], overswirled flow (co-rotating with disks) with $\beta_i = 2.01$ was axially supplied into the cavity at the radial location $x_i = 0.44$ and impinged onto the downstream disk at $x = 0.5$. For $x < 0.5$, recirculation flow developed being radially inward on the disks; for $x > 0.5$, flow over the disks was radially outward. For $x < 0.6$, predicted Nu numbers exhibit the trends peculiar to impinging jets (with a Nu number maximum at $x = 0.5$). Further downstream for $0.6 < x < 0.95$, the Nu number curves show the behaviour typical for boundary layer flows.

As well known, results of numerical simulations of the two- and three-dimensional flows are rather sensitive with respect to a choice of turbulence model, in particular, in singularity points. In simulations [145], two turbulence models were used, namely the model of Morse (1) [128, 129] and the model of Launder-Sharma (2) [101]. Details of the numerical solution of the Navier–Stokes and energy equations in a rotating cavity using an original in-house elliptic solver are documented in work [145] and references therein. Curves of the Nusselt number predicted using the turbulence model of Morse [128, 129] agree with the experimental data much better than those obtained using the Launder-Sharma model [101].

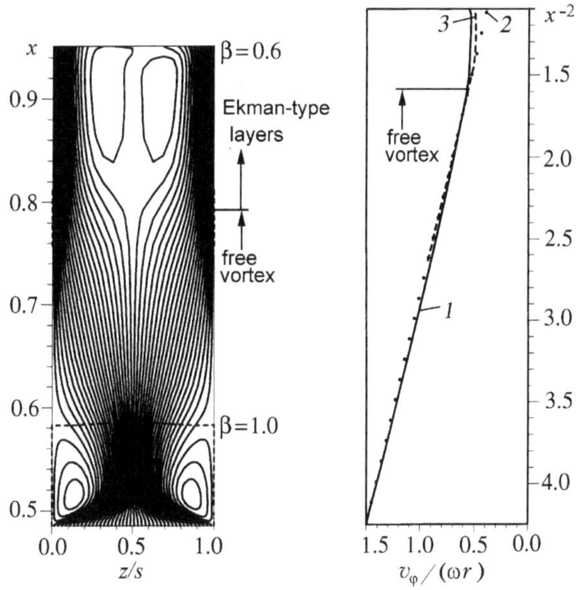

Fig. 6.15 Streamlines (**a**) and swirl parameter $\beta = v_\varphi/\omega r$ variation (**b**) in the rotating cavity with flow overswirl at the inlet ($\beta_i>1$) [79, 145]. *1* – local value of β at $z=s/2$ [145]; *2* – free vortex (6.9); *3* – present integral method [166]

This section contains results of simulations of fluid flow and heat transfer in rotating cavities where air enters radially outwards being overswirled at the inlet with $\beta_i>1$ (Fig. 6.15). The simulations were performed using the present integral method, Eqs. (2.77), (2.78), (2.79) and (2.80) [166]. In the recirculation and source regions, $v_{\varphi,\infty}$ follows the free-vortex law (6.9). On the outer boundary of the boundary layer in the recirculation region and in the Ekman-type layers, boundary condition (6.22) for $v_{r,\infty}$ is used, while in the source region, the value $v_{r,\infty}$ is specified using both Eqs. (6.22) and (6.23). Results of computations are compared with experimental and computational data obtained in the work [145]. The region of fluid flow near the outlet opening from the cavity located in the downstream disk was not considered for computations.

Predictions were carried out for $\beta_i=1.5$; $Re_\varphi=1.5\times10^6$; $C_w=8730$; $\lambda_T=0.1$; $r_i=0.1$ m, $b=0.206$ m; $s=0.005$ m. The wall temperature T_w was described by the seventh-oder polynomial $T_w = \sum_0^7 C_i r^i$, where coefficients C_i for $i=0$–7 were equal to $C_i = 3103.5$; 188545; –4362370; 54725200; –402333000; 1735320000; –4067840000; 4000000000, respectively; $T_i = 293$ K.

As said above, the point $\beta=1$ is the origin of two boundary layers developing oppositely. In view of the features of the integral method, each of these layers was calculated separately. Data for the parameter β in the Ekman-type layers obtained with the help of the present integral method and in computations of [145] correlate well (Fig. 6.15).

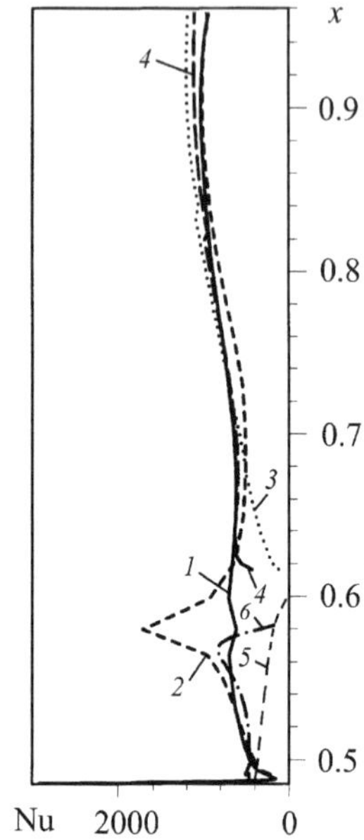

Fig. 6.16 Nusselt numbers in the cavity. *1*, *2* – computations of [145]; *3–6* – present integral method. *1* – turbulence model 1 [128, 129]; *2* – turbulence model 2 [101]; *3* – underswirled flow, α_1=0.7; *4* – underswirled flow, α_1=5.5; *5* – overswirled flow, α_1=0.7; *6* – overswirled flow, α_1=15 [166]

Heat transfer in the region of underswirled flow (β<1) is characterized by the following features (Fig. 6.16). Near the point β=1, the Nusselt number predicted using the integral method for α_1=0.7 (subscript "1" here means conditions for $\beta \to 1$) is noticeably lower than the results computed in [145]. This zone is fairly short, and afterwards the Nusselt numbers obtained using both methods practically coincide up to the outlet opening region, where the integral method is inapplicable. An increase in the initial value of α_1 brings results of the integral method noticeably closer to those of [145]. For α_1=5.5, the curves for both methods practically coincide.

The improving effect of increased values of α_1 can be understood via considering the definition of this quantity $\alpha = -\tau_{wr}/\tau_{w\varphi}$. In the region of the singularity point β=1, the shear stresses $\tau_{w\varphi}$ tend to zero, whereas τ_{wr} remain non-zero as a result of the persistent radial flow. An analogy to this phenomenon can be found in Chap. 5, where fluid flow in the stagnation region of a jet impinging onto an orthogonal surface is considered. It seems rather difficult to propose a justifiable method of determining α_1. Clearly, the magnitude of α_1 should not be smaller than the values for the Ekman layers at β=*idem* [5] and, at the same time, should not generate mass flowrates in the boundary layer (Eq. (2.76)) exceeding 0.5 $\dot{m}$. It is worth noting that α distributions (Fig. 6.17) for various α_1 merge at a certain value of the radial coordinate, the same coordinate where the relations for the Nusselt number merge. The value of α predicted in [145] does not have such a pronounced maximum in the vicinity of the point where β=1, and agrees well with the results obtained using the present integral method starting approximately from the point where the different curves for the Nusselt number merge.

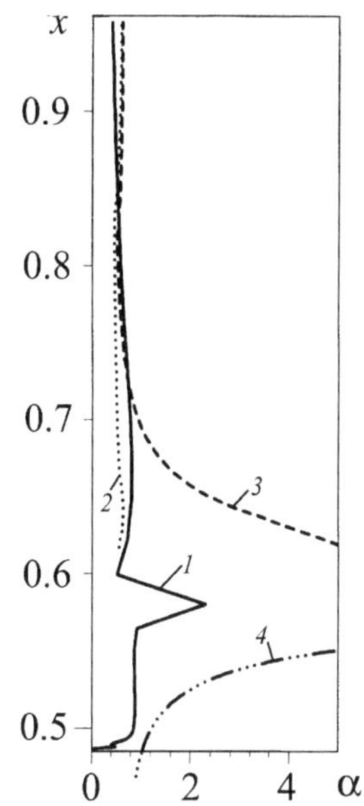

Fig. 6.17 Tangent of the flow swirl angle on the wall. *1* – computations of [145]; *2–4* – present integral method; *2* – underswirled flow, α_1=0.7; *3* – underswirled flow, α_1=5.5; *4* – overswirled flow, α_1=15 [166]

To remind, the effects of the initial value of α on the Nusselt number over a certain span of the radial length near the inlet to the cavity were already discussed above while considering initially underswirled flows (Figs. 6.8 and 6.9).

Heat transfer in the region of overswirled flow has the following characteristic properties. Whilst moving from the point where β=1 towards decreasing r, the Nu numbers predicted using the integral method for α_1=0.7 increase monotonically from zero to finite values for β=1 at the cavity inlet. These values are in fair agreement with the results of [145] near the inlet to the cavity. However, the coincidence ends here. When moving towards increasing r from the cavity inlet, the Nusselt numbers computed in [145] increase rather than decrease. Apparently, the computations of [145] are closer to reality, since the point where β=1 represents a stagnation region where a kind of a ring-like jet, which recirculation flow look like at the point of its reattachment to the wall, impinges onto the disk. As shown in Chap. 5, the Nu numbers for impinging jets have a maximum at the stagnation point and further decrease in its vicinity. A relative weakness of the methodology used in [145] as well as in other kinds of currently existing CFD software is that the results of simulations depend on turbulence model chosen, especially at the singularity points. In the case under consideration, turbulence model 1 [128, 129] appears to be more preferable. However, a local minimum at the point where β=1 instead of a maximum calls for an explanation.

As for the integral method, the condition Nu=0 at β=1 is incorporated in its mathematical model by its very nature. This indicates inadequacy of some model assumptions of the method as applied to recirculating flows. It was necessary to use the value α_1=15 to obtain Nu number distributions more or less consistent with the computations [145] (Fig. 6.16). However, this success is restricted, since, again, recommendations for selection of α_1 in this case are unclear. For example, use of the value α_1=15 for other values of β and C_w lead to results that were in worse agreement with the computations based on the methodology [145] and described in work [78].

Thus, the above investigation of the fluid mechanics and heat transfer for the cases where overswirled flow enters a cavity allowed making the following conclusions.

The flow overswirl at the cavity inlet results in development of a recirculation flow region for $\beta_i \geq \beta \geq 1$.

Modelling of recirculation flows in rotating cavities necessitates use of methodology involving solutions of the complete elliptic Navier–Stokes equations. In addition to this, careful validations are needed in order to select an optimal turbulence model for this class of flows.

Integral methods are effective in the underswirled flow region even when β_i>1 at the inlet to the cavity. However, in the neighbourhood of the point β=1 and in the recirculation region, the present integral method provides markedly lower *Nu* numbers for traditional initial values α_1=0.5–0.7 or requires high values α_1=5.5–15 to improve accuracy of predictions. Also, the problem of the zero values of the *Nu* number obtained using the integral method at the location where β=1 remains unresolved.

6.4 Aerodynamics and Heat Transfer in a Rotating-Disk Air Cleaner

6.4.1 General Characteristics of the Problem

Radially inward flow in rotating-disk cavities takes place in compressors, rotating-disk air cleaners, etc. [139, 145, 188, 202]. If the vector of the radial velocity component of flow fed into the cavity in its periphery is directed towards the axis of rotation of the disks (i.e. radially inwards), the initial underswirl of flow (β_i<1) results in emergence of recirculation flow in a region of the cavity where $\beta_i \leq \beta \leq 1$. In this region, flow moves radially inwards near the centreplane of the cavity, while over the disks radially outward flow develops. As the main flow moves further radially inwards, swirling effect of the disks results in increase in the tangential velocity component of flow, so that the parameter β reaches a value of 1 at a certain radial location, which means the end of the recirculation flow region. Further streamwise (i.e. radially inwards), the Ekman-type layers develop on the disks at a radial location r_e. Finally, flow leaves the cavity at lower radii [139].

Experiments for radially inward flows in rotating cavities of gas turbines with a perforated peripheral shroud elucidating these phenomena were analysed in detail in [139].

The present section describes data of numerical simulations of fluid mechanics and heat transfer in radially inward fluid flow in a rotating-disk air cleaner based on

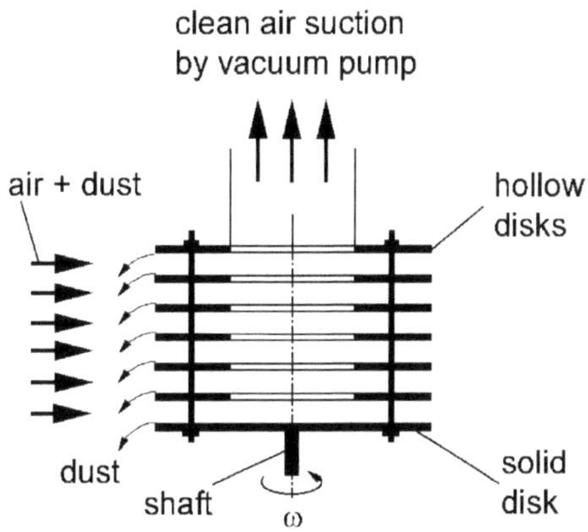

Fig. 6.18 Rotating air cleaner with radially inward air flow between the disks [202]

results obtained in [188]. Modelling of recirculation flow required to solve numerically the full elliptic Navier–Stokes and energy equations, which was performed with the help of the commercial CFD code Phoenics.

6.4.2 Geometrical and Regime Parameters of the Air Cleaner

Rotating air cleaners are widely used in industrial and technological processes [88, 188, 202]. In rotating air cleaners, separation of dust particles takes place due to effects of centrifugal forces in two-phase flow in channels of a rotor. By the direction of fluid flow in the rotor, the entire variety of design of rotating air cleaners can be divided into three groups. The first group includes air cleaners where air flows along axial channels of the rotor (i.e. parallel to the axis of rotation). The second group includes air cleaners in which main air flow and dust move in the radially outward direction. The third group includes air cleaners in which air is sucked by a fan into the rotor channels and moves radially inwards from periphery towards the axis of rotation (Fig. 6.18). Centrifugal forces push dust particles to a casing and then towards a dust collector. Devices of this type require high specific energy for driving the rotor, as compared to devices of the first two groups [88, 202]. However, none of the known works, except for work [188], attempted to model aerodynamics in the third-group devices to elucidate the reasons for the increased energy use and identify methods for optimization of such devices. In work [188], simulations of aerodynamics of a carrying phase (air) in absence of solid particles using a commercial code Phoenics (version 3.5.1) were carried out. Since incoming air flow is often more (or less) heated compared to the air cleaner, heat transfer in this device has been also simulated.

Figure 6.19 shows a segment of the air cleaner used as a model for numerical computations by means of the Phoenics code. This model included three disks with an identical outer radius b and height h between disks rotating with frequency n (r.p.m.). One of the disks was solid, while the other two had an internal hole with

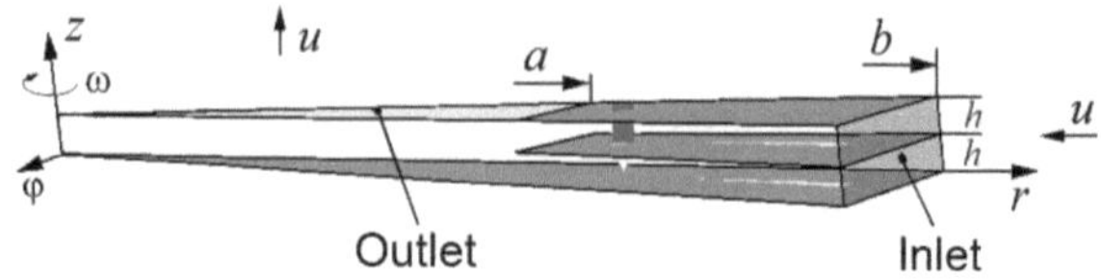

Fig. 6.19 Model of the rotating air cleaner used in simulations with the help of a commercial Phoenics code [188]

a radius a. Since the problem is axisymmetric, modelling was performed using a segment with the angle of 0.1 rad.

Air was fed into the cavities between the disks with the volumetric flowrate Q (m^3/h). The number of cavities between the disks in a real industrial prototype was 39; one of them was solid and the rest 39 had internal holes in the centre. In simulations, the number of cavities m between the disks, through which the flowrate Q was pumped, varied that affected the magnitude of the air velocity at the inlet to the model air cleaner (see Table 6.2).

The radial flow velocity at the inlet to the cavity v_{ri} was calculated based on the formula

$$v_{ri} = \frac{Q}{3600 \cdot 2\pi bhm}. \tag{6.25}$$

Tangential $v_{\varphi i}$ and axial v_{zi} velocity components at the inlet to the cavities were also specified as regime parameters.

Angular velocity of rotation of the disks ω (rad/s) was determined from the relation

$$\omega = 2\pi n/60. \tag{6.26}$$

In simulations using the Phoenics code [188], an excess static pressure field P was computed relative to the atmospheric pressure. It was assumed that the excess pressure was equal to zero at the outlet from the device, while the magnitude of P at the inlet was determined from computations. The physical meaning of the value P at the inlet is therefore the pressure needed to pump air through the chosen model of the air cleaner or, which is the same, the pressure losses in this model.

Values of the geometric and regime parameters of the rotating air cleaner used in simulations are documented in Table 6.2. Variant 1 was taken as the baseline, while the others were selected in such a way that one or two parameters varied in comparison with variant 1. Variants 13 and 14 stand aside from the common sequence, because in these variants external and internal radii of the cleaner varied, with the ratio between them remaining approximately the same.

Heat transfer was simulated for the baseline variant 1. In investigations fulfilled in work [188], only modelling of temperature field in the cleaner was performed.

Table 6.2 Regime and geometric parameters of the rotating air cleaner in simulations [188]

No	Geometric parameters				Regime parameters					
	b, m	a, m	m	h, m	n, r.p.m.	ω, rad/s	Q, m^3/h	v_{ri}, m/s	$v_{\varphi i}$, rad/s	v_{zi}, m/s
1	0.4	0.3	39	0.008	1500	157.08	3000	–1.063	0	0
2	0.4	0.3	39	0.008	1500	157.08	1500	–0.531	0	0
3	0.4	0.3	39	0.008	3000	314.16	3000	–1.063	0	0
4	0.4	0.3	39	0.008	750	76.214	3000	–1.063	0	0
5	0.4	0.3	39	0.012	1500	157.08	3000	–0.709	0	0
6	0.4	0.3	39	0.004	1500	157.08	3000	–2.126	0	0
7	0.4	0.2	39	0.008	1500	157.08	3000	–1.063	0	0
8	0.4	0.3	39	0.008	1500	157.08	3000	–1.063	31.4	0
9	0.4	0.3	39	0.008	1500	157.08	3000	–1.063	31.4	–2
10	0.4	0.3	78	0.004	1500	157.08	3000	–1.063	0	0
11	0.4	0.3	26	0.012	1500	157.08	3000	–1.063	0	0
12	0.4	0.3	19.5	0.016	1500	157.08	3000	–1.063	0	0
13	0.3	0.22	39	0.008	1500	157.08	2250.6	–1.063	0	0
14	0.2	0.12	39	0.008	1500	157.08	1500	–1.063	0	0

6.4.3 Parameters of the Computational Scheme

The total height of computational domain (coordinate z) was equal to $2h$, angular segment (coordinate φ) was limited by 0.1 rad and the maximum external radius of the domain equalled to b.

Parameters of the computational mesh for the model domain used in simulations were as follows:

(a) 80 cells in the radial direction, with an even distribution in each region; of them, 60 cells in the region $0 \le r \le a$, and 20 cells in the region $a \le r \le b$ for all variants 1–6 and 8–14; 40 cells in the region $0 \le r \le a$ and 40 cells in the region $a \le r \le b$ for variant 7;
(b) 120 cells in the axial direction (coordinate z); of them, 30 cells in each region $0 \le z \le h/2$, $h/2 \le z \le h$, $h \le z \le 3h/2$ and $3h/2 \le z \le 2$, with the mesh being refined in the direction from the centreplanes of the cavities (values $z = h/2$ and $3h/2$) towards the surfaces of the disks (stretching factor 2.0);
(c) one cell along the coordinate φ, because the problem is axisymmetrical.

Turbulent flow parameters were calculated using the k–ε turbulence model [100], whereas the solver is based on the finite-difference method.

6.4.4 Results of Simulations

Geometric division of the air cleaner domain into the hole region in the disks for $0 \le r \le a$ and the region of cavities between the disks for $a \le r \le b$ determines principal

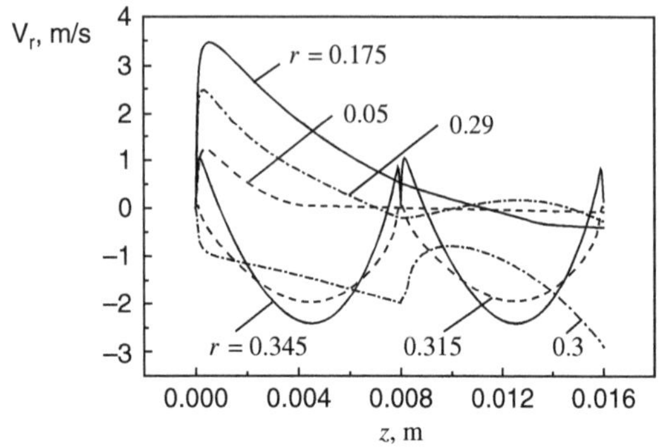

Fig. 6.20 Variant 1: radial velocity variation over the height of the air cleaner

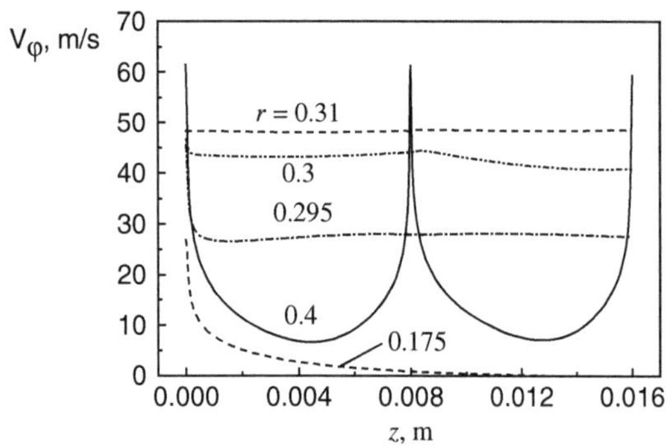

Fig. 6.21 Variant 1: tangential velocity variation over the height of the air cleaner

features of flow aerodynamics in these regions. Results of computations of the velocity components and static pressure in the air cleaner partially published in [188] are presented below in Figs. 6.20, 6.21, 6.22, 6.23, 6.24, 6.25, 6.26 and 6.27.

Flow of air in the hole region is radial outward, close by its features to the flow near a free rotating disk (Chap. 3), whose role is played in this case by the lower solid disk. The tangential velocity component v_φ varies from the value ωr on the solid disk at z=0 to zero at $z \to 2h$. Radial velocity component v_r has a form typical for wall jets (similar to flow near a free rotating disk) directed radially outwards. The axial velocity component v_z is directed downwards (towards the solid disk), which was caused by the need to compensate the volume of air pushed due to pumping effect of the solid disk radially outwards to the region with $r \to a$.

Main features of aerodynamics in the cavities between the disks are similar to those described above for rotating-disk cavities in gas turbines.

Variant 1. For the baseline variant 1, radial flow of air near the centreplane of the cavity between the disks at the inlet to it (r is somewhat smaller than b) is radially inward with $v_r \approx$const, whereas on the disks boundary layers exist with radially outward flow within them v_r>0 (Fig. 6.20).

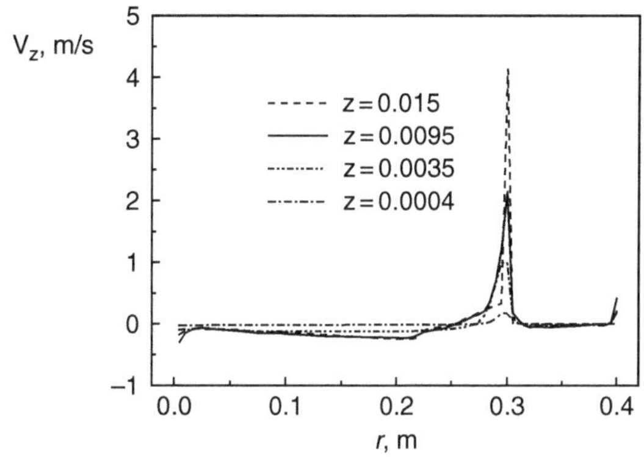

Fig. 6.22 Variant 2: axial velocity variation over the height of the air cleaner

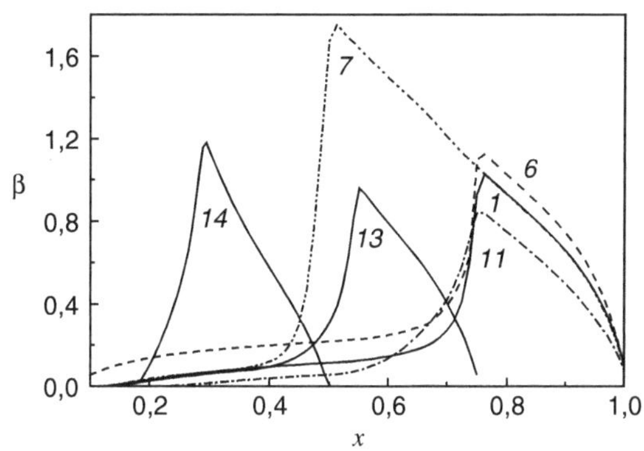

Fig. 6.23 Radial variation of the swirl parameter in the air cleaner. The number of the curve corresponds to the number of variant

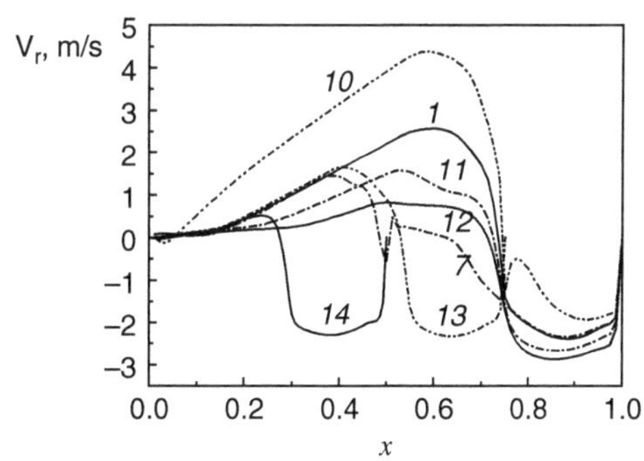

Fig. 6.24 Radial variation of v_{rC} in the air cleaner. The number of the curve corresponds to the number of variant

Tangential velocity component v_φ at the inlet to the cavity for variant 1 (and all the other variants, where v_φ was specified zero at the inlet), in fact, acquired some small non-zero value. This agrees with experimental results of [138], on the basis of which the authors of [138] introduced an idea of an efficient swirl parameter at the inlet into the cavity. This swirl is formed practically instantly and is related to the

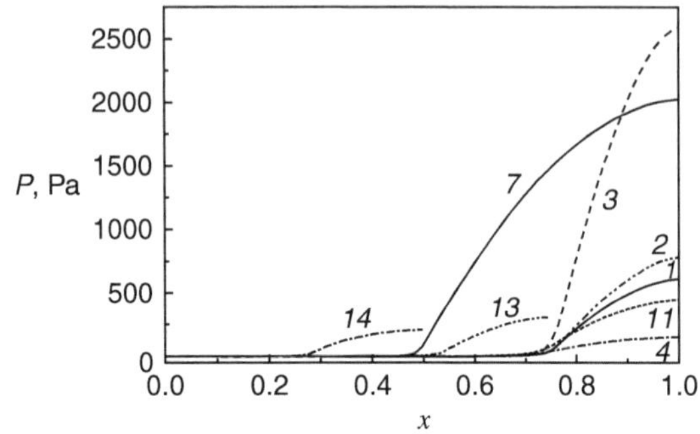

Fig. 6.25 Radial variation of the excessive static pressure in the air cleaner. The number of the curve corresponds to the number of variant

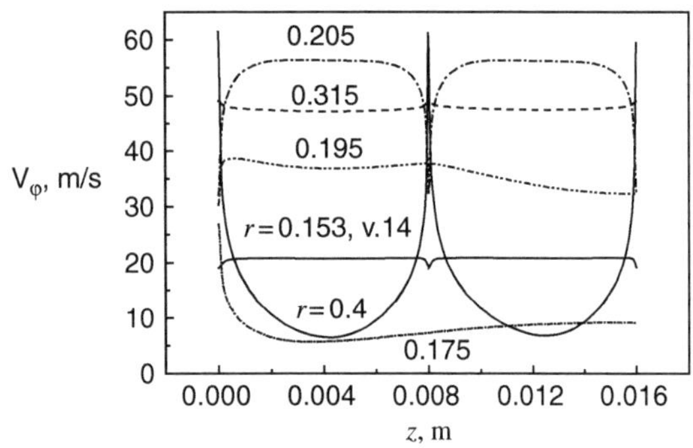

Fig. 6.26 Variants 7 and 14 (one curve): tangential velocity variation over the height of the air cleaner

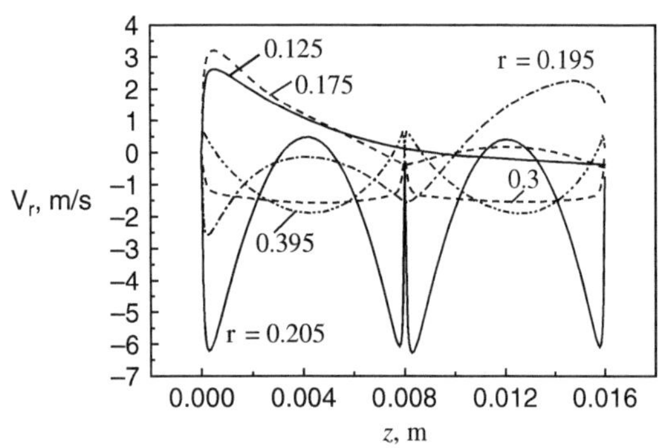

Fig. 6.27 Variant 7: radial velocity variation over the height of the air cleaner

process of flow restructuring. Further, with a decrease in the coordinate r, velocity component v_φ was virtually constant $v_\varphi = v_{\varphi C}$ over the height of the cavity forming a boundary layer in the vicinity of the disks, whereas v_φ varied from the value $v_{\varphi C}$ to ωr on the surfaces of the disks (Fig. 6.21).

Within a narrow radial region $r<a$, radially outward flow from the hole region and radially inward flow leaving the space between the disks collided, turned and moved upwards along the axis of rotation of the air cleaner, and finally left the hole region at $z=2h$ (Fig. 6.22). Thus, two opposite air flows were observed in the outlet cross-section of the cleaner, with one of them (non-swirling) moving into the air cleaner towards the solid disk and occupying almost the entire hole region. The second flow (with the tangential velocity $v_\varphi \approx 0.5\omega r$) moved outwards from the solid disk, being confined within a narrow region at $r \leq a$. Radial variation of the axial velocity component is qualitatively the same for all variants 1–14.

Effective parameter of swirl β at the inlet into the cavity was equal to 0.11. Then, the value of β gradually increased with a decrease of r due to the swirling effect of the disks (Fig. 6.23). In the radial location $r \approx 0.315$, the value of β near the centreplane of the cavity became equal to unity; and with the further movement of flow towards smaller r, the flow became overswirled relative to the disks ($v_\varphi > \omega r$) over the entire height of the cavity.

Radial variation of v_{rC} is similar for all variants: a minimum point was observed at a certain radial location in the cavity (which corresponded to maximum strength of inward flow between the disks), whereas in the hole region, a maximum of radially outward flow due to pumping effect of the bottom solid disk was observed. The values v_{rC} are close to zero in the region where radially inward and outward flows collide, i.e. where $r \rightarrow a$ (Fig. 6.24). In the region where $\beta>1$, flow becomes radially inward over the entire height of the cavity (Fig. 6.24).

Total pressure drop in the air cleaner occurred almost in full over the cavities region between the disks, while in the hole region the pressure was practically constant (Fig. 6.25).

Compared to variant 1, the swirling effect of the disks on forced radial flow is amplified in the following cases: with a decrease of the air flowrate (and v_{ri}) in variant 2; with a decrease of the disk rotational velocity in variant 3; with an increase in the height of the inter-disk spacing in variants 6 and 10; with shifting the disk hole radius inside the air cleaner to the value $a=0.2$ m in variant 7; with an increase in the tangential velocity of flow at the inlet to the cavity in variants 8 and 9; with a relative enlarging cavities region between the disks in variant 14. This effect is most prominent in variant 7, where an extended region of overswirled flow is observed for $r=0.2$–0.315 m, with a maximum value of the swirl parameter $\beta=1.67$ being located at $r=0.215$ m (Figs. 6.23 and 6.26). In the region of underswirled flow at the inlet into the cavities, there is a substantial radially outward flow in the boundary layers on the disks, whereas in the region of overswirled flow radially inwards developing boundary layers emerge with pronounced radial velocity maxima near the disks (Fig. 6.27).

Enhancement of the overswirled flow region and an increase in the rotational velocity of the disks increase the pressure drop required to pump flow through the air cleaner (Fig. 6.25).

Compared to variant 1, the swirling effect of the disks on the forced radial flow weakened, the overswirled flow did not develop and the pressure drop across the cavity decreased in the following cases: with a decrease in the angular speed of disk rotation in variant 4 and an increase in the inter-disk spacing in variants 5, 11 and 12 (Figs. 6.23 and 6.25).

An increase in the spacing between the disks resulted in some additional effects. For instance, if the mass flowrate or the radial velocity at the inlet into the cavity remained constant in variants 5, 11 and 12, the radially outwards developing boundary layers on the surfaces of the disks became stronger.

Scaling of the air cleaner and keeping the same ratio between the external and internal radii of the disks and a constant value of the radial velocity at the inlet to the cavities (variants 13 and 14) in fact retained the unchanged basic features of flow aerodynamics in the cavities and resulted in decrease in the pressure drop due to decrease in the mass flowrate through the air cleaner (Fig. 6.25).

In work [188], a case of heat transfer was simulated where air with T_i=20°C was fed into the air cleaner, which was heated up to a constant temperature T_w=100°C. This corresponds to one of non-optimum operating modes of the air cleaner, in which the surface friction was so high that the air cleaner was heated up to such a high temperature. The temperature field in the air cleaner is similar to the field of the tangential velocity component (Fig. 6.28).

A substantial difference of the temperature profiles in Fig. 6.28 from the profiles of v_φ consists in that the local temperature in the centre of the gap between the disks cannot exceed the value of the wall temperature T_w. Maximum air temperatures were observed near the outlet from the inter-disk cavity ($r \rightarrow a$). Cold air was constantly sucked into the outlet region from the outside; this air mixed with heated air flowing towards the outlet region from the cavities between the disks. More detailed investigations of heat transfer in the air cleaner were not performed in [188].

Thus, based on the parametric studies of the aerodynamics of turbulent flow in the rotating-disk air cleaner, the following conclusions were made. It is possible to improve performance of the air cleaner in terms of reduction of pressure losses in it, if regimes with flow overswirl are avoided. It is obvious that particle separation can be improved, provided that radially outwards developing boundary layers on the disks are enhanced, which is possible both if flow overswirl conditions are avoided

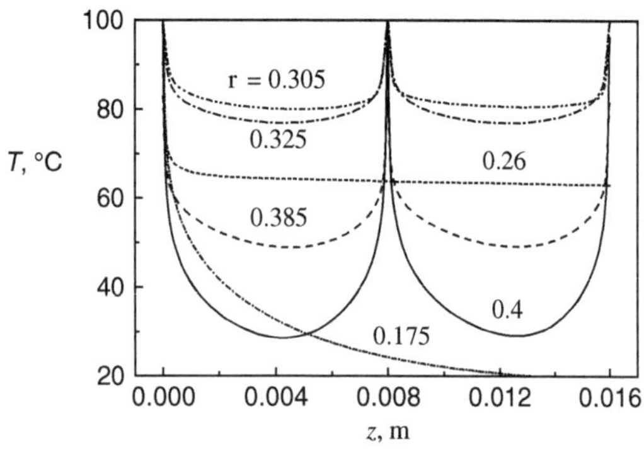

Fig. 6.28 Variant 1: local temperature variation over the height of the air cleaner for T_w=100°C and T_i=20°C

and if the spacing between the disks is increased to some extent, compared to the baseline variant 1, with all the other conditions being equal.

The investigation of the temperature field demonstrated that maximum air temperatures are observed in the region of the outlet from the cavities between the disks ($r \rightarrow a$), while cold air is constantly sucked into the outlet region from outside.

Chapter 7
Laminar Fluid Flow and Heat Transfer in a Gap Between a Disk and a Cone that Touches the Disk with Its Apex

7.1 General Characterization of the Problem

Conical diffusers with stationary walls (Fig. 7.1) are known in engineering [158]. For modelling fluid flow in such devices, with flow swirl at the inlet being absent, simplified Navier–Stokes equations were used in [158], while heat transfer has never been simulated in any of the known investigations.

Devices in which fluid flow occurs in a gap with small solid angles γ=1...5 degree between a rotating cone and a stationary flat surface (Fig. 7.2) are used in viscosimetry [50, 127, 161]. In medicine, such a device is used for growing endothelium cells, which are placed in a form of a monolayer on the stationary surface, while the slowly rotating cone makes possible circulation of feeding culture medium in the gap [16, 17].

Fig. 7.1 Schematic of swirling flow in a stationary conical diffuser

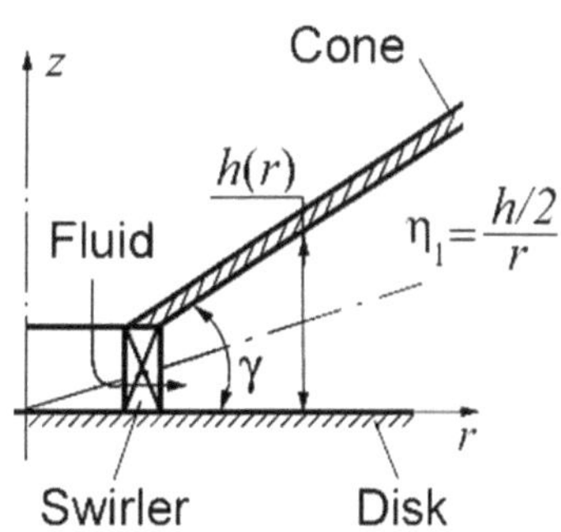

For the cases of small angles γ, the Navier–Stokes equations (2.1), (2.2) and (2.3) can be simplified as follows [16, 17, 161]:

$$-\frac{v_\varphi^2}{r} = -\frac{1}{\rho}\frac{\partial p}{\partial r} + \nu\frac{\partial^2 v_r}{\partial z^2}, \tag{7.1}$$

$$v_r\frac{\partial v_\varphi}{\partial r} + v_z\frac{\partial v_\varphi}{\partial z} + \frac{v_r v_\varphi}{r} = \nu\frac{\partial^2 v_\varphi}{\partial z^2}, \tag{7.2}$$

I.V. Shevchuk, *Convective Heat and Mass Transfer in Rotating Disk Systems*, Lecture Notes in Applied and Computational Mechanics 45, DOI 10.1007/978-3-642-00718-7_7,

© Springer-Verlag Berlin Heidelberg 2009

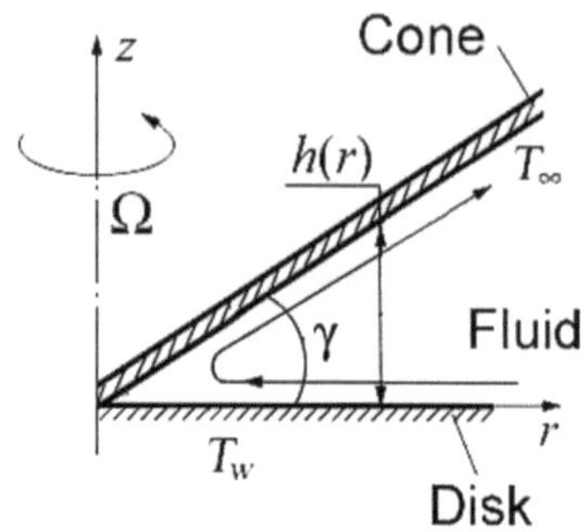

Fig. 7.2 Schematic of fluid flow in a conical gap with a rotating cone and a stationary disk

$$-\frac{1}{\rho}\frac{\partial p}{\partial z} + \nu\frac{\partial^2 v_z}{\partial z^2} = 0. \quad (7.3)$$

In work [161], Eqs. (7.1), (7.2) and (7.3) were solved by a method of expansion in the small parameter $Re = Re_\Omega \eta_1^2/12$. Here $\eta_1 = h/r$, whereas $h{=}r{\cdot}\mathrm{tg}\gamma$ is a height of the gap. As a result, the following relations for the velocity components were obtained:

$$v_r/(\Omega\, r) = Re\,(1.8\tilde{z}^2 - \tilde{z}^4 - 0.8\tilde{z}), \quad (7.4)$$

$$v_\varphi/(\Omega\, r) = \tilde{z} + Re^2\,(-83\tilde{z} + 70\tilde{z}^4 + 63\tilde{z}^5 - 50\tilde{z}^7)/175, \quad (7.5)$$

$$v_z/(\Omega\, r) = Re\eta_1(\tilde{z}^2 - \tilde{z}^3), \quad (7.6)$$

where $\tilde{z} = z/h$.

Analysis of Eqs. (7.4), (7.5) and (7.6) in comparison with data of a self-similar solution of the full Navier–Stokes equations obtained by the author [185] is given below. Using expressions (7.4) and (7.5) for the velocity components v_r and v_φ, one can obtain the following relations for the flow swirl angle on the surface of the disk φ_w:

$$\varphi_w = \arctan[0.8Re/(1 - 83Re^2/175)] \quad \text{for } Re = 1 - 1.452, \quad (7.7)$$

$$\varphi_w = \arctan(0.8Re) \quad \text{for } Re << 1. \quad (7.8)$$

To remind, the flow swirl angle is defined as $\varphi = \arctan[v_r/(\Omega\, r - v_\varphi)]$.

Equation (7.7) agrees with experiments [161] and Eq. (7.8) only for $Re \leq 0.5$ (Fig. 7.3). Equation (7.8), being formally valid only for $Re << 1$, agrees well with the experiments up to $Re = 2$. It should be noticed that authors [161] derived and compared with experiments only Eq. (7.8), while Eq. (7.7) logically resulting from Eqs. (7.4) and (7.5) was not obtained in [161]. It should be pointed out that Eq. (7.7) holds only for $Re \leq 1.452$. At $Re = 1.452$, the value of φ_w according to Eq. (7.7) is equal to 90 degree, while the argument of the function *arctan* in the right-hand side of Eq. (7.7) tends to infinity that does not agree with physics of the process considered.

Authors [16, 17], similar to work [161], solved the system of Eqs. (7.1), (7.2) and (7.3) using a method of series expansion in a small parameter *Re*. However, while the solutions (7.4), (7.5) and (7.6) [161] are restricted only to two terms of the power series in *Re*, the authors of [16, 17] used up to 70 terms of the expansion. In doing so, the calculated parameter φ_w [16, 17] agreed with the experiments [161] for *Re*=0.5–1 somewhat worse than Eq. (7.8), while at $Re = 1.2928$ the value of φ_w computed by [16, 17] tends to infinity that contradicts to the physics of the process.

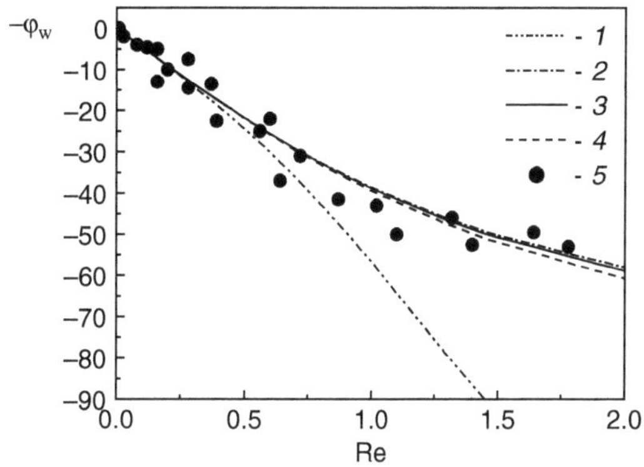

Fig. 7.3 Flow swirl angle on the surface of a stationary disk with a cone rotating. Computations: *1* – Eq. (7.7); *2* – Eq. (7.8) [161]; *3* – self-similar Eqs. (7.19), (7.20), (7.21) and (7.22) [185]; *4* – self-similar Eqs. (7.24), (7.25) and (7.26) [185]. Data *5* – experiments [161]

Full Navier–Stokes equations for small values of γ were solved numerically in [50].

None of the known investigations has dealt with the cases of (a) a rotating disk at a stationary cone and (b) simultaneous rotation of a disk and a cone with different angular speeds, which could allow determining optimal parameters of the entire system. Heat transfer in the cone–disk systems has never been simulated as well. Besides, no attempt has been made to derive self-similar forms of the full Navier–Stokes equations together with the thermal boundary layer equation for a conical diffuser for any value of the conical angle γ.

This chapter represents results for the cases of swirl flow in the gap between a stationary cone and a disk (Fig. 7.1) and a rotating disk and/or a cone with no initial swirl of the flow (Fig. 7.2) [182, 185, 186].

7.2 Navier–Stokes and Energy Equations in the Self-similar Form

In this study, full Navier–Stokes equations in cylindrical coordinates for a steady-state axisymmetric laminar flow (2.1), (2.2) and (2.3) and simplified Eqs. (7.1), (7.2) and (7.3) will be solved. The energy equation is used in a form of the thermal boundary layer equation (2.20) for laminar flow. Boundary conditions for the case of a rotating disk and/or a cone with no initial swirl of flow look as follows:

$$z = 0:\quad \mathrm{v}_r = 0,\quad \mathrm{v}_z = 0,\quad \mathrm{v}_\varphi = \omega r,\quad T_w - T_\infty = c_0 r^{n_*}, \tag{7.9}$$

$$z = h:\quad \mathrm{v}_r = 0,\quad \mathrm{v}_z = 0,\quad \mathrm{v}_\varphi = \Omega r,\quad T = T_\infty. \tag{7.10}$$

For the case of swirl flow in a gap between a stationary cone and a disk

$$z = 0:\quad \mathrm{v}_r = \mathrm{v}_\varphi = \mathrm{v}_z = 0,\quad T_w - T_\infty = c_0 r^{n_*}, \tag{7.11}$$

$$z = z_1 = h/2:\quad \mathrm{v}_r = \mathrm{v}_{r1},\quad \mathrm{v}_\varphi = \mathrm{v}_{\varphi 1},\quad d\mathrm{v}_r/dz = 0,\quad T = T_\infty. \tag{7.12}$$

Here c_0 and n_* are constants, and subscript "1" relates to boundary conditions at $z = z_1 = h/2$. In the present work, heat transfer of a *disk* with air is studied for boundary conditions (7.9) and (7.11) equivalent to boundary condition (2.35). Therefore, for the sake of simplification, it is accepted that the temperature of a cone is constant, $T_\infty = \text{const}$.

The thermal boundary layer equation is used instead of the full energy equation, because in the case of laminar flow viscous dissipation effects are minor, and thermal conductivity in the radial (i.e. marching) direction should be taken into account, as a rule, only for Pr→0 [22]. In addition, a self-similar form of the full energy equation imposes restrictions on the boundary conditions for T_w given by Eqs. (7.9) or (7.11): for rotating disks, for example, one can use only the value of $n_* = 2$ [41, 138, 139].

The use of variables and functions that are self-similar relative to coordinate r enables, in a number of physical problems, to reduce partial differential equations (2.1), (2.2), (2.3) and (2.4), (7.1), (7.2) and (7.3) and (2.20) to a set of ordinary differential equations, which can be solved numerically using applied mathematical computer codes (Mathcad, etc.) [22, 41, 138, 158, 165, 174, 180].

A form of self-similar variables and functions for the problem under consideration can be found by the method of group analysis of differential equations already used above in Sect. 4.2 [185, 186]. Let us introduce the so-called linear transformation:

$$r = A^{\alpha_1}\bar{r},\quad z = A^{\alpha_2}\bar{z},\quad \mathrm{v}_r = A^{\alpha_3}\bar{\mathrm{v}}_r,\quad \mathrm{v}_\varphi = A^{\alpha_4}\bar{\mathrm{v}}_\varphi,\quad \mathrm{v}_z = A^{\alpha_5}\bar{\mathrm{v}}_z,\quad p = A^{\alpha_6}\bar{p}, \tag{7.13}$$

where α_k ($k = 1, \ldots, 6$) and a coefficient of transformation A are constants [22]. Relations (7.13) are substituted into Eqs. (2.1), (2.2), (2.3) and (2.4), (2.20) and (7.1), (7.2) and (7.3). The initial and transformed forms of each equation are invariant, if the overall exponents of the constant A are the same for each term of the transformed equation (so that this constant can be eliminated). As a result, one can obtain the following relations between constants α_k:

$$\alpha_1 = \alpha_2 = \alpha,\quad \alpha_3 = \alpha_4 = \alpha_5 = -\alpha,\quad \alpha_6 = -2\alpha. \tag{7.14}$$

Having obtained Eqs. (7.14) for the constants α_k, one can find relations between parameters with overbars and without them in Eqs. (7.13). This results in the following:

$$A=\left(\frac{r}{\bar{r}}\right)^{1/\alpha}=\left(\frac{z}{\bar{z}}\right)^{1/\alpha}=\left(\frac{\bar{\mathrm{v}}_r}{\mathrm{v}_r}\right)^{1/\alpha}=\left(\frac{\bar{\mathrm{v}}_\varphi}{\mathrm{v}_\varphi}\right)^{1/\alpha}=\left(\frac{\bar{\mathrm{v}}_z}{\mathrm{v}_z}\right)^{1/\alpha}=\left(\frac{\bar{p}}{p}\right)^{1/(2\alpha)}, \tag{7.15}$$

$$\bar{z}/z=\bar{r}/r,\quad \bar{\mathrm{v}}_r\bar{r}=\mathrm{v}_r r,\quad \bar{\mathrm{v}}_\varphi\bar{r}=\mathrm{v}_\varphi r,\quad \bar{\mathrm{v}}_z\bar{r}=\mathrm{v}_z r,\quad \bar{p}\cdot\bar{r}^2=p\cdot r^2. \tag{7.16}$$

Combinations of variables (7.16) are invariant with respect to the linear transformations (7.13) and are named absolute variables. In accordance with the Morgan theorem [22], they are similar variables provided that the boundary conditions of a particular problem can be expressed in an r-independent form.

Based on Eqs. (7.15) and (7.16), one can derive the following self-similar variables and functions:

$$\eta=z/r,\quad F=\mathrm{v}_r r/\nu,\quad G=\mathrm{v}_\varphi r/\nu,\quad H=\mathrm{v}_z r/\nu, \tag{7.17}$$

$$P=pr^2/(\rho\nu^2),\quad \theta=(T-T_\infty)/(T_w-T_\infty). \tag{7.18}$$

The function θ remains therefore invariable.

Substituting formulas (7.17) and (7.18) into the full Navier–Stokes equations (2.1), (2.2) and (2.3), as well as into Eqs. (2.4) and (2.20), and omitting the terms that include derivatives with respect to r, one can obtain a final system of ordinary differential equations:

$$F^2+G^2+2P+F'L+\eta P'+F''M=0\,, \tag{7.19}$$

$$G'L+G''M=0\,, \tag{7.20}$$

$$P'-H(1+F)-H'L-H''M=0\,, \tag{7.21}$$

$$H'-\eta F'=0\,, \tag{7.22}$$

$$\theta''=\mathrm{Pr}\left[n_*F\theta+\theta'(H-\eta F)\right], \tag{7.23}$$

where $M=1+\eta^2$, $L=3\eta+\eta F-H$. Here, the primes denote differentiation with respect to coordinate η.

Simplified Eqs. (7.1), (7.2) and (7.3) take the following form:

$$G^2+2P+\eta P'+F''=0\,, \tag{7.24}$$

$$G'\,(\eta F-H)+G''=0\,, \tag{7.25}$$

$$P'=H''\,, \tag{7.26}$$

Transformed boundary conditions (7.9) and (7.10) look as follows:

$$\eta = 0: \quad F = H = 0, \quad G = G_0, \quad \theta = 1, \tag{7.27}$$

$$\eta = \eta_1: \quad F = H = 0, \quad G = G_1, \quad \theta = 0, \tag{7.28}$$

where $\eta_1 = h/r$, $G_0 = Re_\omega$, $G_1 = Re_\Omega$. Subscripts "0" and "1" denote conditions at $\eta = 0$ and $\eta = \eta_1$ respectively.

Transformed boundary conditions (7.11) and (7.12) look as follows:

$$\eta = 0: \quad F = G = H = 0, \quad \theta = 1, \tag{7.29}$$

$$\eta = \eta_1: \quad G = G_1, \quad F = F_1, \quad F' = 0, \quad \theta = 0, \tag{7.30}$$

where $\eta_1 = 0.5\, h/r$.

Strictly speaking, boundary conditions (7.27) and (7.28) for functions $G_0 = Re_\omega$ and $G_1 = Re_\Omega$ are not self-similar, since they depend on the coordinate r. The boundary conditions are self-similar in the case of $G_0 = 0$, i.e. Eq. (7.29) (stationary disk), and $G_1 = \text{const}$, $F_1 = \text{const}$, Eqs. (7.30), i.e. for the case where the tangential and radial velocities at the point $\eta = \eta_1$ develop as a free vortex:

$$(v_\varphi)_{\eta=\eta_1} = G_1 \nu/r, \quad (v_r)_{\eta=\eta_1} = F_1 \nu/r. \tag{7.31}$$

However, the boundary conditions (7.27) and (7.28) may be regarded as locally self-similar, and it may be assumed that G_0 and G_1 are parameters of the problem at each particular point r [185, 186]. As will be shown below, this approach enables to obtain results, which agree well with the known experimental and computational data.

7.3 Rotating Disk and/or Cone

7.3.1 Numerical Values of Parameters in the Computations

Equations (7.19)–(7.26) were numerically solved using the Mathcad software with its built-in sbval function (shooting method). Use of this method in the above chapters, in particular, to solve the self-similar set of Eqs. (2.37), (2.38), (2.39), (2.40) and (2.41) for rotating-disk systems (Chaps. 3 and 5) made it possible to calculate profiles of the velocity components and temperature, together with their derivatives, which coincide with the known data [41, 138, 139] up to the fourth significant digit.

The full self-similar set of Eqs. (7.19), (7.20), (7.21), (7.22) and (7.23), which is valid for any value of the angle γ, was used to solve all of the problems investigated below. The simplified set of Eqs. (7.24), (7.25) and (7.26) (simultaneously with Eqs. (7.22) and (7.23)) was used to compute velocity components in the problem of rotation of a cone with a stationary disk for small angles $\gamma \leq 5$ degree. The cases of small angle $\gamma = 4$ degree ($\eta_1 = 0.0698$) and rather large cone angles $\gamma = 45$ degree (η_1=1) were investigated. As was revealed by the computations, the value of η_1 in the range of γ=1–5 degree has no effect on the parameters of flow and heat transfer.

The value of $Pr = 0.71$ (air) was chosen for the heat transfer investigation; this enables one to make comparisons with the cases of air cooling in rotating-disk systems, which were well investigated both theoretically and experimentally (see Chap. 3). An increase in the Prandtl number causes an increase in the Nusselt number (see Chaps. 3 and 5 for the cases of $Pr \leq 1$, as well as Chap. 8 for the cases of $Pr \geq 1$). However, in the present chapter, effects of the Prandtl number on heat transfer have not been studied. The cases of decreasing ($n_* = -1$), constant ($n_* = 0$), and square-law increasing ($n_* = 2$) disk temperatures were investigated, which incorporate the range of thermal boundary conditions on a disk observed in engineering [41, 138, 139, 196]. The calculations were performed for the value of $Re = Re_\omega \eta_1^2/12$ (or $Re = Re_\Omega \eta_1^2/12$) equal to unity, which corresponds to $Re_\omega = 12$, $Re_\Omega = 12$ at $\eta_1 = 1$, and $Re_\omega = 2463$, $Re_\Omega = 2463$ at $\eta_1 = 0.0698$.

7.3.2 Cone Rotation at a Stationary Disk

The components of the velocity profile in the gap calculated by Eqs. (7.19), (7.20), (7.21), (7.22), (7.23), (7.24), (7.25) and (7.26) and obtained in [161] by the method of expansion in the small parameter Re are given in Figs. 7.4, 7.5 and 7.6. Here, $Re = 1$ ($Re_\omega = 2463$). It follows from these figures that the tangential velocity component (Fig. 7.6) is an order of magnitude higher than the radial component (Fig. 7.4) which, in turn, is an order higher than the axial component (Fig. 7.5). The results of calculations by Eqs. (7.19), (7.20), (7.21) and (7.22) and (7.24), (7.25) and (7.26) almost coincide, and this confirms the possibility of using the simplified Eqs. (7.24), (7.25) and (7.26) for conical gaps with small angles γ. The results of calculations by Eqs. (7.19), (7.20), (7.21) and (7.22) and (7.24), (7.25) and (7.26) and the data of [161] for the radial velocity component v_r are in a quite close agreement (Fig. 7.4).

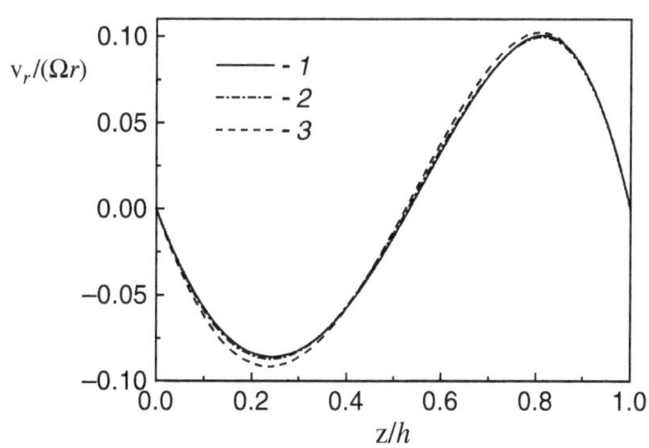

Fig. 7.4 Profiles of the radial velocity component in a gap between a rotating cone and a stationary disk. *1* – Eqs. (7.19), (7.20), (7.21) and (7.22); *2* – Eqs. (7.24), (7.25) and (7.26); *3* – Eq. (7.4) [161]

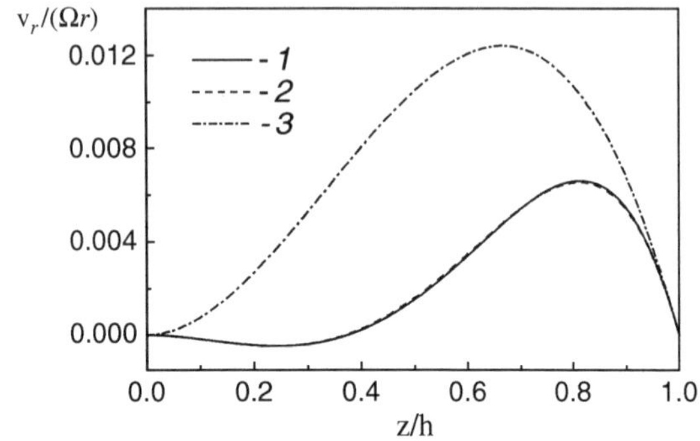

Fig. 7.5 Profiles of the axial velocity component in a gap between a rotating cone and a stationary disk. *1* – Eqs. (7.19), (7.20), (7.21) and (7.22); *2* – Eqs. (7.24), (7.25) and (7.26); *3* – Eq. (7.6) [161]

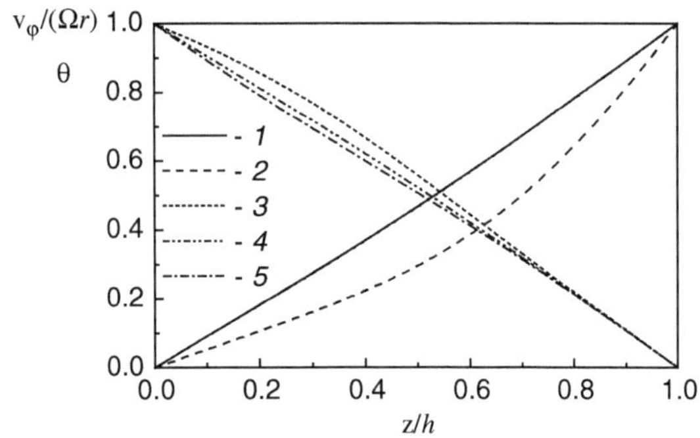

Fig. 7.6 Profiles of the axial velocity component (*1, 2*) and dimensionless temperature θ (*3–5*) in a gap between a rotating cone and a stationary disk. *1* – Eqs. (7.19), (7.20), (7.21) and (7.22) and (7.24), (7.25) and (7.26); *2* – Eq. (7.5) [161]; *3* – Eqs. (7.19), (7.20), (7.21), (7.22) and (7.23) at $n_*=2$; *4* – (7.19), (7.20), (7.21), (7.22) and (7.23) at $n_*=0$; *5* – (7.19), (7.20), (7.21), (7.22) and (7.23) at $n_*=-1$

The differences between the two sets of equations under consideration with respect to the components v_z and v_φ are more pronounced (Figs. 7.5 and 7.6)

The adequacy of simulation of the tangential velocity component is illustrated by comparison of the computed results and experiments of [161] for the flow swirl angle on the disk surface φ_w. On a disk at $\eta=0$, we have $\varphi_w = \arctan[v_r/(\Omega\, r - v_\varphi)]_{z=0} = \arctan(-F'_w/G'_w)$. The prediction data for φ_w agree well with the experimental results of [161] over the entire range of variation of the Reynolds number *Re* (Fig. 7.3). It follows from this that the profiles of the velocity components in Figs. 7.4, 7.5 and 7.6 obtained as a result of calculations by Eqs. (7.19), (7.20), (7.21) and (7.22) and (7.24), (7.25), (7.26), (7.27) and (7.28) describe real flow in the gap more adequately than the solutions (7.4), (7.5), (7.6), (7.7) and (7.8) [161].

The dimensionless temperature in the gap, which decreases from unity on the disk to zero on the cone, depends on the value of n_*; it should be pointed out that in the vicinity of the disk the θ curve becomes smoother with increasing n_* (Fig. 7.6).

The local Nusselt number is calculated by the relation:

$$Nu = -\theta'_{\eta=0}, \tag{7.32}$$

where differentiation in the right-hand part of Eq. (7.32) is done with respect to the variable η.

In analogy with the relations (3.4) and (3.5) for the rotating-disk systems, the Nusselt number may be represented as

$$Nu = K_1 Re_{\Omega}^{1/2} \tag{7.33}$$

$$K_1 = \frac{-(d\theta/d\eta)_{\eta=0}}{Re_{\Omega}^{1/2}} = -\left(\frac{d\theta}{d\zeta}\right)_{\zeta=0} \tag{7.34}$$

where $\zeta = z\sqrt{\Omega/\nu}$.

As a result, for the case where $\eta_1 = 0.0698$ (or $Re_{\Omega} = 2463$), it was obtained that $Nu = 15.28$, 13.40, 9.35 and $K_1 = 0.308$, 0.270, 0.188 for $n_* = -1$, 0, 2, respectively. The order of magnitude of the constant K_1 is comparable with the data for a free rotating disk, where $K_1 = 0.189$, 0.326, 0.519 for the same values of n_* (see Table 3.8). It should be pointed out that, following an increase in n_*, the constant K_1 decreases for the problem under consideration, while for a free rotating disk K_1 increases. It can be thus concluded that the decrease in the constant K_1 with increasing n_* observed in the currently considered case occurs under conditions of radially inward flow over a stationary disk caused by a rotating cone. And vice versa, an increase in the constant K_1 with increasing n_* is observed under conditions of radially outward flow over a rotating disk in a stationary or co-rotating fluid (see Chaps. 3, 4, 5 and 6). At $\eta_1 = 1$ and $Re = 1$ (or $Re_{\Omega} = 12$), it was obtained that $Nu = 1.047$, 0.954, 0.760 and $K_1 = 0.302$, 0.275, 0.219 for the same values of n_*. It is thus obvious that the constant K_1 is a conservative parameter and varies rather weakly with an increasing angle of the gap γ.

7.3.3 Disk Rotation at a Stationary Cone

The direction of radial flows for the problem at hand is opposite to that in the case of cone rotation at a stationary disk: the flow over the cone is radially inward, and that over the disk is radially outward (Fig. 7.7). The tangential velocity v_{φ} linearly decreases with increasing coordinate z, and the profile of the axial component v_z is mirror-symmetrical (relative to the abscissa axis) to the v_z curve given in Fig. 7.5. The profile of the dimensionless temperature θ in Fig. 7.7 is close to the $v_{\varphi}/(\omega r)$ distribution.

The Nusselt number on the disk is calculated by Eqs. (7.32), (7.33) and (7.34) substituting ω instead of Ω (in the definition of both the Reynolds number Re and the coordinate ζ) or, that is in fact the same, using Eqs. (3.4) (for $n_R = 1/2$) and (3.5). As a result, for $Re = 1$ ($Re_{\omega} = 2463$) and $\eta_1 = 0.0698$, it was obtained that $Nu = 13.33$, 15.35, 19.13 and $K_1 = 0.269$, 0.309, 0.386 at $n_* = -1$, 0, 2, respectively.

One can see that, as n_* increases, the constant K_1 also increases, though at a slower rate than in the case of a free rotating disk (where, just to remind, K_1=0.189,

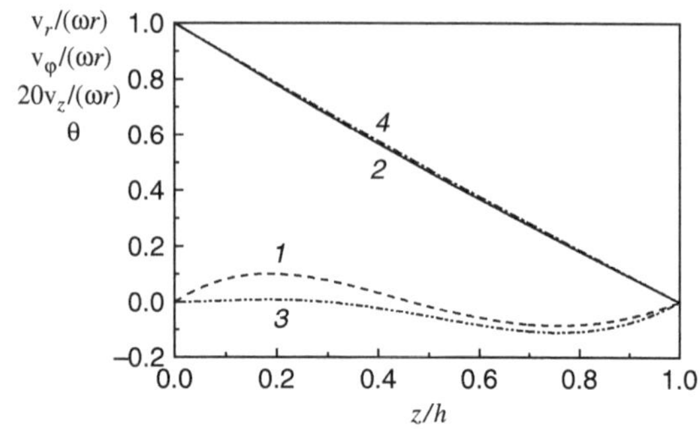

Fig. 7.7 Velocity and temperature profiles in the gap between a rotating disk and a stationary cone for $Re = 1$ ($Re_\omega = 2463$) and $\eta_1 = 0.0698$. *1* – $v_r/(\omega r)$; *2* – $v_\varphi/(\omega r)$; *3* – 20 $v_z/(\omega r)$; *4* – θ ($Pr = 0.71$, $n_* = 0$)

0.326, 0.519 for the same n_*). With η_1=1 and Re=1 (Re_ω=12) one can obtain Nu=0.96, 1.041, 1.197 and K_1=0.277, 0.301, 0.345 for the same values of n_*, i.e., the constant K_1 is still conservative with respect to the cone angle γ.

7.3.4 Co-rotating Disk and Cone

In this case, the flow pattern is determined by the ratio between the Re_Ω and Re_ω numbers. If a cone rotates faster than a disk, i.e. for $Re_\Omega > Re_\omega$, the fluid flows radially outward over the cone and radially inward over the disk. If a disk rotates faster than a cone, i.e. for $Re_\Omega < Re_\omega$, the flow pattern is reversed. The Nusselt number was calculated by Eqs. (7.32), (7.33) and (7.34), where Ω is the reference angular velocity. It is interesting to analyse the case of approximately the same angular speed of rotation of the disk and cone. With $Re_\omega = 1.01\ Re_\Omega$, the calculations give Nu = 14.31, 14.35, 14.43 and K_1 = 0.288, 0.289, 0.291, while condition Re_ω = 0.99 Re_Ω results in Nu = 14.35, 14.31, 14.23 and K_1 = 0.289, 0.288, 0.287 at n_* = −1, 0, 2, respectively. For both examples, the conditions $Re = 1$ and $\eta_1 = 0.0698$ ($Re_\Omega = 2463$) were held. With $Re_\omega = 1.01\ Re_\Omega$, $Re = 1$ and $\eta_1 = 1$ ($Re_\Omega = 12$), the calculated values were Nu = 0.999, 1.001, 1.004 and K_1 = 0.288, 0.289, 0.290.

It is evident that the constant K_1 is almost invariable for all variants. The effect of the flow direction on the dependence of the Nusselt number on the exponent n_* exhibits the same pattern (in a weak form) as that mentioned above in Sect. 7.3.2. For the radially outward flow over the disk Nu numbers increase, and for the radially inward flow the values of Nu decrease with increasing n_*. An increase in the gap height has almost no effect on the value of K_1.

7.3.5 Counter-Rotating Disk and Cone

In this case, one can observe the most complex pattern of radial flow, which is radially outward over the disk and the cone and radially inward in the middle of the gap (Fig. 7.8). The axial velocity component v_z changes its sign twice, being negative

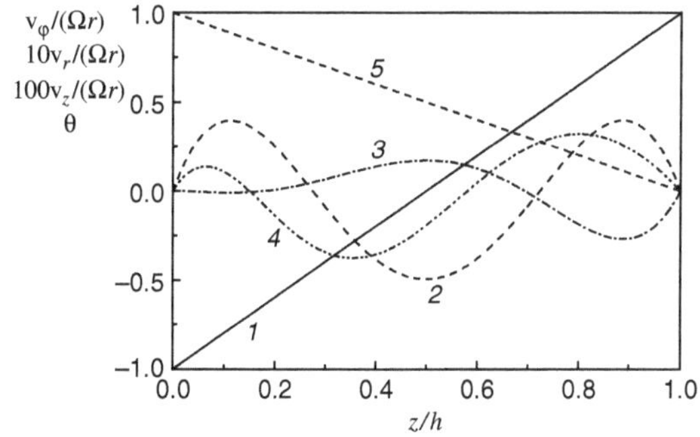

Fig. 7.8 Velocity and temperature profiles for a counter-rotating disk and a cone at $Re = 1$. *1–3* – $Re_\omega = -Re_\Omega = 2463$ and $\eta_1 = 0.0698$; *4* – $Re_\omega = -Re_\Omega = 12$ and $\eta_1 = 1$. *1* – $v_\varphi/(\omega r)$; *2, 4* – $10\, v_r/(\omega r)$; *3* – $100\, v_z/(\omega r)$; *5* – θ ($Pr = 0.71$, $n_* = 0$)

in the vicinity of the surfaces and positive in the middle of the gap; the tangential velocity $v_\varphi/(\Omega r)$ linearly increases from −1 to 1 with increasing coordinate z, and the temperature θ monotonically decreases from unity to zero. With $\eta_1 = 0.0698$, $Re = 1$ and $Re_\omega = -Re_\Omega = 2463$, the profile of the radial velocity v_r becomes symmetrical about the middle of the gap (curve *2*, Fig. 7.8). The Nusselt number was calculated by Eqs. (7.32), (7.33) and (7.34), where Ω was a reference velocity. On the disk, one can obtain $Nu = 14.21$, 14.44, 14.85 and $K_1 = 0.286$, 0.201, 0.299 at $n_* = -1, 0, 2$, i.e., the Nusselt number increases with n_*. With $\eta_1 = 1$, $Re = 1$, $Re_\omega = 12$ and $Re_\Omega = -12$, the profile of the radial velocity v_r becomes non-symmetrical, because radial flow is more pronounced over the cone (curve *4*, Fig. 7.8). In view of this, the Nusselt number on the disk decreases with increasing n_*: $Nu = 1.011$, 0.989, 0.942 and $K_1 = 0.292$, 0.285, 0.272 for the same values of n_* as above.

Therefore, in this case, heat transfer is weakly sensitive to the variation of the value and sign of the disk temperature gradient. The variation of the gap height and of the angular speeds of disk and cone rotation affect the v_r profile and the qualitative character of the influence of n_* on Nu.

7.4 Radially Outward Swirling Flow in a Stationary Conical Diffuser

Computations were performed for a stationary diffuser with an angle of $g = 35$ degree or $h_1 = 0.35$. From a physical point of view, boundary conditions (7.30) and (7.31) mean that swirling flow develops as a free vortex along the gap angle bisector. In reality, the free vortex may occupy a considerable part of the gap height, while the boundary layer is formed only near the walls. In this study, we consider a somewhat idealized case in which conditions (7.30) and (7.31) hold only for $h = h_1$.

Computations showed that non-swirling radial flow ($G_1 = 0$) does not separate from the walls of the diffuser for $F_1 < 63$. Separation occurs for $F_1 \approx 63$, while for $F_1 > 63$ one can observe a recirculation flow zone near the disk surface (Fig. 7.9). The reason for the flow separation is the large angle of conicity: reducing it to $\eta_1 = 0.035$ results in the separation value of F_1 increasing to about 7500.

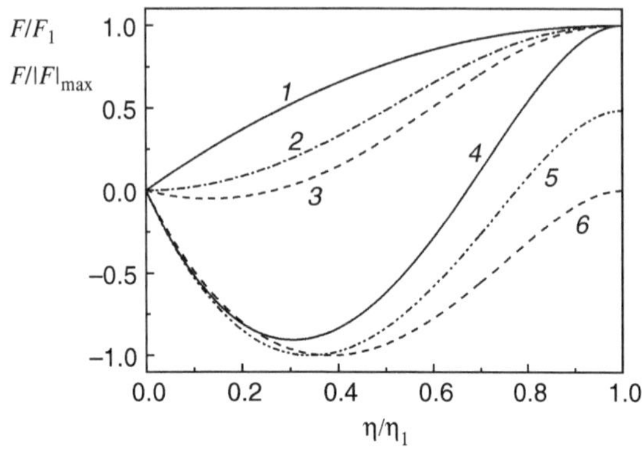

Fig. 7.9 Profiles of the radial velocity F/F_1 (*1–4*) or $F/|F|_{max}$ (*5, 6*) in a gap between a disk and a cone. $\mathbf{G_1 = 0}$: *1* – $F_1 = 2$; *2* – $F_1 = 63$; *3* – $F_1 = 90$. $\mathbf{G_1 = 97.96}$: *4* – $F_1 = 20$; *5* – $F_1 = 10$, $|F|_{max} = 20.66$; *6* – $F_1 = 0$, $|F|_{max} = 24.28$

Flow swirl ($G_1 = 97.96$, $Re = G_1\eta_1^2/12 = 1$) inevitably leads to formation of pronounced recirculation flow near the disk (Fig. 7.9). In the limiting case of zero radial velocity $F_1 = 0$ at the swirler outlet, the velocity v_r is negative for any η except for $\eta = \eta_1$ (curve *6*; the quantity $|F|_{max}$ is the maximum absolute magnitude of F at the minimum of the curve $F/|F|_{max}$ at $\eta/\eta_1 \approx 0.4$, that is, $F_{max} = -24.28$). As F_1 increases, the recirculation flow zone shrinks, while near the axis of the gap the radially outward flow region expands. In the gap, the tangential velocity component G/G_1 increases almost linearly from zero at $\eta = 0$–1 for $\eta = \eta_1$ (curve *1* in Fig. 7.10).

The diffuser serves to restore static pressure: the pressure increases with the r coordinate due to reduction in the velocity components $(v_r)_{\eta=\eta_1}$ and $(v_\varphi)_{\eta=\eta_1}$ with increasing r. In order for the function P to be self-similar, it is necessary that the quantity p in Eq. (7.18) be understood to represent the excess pressure $p - p_\infty$, which decreases as r increases ($p = p_\infty = \text{const}$ for $r \to r_\infty$). In non-swirling flow ($G_1 = 0$), the parameter P characterizing the pressure recovery rate increases with F_1 (Fig. 7.11). For $G_1 = 97.96$, the flow swirl leads to a noticeable increase in the pressure recovery rate P, while the contribution of the parameter F_1 over its range of variation from $F_1 = 0$–20 becomes weak.

For F_1 ranging from 50 to 63, the curves of the Nusselt number (calculated by Eq. (7.32)) versus F_1 exhibit points of maxima for $n_* = 2$ and 0 and minima for $n_* = -1$ (Fig. 7.12).

If the inlet flow swirl is absent ($G_1 = 0$) and values of the radial velocity at the inlet F_1 are non-zero, an increase in the exponent n_* leads to an increase in the Nusselt number (curves *1–3*, Fig. 7.12). Over the range of variation of $n_* = 0$–2, the Nusselt numbers following an increase in F_1 also increase for non-separating radially outward flow, remain approximately constant for the values of F_1 close to separating flow and decrease for radially inward recirculation flow near the disk surface (curves *1* and *2* in Fig. 7.12). For $n_* = 0$, these tendencies are very weak, while the maximum of the *Nu* number curve shifts towards larger values of F_1.

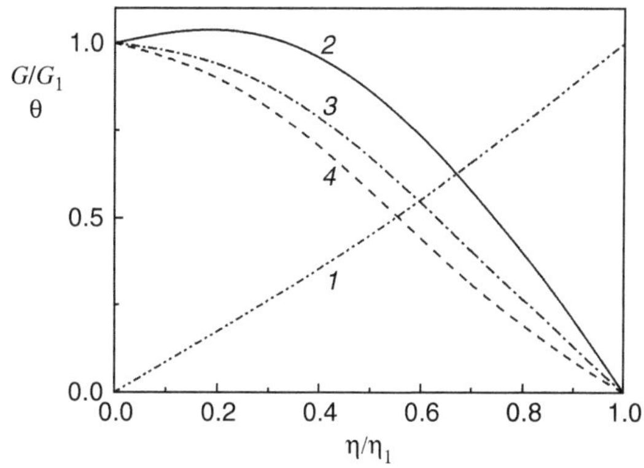

Fig. 7.10 Profiles of the tangential velocity component G/G_1 and temperature θ in the gap between a cone and a disk. *1* – G/G_1 for $F_1 = 30$, $G_1 = 97.96$; *2* – θ for $G_1 = 97.96$ and $F_1 = 10$; *3* – θ for $G_1 = 97.96$ and $F_1 = 30$; *4* – θ for $G_1 = 97.96$ and $F_1 = 60$ ($Pr = 0.71$, $n_* = 2$)

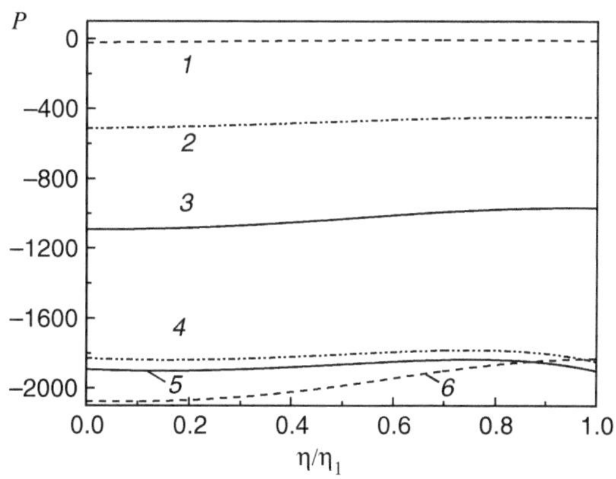

Fig. 7.11 Static pressure drop in the gap for $G_1 = 0$ (curves *1–4*) and $G_1 = 97.96$ (curves *5, 6*). *1, 6* – $F_1 = 20$; *2* – $F_1 = 45$; *3* – $F_1 = 63$; *4* – $F_1 = 90$; *5* – $F_1 = 10$

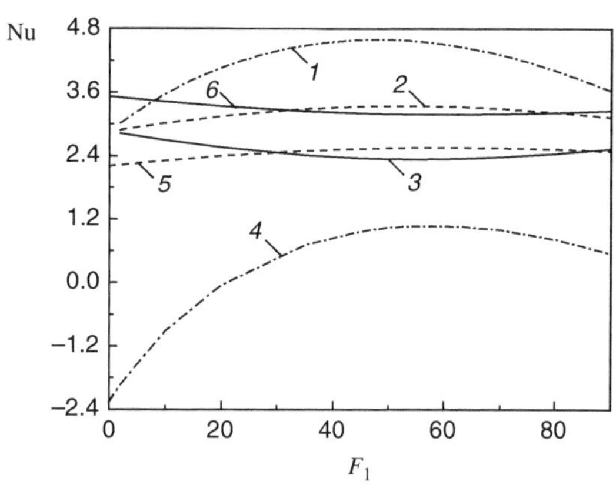

Fig. 7.12 Nusselt numbers in the gap for $G_1 = 0$ (curves *1–3*) and $G_1 = 97.96$ (curves *4–6*). *1, 4* – $n_* = 2$; *2, 5* – $n_* = 0$; *3, 6* – $n_* = -1$

Imposing initial swirl with $G_1 = 97.96$ leads to different signs of v_r and dT_w/dr and a decrease in the Nusselt numbers for $n_* = 2$ and 0 (curves *4* and *5* in Fig. 7.12) in comparison with non-swirling flow. At the same time, for $n_* = -1$, the signs of v_r and dT_w/dr are the same, and the Nusselt number of swirling flow (curve *6* in Fig. 7.12) increases in comparison with non-swirling flow. It should be pointed out that though for $G_1 = 97.96$ the radial velocity component v_r does not change its sign near the disk as F_1 increases (i.e. near-disk flow remains always radially inward), the shapes of curves *4*, *5* and *6* of the *Nu* number dependency on F_1 are qualitatively similar to curves *1*, *2* and *3* for non-swirling flow for $n_* = idem$ (see Fig. 7.12).

For swirling flow ($G_1 = 97.96$) and $n_* = 2$, temperature profiles decrease monotonically along the height of the gap under condition $F_1 > 21$ (curves *2* and *3* in Fig. 7.10). Under condition of $F_1 \leq 21$, the temperature profile in a fluid flowing in the direction of simultaneous decrease in r and T_w has a maximum near the wall (curve *4* in Fig. 7.10). Because of this, the Nusselt number becomes negative, which indicates a change in the heat flux sign: the fluid heats the disk surface, rather than the reverse (curve *4* in Fig. 7.12).

Thus, in the present chapter, self-similar forms of the full and simplified Navier–Stokes equations and the energy equation applicable to the fluid flow in a gap between a disk and a cone that touches the disk with its apex were derived for any angles of conicity. Exact solutions of the self-similar sets of equations were obtained for the problems of cone rotation at a stationary disk, disk rotation at a stationary cone, simultaneous co-rotation or contra-rotation of a disk and a cone, as well as for radially outward swirling flow in a stationary conical diffuser. Peculiarities of fluid flow and heat transfer in these conditions were analysed. It was shown that computations agree well with the experimental data available in the literature. Effects of the parameters of the boundary conditions on the pressure, velocity and temperature profiles, as well as on the Nusselt numbers, were investigated.

Chapter 8
Heat and Mass Transfer of a Free Rotating Disk for the Prandtl and Schmidt Numbers Larger than Unity

8.1 Laminar Flow

Problems of heat and mass transfer in fluid flow over a rotating disk for the Prandtl or Schmidt numbers larger than unity are of great importance in a number of industrial and scientific applications. To mention a few, one should count electrochemistry, where the rotating-disk electrode technique is widely used for experimental determination of the diffusion coefficient at the Schmidt numbers much larger than unity [3, 6, 8, 37, 39, 40, 42, 48, 81, 102, 105, 126, 131, 157]. The second area of application actual for the present research is the naphthalene sublimation technique often employed in experimental measurements of the mass transfer coefficients α_m [24, 28, 29, 59, 64, 96, 97, 117, 193, 195, 201, 208].

Differential equations of convective diffusion for axisymmetric flow in the cylindrical polar coordinate system for laminar and turbulent flow regimes, Eqs. (1.44) and (1.45), are analogous to the respective forms of the energy equation, Eqs. (2.5) and (2.9), in case of reciprocal substitution of local temperatures T by local concentrations C, as well as substitution of thermal diffusivity a by the diffusion coefficient D_m. The Navier–Stokes equations for laminar flow, Eqs. (2.1), (2.2) and (2.3), and those for turbulent flow, Eqs. (2.6), (2.7) and (2.8), as well as continuity equation (2.4) retain the very same form as stated in Chap. 2.

The system of Eqs. (2.37), (2.38), (2.39), (2.40) and (2.41) for steady-state axisymmetric laminar flow in the self-similar form can be also used to model convective mass transfer, if the non-dimensional function θ is defined as follows:

$$\theta = (C - C_\infty)/(C_w - C_\infty), \tag{8.1}$$

and the Prandtl number in Eq. (2.41) is substituted by its analogue for convective mass transfer, i.e. Schmidt number Sc (called sometimes Prandtl–Schmidt number).

The boundary condition for the concentration on the disk surface C_w is simplified in comparison with the boundary conditions (2.34), (2.35) and (2.36), because the surface concentration of a substance is constant in the radial (as well as circumferential) direction:

I.V. Shevchuk, *Convective Heat and Mass Transfer in Rotating Disk Systems*, Lecture Notes in Applied and Computational Mechanics 45, DOI 10.1007/978-3-642-00718-7_8,
© Springer-Verlag Berlin Heidelberg 2009

$$C_w = const. \tag{8.2}$$

The self-similar convective diffusion equation (i.e. transformed equation (2.41)) takes the following form:

$$\theta'' - ScH\theta' = 0. \tag{8.3}$$

Local and average Nusselt numbers are, as before, calculated by the generalized equations (3.4), which hold both for laminar and for turbulent flow regimes. The Sherwood number Sh, which is often referred to as the mass transfer Nusselt number Nu_m, is calculated by the relations that formally coincide with Eqs. (3.4):

$$Sh \equiv Nu_m = K_1 Re_\omega^{n_R}, \quad Sh_{\text{av}} \equiv Nu_{m,\text{av}} = K_2 Re_\varphi^{n_R}. \tag{8.4}$$

In Eqs (8.4), coefficients K_1 and K_2 depend on the boundary conditions, flow regime and Sc numbers, while the constant n_R is, generally speaking, dependent only on the flow regime; for laminar flow $K_1=K_2$ and $n_R=1/2$.

Thus, the energy and convective diffusion equations, along with their boundary conditions $T_w = const$ or $C_w = const$, are analogous, so that one can switch between their solutions (or equations for heat/mass transfer parameters obtained experimentally) simply substituting C, Sc and Sh instead of T, Pr and Nu (and vice versa), respectively.

For laminar flow, the problem under consideration can be rather easily solved numerically using the self-similar system of Eqs. (2.37), (2.38), (2.39), (2.40) and (2.41). To remind, for the case of a free rotating disk, parameters N and β in Eq. (2.37) are equal to zero. Numerical solution of Eqs. (2.37), (2.38), (2.39), (2.40) and (2.41) for the Prandtl and Schmidt numbers larger than unity has been performed using MathCAD software similar to the case of $Pr \leq 1$ analysed in Chap. 3. Computed values of the constant K_1 depending on the Pr (or Sc) number are given in Table 8.1.

Following increase in the Prandtl numbers, numerical values of the constant K_1 also increase: by 6.73 times for $Pr = 100$ and by 155 times for $Pr = 10^6$ in comparison with the case of $Pr = 1$ (data for $n_* = 0$, Table 8.1). For a constant value of the Prandtl number, the coefficient K_1 increases also with the exponent n_*. For instance, for $Pr = 2$ the value of K_1 increases by 2.73 times, while for $Pr = 10^6$ the increase amounts up to 2.20 times with the increase in the exponent n_* from –1 to 3. As mentioned in Chap. 3, for $Pr = 0.71$ the constant K_1 increases by 3.3 times over the same range of variation of the exponent n_*. This means that the effect of the exponent n_* on the coefficient K_1 becomes weaker at increased Prandtl numbers.

Use of a tabulated exact solution is often inconvenient in practice, if there exists a need to express a relation between parameters in an explicit form.

The only known approximate analytical solution for the constant K_1 under the boundary condition (2.35) at $Pr \geq 1$ and arbitrary non-zero values of the exponent n_* was found by Dorfman in the form of Eq. (3.6) [41]. As it is evident from Table 8.2, the values of K_1 calculated by Eq. (3.6) systematically exceed data of the exact

Table 8.1 Values of the constant K_1 according to the exact solution of Eqs. (2.37), (2.38), (2.39), (2.40) and (2.41) for the Prandtl or Schmidt numbers larger than unity

Pr (Sc)	n_*=–2	n_*=–1.5	n_*=–1	n_*=–0.5	n_*=0	n_*=1	n_*=2	n_*=3	n_*=4
1.0	0.0	0.1305	0.2352	0.3221	0.3963	0.5180	0.6159	0.6982	0.7693
1.05	0.0	0.1347	0.2422	0.3312	0.4070	0.5310	0.6306	0.7142	0.7865
1.1	0.0	0.1388	0.2491	0.3401	0.4173	0.5436	0.6449	0.7297	0.8031
1.5	0.0	0.1682	0.2979	0.4028	0.4906	0.6324	0.7450	0.8389	0.9199
2.0	0.0	0.1989	0.3482	0.4669	0.5653	0.7226	0.8466	0.9498	1.0386
2.28	0.0	0.2140	0.3728	0.4982	0.6016	0.7663	0.8960	1.0036	1.0963
2.4	0.0	0.2201	0.3827	0.5108	0.6162	0.7840	0.9158	1.0253	1.1195
2.5	0.0	0.2251	0.3907	0.5209	0.6280	0.7982	0.9319	1.0428	1.1383
3.0	0.0	0.2480	0.4279	0.5680	0.6826	0.8640	1.0061	1.1238	1.2251
4.0	0.0	0.2873	0.4912	0.6480	0.7753	0.9758	1.1321	1.2615	1.3728
5.0	0.0	0.3206	0.5445	0.7153	0.8533	1.0697	1.2382	1.3774	1.4971
7.0	0.0	0.3756	0.6325	0.8262	0.9818	1.2247	1.4131	1.5687	1.7024
9.0	0.0	0.4209	0.7047	0.9173	1.0873	1.3518	1.5568	1.7259	1.8711
10.0	0.0	0.4410	0.7368	0.9577	1.1341	1.4083	1.6206	1.7957	1.9460
11.0	0.0	0.4599	0.7668	0.9955	1.1779	1.4611	1.6802	1.8609	2.0161
13.0	0.0	0.4943	0.8216	1.0646	1.2579	1.5576	1.7893	1.9804	2.1444
15.0	0.0	0.5254	0.8710	1.1268	1.3300	1.6446	1.8877	2.0880	2.2599
17.0	0.0	0.5537	0.9162	1.1836	1.3958	1.7241	1.9775	2.1864	2.3656
19.0	0.0	0.5800	0.9579	1.2361	1.4567	1.7975	2.0605	2.2773	2.4632
20.0	0.0	0.5924	0.9776	1.2610	1.4855	1.8323	2.0999	2.3203	2.5095
50	0.0	0.8536	1.3925	1.7835	2.0909	2.5635	2.9269	3.2260	3.4825
100	0.0	1.1108	1.8009	2.2979	2.6871	3.2840	3.7422	4.1190	4.4421
500	0.0	1.9943	3.2033	4.0644	4.7351	5.7596	6.5442	7.1888	7.7413
1000	0.0	2.5467	4.0802	5.1691	6.0162	7.3083	8.2972	9.1096	9.8057
5000	0.0	4.4470	7.0967	8.9694	10.423	12.636	14.329	15.718	16.909
10^4	0.0	5.6363	8.9846	11.348	13.181	15.971	18.104	19.855	21.356
10^5	0.0	12.291	19.548	24.657	28.613	34.632	39.230	43.003	46.236
10^6	0.0	26.626	42.304	53.328	61.860	74.834	84.742	92.873	99.838

solution. Inaccuracies of Eq. (3.6) at $n_*\leq 0$ are unacceptably high and reach 16–40% already for $Pr=1$. For the case of $n_*=0$, the inaccuracies of Eq. (3.6) for $Pr=1$–3 reach 10–11%, respectively. At higher values of the exponent n_* and $Pr=1$–3, inaccuracies of Eq. (3.6) are relatively low (1–6%); however, for further increased Prandtl numbers the inaccuracies of Eq. (3.6) become approximately the same for any n_* and abruptly increase for $Pr\rightarrow\infty$.

The effect of the Schmidt (or Prandtl) number on the constant K_1 and the rate of increase in the divergence of the Dorfman's equation (3.6) and the exact solution at $n_*=0$ ($T_w=const$ or C_w=const) is illustrated in Fig. 8.1. It is evident that inaccuracies of Eq. (3.6) make it unacceptable for problems of mass transfer at $C_w=const$ even over the range of the Prandtl or Schmidt numbers moderately different from unity, e.g. for Sc=1–3.

As mentioned in Chap. 3, for a disk with a constant surface temperature $T_w=const$ the authors of [26, 106] developed approximate equations (3.7) and (3.8) for the constant K_1 over the range of the Prandtl numbers Pr=0–∞. As applied to

Table 8.2 Values of the constant K_1 calculated by Eq. (3.6) for the Prandtl or Schmidt numbers larger than unity

$Pr\,(Sc)$	n_*=–2	n_*=–1.5	n_*=–1	n_*=–0.5	n_*=0	n_*=1	n_*=2	n_*=3	n_*=4
1.0	0.0	0.2178	0.3080	0.3772	0.4356	0.5335	0.616	0.6887	0.7544
2.0	0.0	0.308	0.4356	0.5335	0.616	0.7544	0.8712	0.9740	1.0669
2.28	0.0	0.3289	0.4651	0.5696	0.6577	0.8055	0.9301	1.0399	1.1392
3.0	0.0	0.3772	0.5335	0.6534	0.7544	0.9240	1.0669	1.1929	1.3067
5.0	0.0	0.4870	0.6887	0.8435	0.9740	1.1929	1.3774	1.5400	1.6870
10.0	0.0	0.6887	0.9740	1.1929	1.3774	1.6870	1.9480	2.1779	2.3858
50	0.0	1.5400	2.1779	2.6674	3.0801	3.7722	4.3558	4.8699	5.3347
500	0.0	4.8699	6.8871	8.4349	9.7398	11.929	6.5442	15.400	16.870
1000	0.0	6.8871	9.7398	11.929	13.775	16.870	19.480	21.779	9.8057
5000	0.0	15.400	21.779	26.674	30.800	37.722	43.558	48.699	53.347
10^4	0.0	21.779	30.800	37.722	43.558	53.347	61.600	68.871	75.444
10^5	0.0	68.871	97.398	119.29	137.74	168.70	194.80	217.79	238.58
10^6	0.0	217.79	308.00	377.22	435.58	533.47	616.00	688.71	754.44

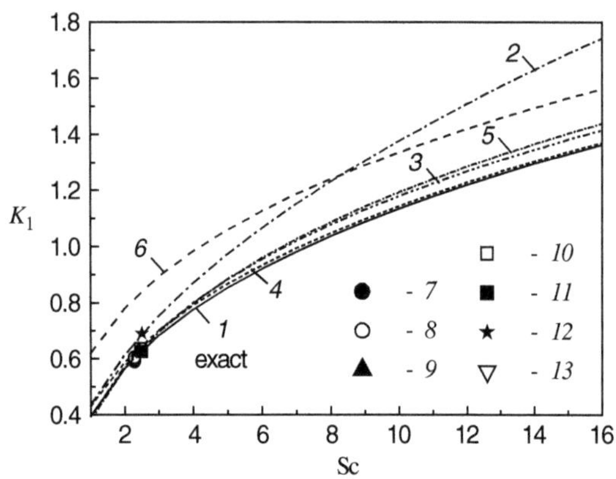

Fig. 8.1 Constant K_1 in Eq. (8.4), laminar flow at C_w=*const*. *1* – exact solution, Eqs. (2.37), (2.38), (2.39), (2.40) and (2.41); *2* – Eq. (3.6) for n_*=0; *3* – Eqs. (8.5), (8.6); *4* – Eq. (8.7); *5* – Eq. (8.8); *6* – Eq. (8.9). Experiments: *7* – K_1=0.59, Sc=2.28 [24]; *8* – K_1=0.604, Sc=2.28 [29]; *9* – K_1=0.625, Sc=2.4 [96, 97, 208]; *10* – K_1=0.636, Sc=2.44 [64]; *11* – K_1=0.625, Sc=2.5 [201]; *12* – K_1=0.69, Sc=2.5 [195]; *13* – K_1=0.628, Sc=2.5 [117]

the problem of mass transfer at the boundary condition C_w=*const*, Eqs. (3.7) and (3.8) take the following form, respectively:

$$K_1 = 0.6109Sc/(0.5301 + 0.3996Sc^{1/2} + Sc)^{2/3}, \tag{8.5}$$

$$K_1 = 0.6Sc/(0.56 + 0.26Sc^{1/2} + Sc)^{2/3}. \tag{8.6}$$

Data computed by Eqs. (8.5) and (8.6) practically coincide. Maximal inaccuracies of these equations with respect to the exact solution reach 4 and 5%, respectively, over the range of Sc=5–20 (Fig. 8.1). For further increased Prandtl or Schmidt numbers, the inaccuracies of Eqs. (8.5) and (8.6) become negligibly small (Table 8.3).

Authors of the work [131] offered another relation

Table 8.3 Values of the constant K_1 given by Eqs. (8.5), (8.6), (8.7), (8.8) and (8.9) for a disk at T_w=*const* or C_w=*const*

Pr (*Sc*)	Exact	Equation (8.5) [26]	Equation (8.6) [106]	Equation (8.7) [131]	Equation (8.8) [6]	Equation (8.9) [105]
1.0	0.3963	0.3941	0.4025	0.4303	0.3827	0.62
2.0	0.5653	0.5753	0.5864	0.5892	0.5664	0.7812
2.28	0.6016	0.6144	0.6257	0.6238	0.6065	0.816
2.4	0.6162	0.6300	0.6415	0.6378	0.6227	0.8301
2.5	0.6280	0.6430	0.6543	0.6491	0.6358	0.8415
5.0	0.8533	0.8839	0.8946	0.8676	0.8855	1.0602
20.0	1.4855	1.5414	1.5414	1.4924	1.5688	1.6829
50	2.0909	2.1552	2.1424	2.0958	2.2019	2.2841
100	2.6871	2.753	2.7278	2.6915	2.8147	2.8778
500	4.7351	4.7885	4.7222	4.7400	4.8890	4.9209
1000	6.0162	6.056	5.9651	6.0218	6.1773	6.2000
10^4	13.181	13.126	12.904	13.192	13.3565	13.358
10^5	28.613	28.332	27.834	28.639	28.7954	28.7779
10^6	61.860	61.074	59.990	61.915	62.0461	62.000

$$K_1 = 0.621Sc/(1 + 0.298Sc^{-1/3} + 0.14514Sc^{-2/3}), \tag{8.7}$$

which at the expense of decreased accuracy at Sc=1–2 (at Sc=1 its inaccuracy with respect to the exact solution is about 8% and decreases to 4% already at the point Sc=2) provides inaccuracy of not more than 1–3% with respect to the exact solution over the remaining range of variation of the Prandtl or Schmidt numbers (Fig. 8.1, Table 8.3).

Equations (8.5), (8.6) and (8.7) agree fairly well with the asymptotic solution K_1/Sc=0.885 [199]. In the limiting case of $Sc\to\infty$, these equations agree with the asymptotic solution K_1=0.62$Sc^{1/3}$ [105, 199].

Author of the work [6] constructed a relation for the constant K_1 over the range Pr=0–∞ by way of combining the asymptotic solutions for the cases $Pr\to 0$ and $Pr\to\infty$. As applied to the mass transfer problems, the relation of [6] has the following form:

$$K_1 = \left[(0.88447Sc)^{-1.077} + (0.62048Sc^{1/3})^{-1.077}\right]^{-1/1.077}. \tag{8.8}$$

Relation (8.8) practically coincides with the exact solution at Sc=2 (inaccuracy 0.2%). Following increase in the Schmidt numbers, inaccuracy of Eq. (8.8) increases up to 3.2% already at Sc=2.5 and reaches its maximum of 5.6% at $Sc\approx 20$, slowly decreasing afterwards with the following increase in the Schmidt numbers (e.g. inaccuracy of Eq. (8.8) is 4.7% at Sc=100; 2.7% at Sc=1000; 0.6% at $Sc=10^5$). If the Schmidt numbers decrease over the range of Sc<2, inaccuracy of Eq. (8.8) changes its sign, while its absolute value steadily increase reaching 3.4% at $Sc=1$ and 8.2% at $Sc=0.1$ (see Table 8.3).

Thus, Eqs. (8.5), (8.6), (8.7) and (8.8) provide, generally speaking, approximately the same accuracy of prediction of the constant K_1. However, in particular applied problems, preference should be given to a relation, which provides the highest accuracy over the particular range of the Schmidt numbers peculiar to the problem at hand.

Application to electrochemistry problems. An asymptotic theoretical solution for mass transfer problems at very high Schmidt numbers $Sc>>1$ was obtained by Levich [105]

$$K_1 = 0.62 Sc^{1/3}. \tag{8.9}$$

This solution coincides with the respective asymptotic solution for heat transfer problems obtained at $Pr>>1$ in the work [199].

As can be seen from Table 8.3, asymptotic equation (8.9) agrees well with the exact solution at $Sc>500$ (inaccuracy of Eq. (8.9) at $Sc=500$ equals to only 3.9% and tends practically to zero at $Sc\to\infty$). Equation (8.9) looses its accuracy for $Sc<500$ and exceeds the exact solution by 7.1% at $Sc=100$ and by 56.7% at $Sc=1$ (see Table 8.3 and Fig. 8.1).

Authors of experimental investigations [37, 39, 42, 48, 126, 157] confirmed validity of the Levich's equation (8.9) for high Schmidt numbers.

Rotating-disk electrodes are widely used in experimental electrochemistry [3, 105]. Due to the pumping effect of a rotating-disk electrode, an electrolyte moves from the outer volume of liquid to the disk surface. Flow of an electrolyte towards the disk surface is proportional to the angular velocity of rotation of the electrode. The process of convective mass transfer, which in this case is represented by intensity of diffusion electrical current on the electrode, is described by Eqs. (1.44) and (1.45).

Solution (8.4) with account for Eq. (8.9) is usually employed in the following form [3, 105]:

$$i_L = 0.62 n F A C_0 D_m^{2/3} \nu^{-1/6} \omega^{1/2}, \tag{8.10}$$

where i_L is limiting diffusion current of electrons to a disk electrode; n is the number of electrons transferred; F is the area of the electrode; A is Faraday constant (96,485 C/mol); C_0 is concentration of diffusion substance in the volume at a sufficiently large distance from the electrode, mol/m^3. It is easy to see that in this case the mass transfer coefficient is equal to $\alpha_m = i_L/(nFAC_0)$, whereas Eq. (8.10) can be reduced to Eq. (8.9).

The following three problems are interesting for practice: (a) finding out dependence of i_L on ω; (b) determination of the diffusion coefficient D_m based on the value of i_L; and (c) analysis of volt–ampere characteristics obtained on a rotating-disk electrode.

Application to experimental determination of mass and heat transfer coefficients with the help of naphthalene sublimation technique. As mentioned above, there exists analogy between the processes of convective heat transfer from a surface with surrounding air and mass transfer for naphthalene sublimation into air from the same surface. In the past, the naphthalene sublimation technique was used only to determine surface-averaged mass transfer of an *entire* disk by means of its weighing before and after the experiment and subsequent determination of the mass of naphthalene sublimated for the period of experiments [64, 96, 97, 201, 208]. At present, exact and reliable techniques are already available for the determination of not only surface-averaged, but also local mass transfer coefficients for laminar, transitional and turbulent flow at any point on the surface [24, 28–30, 59, 117, 195].

The essence of the analogy between the processes of heat and mass transfer from a surface consists in the following. Relations for the coefficients K_1 in the equations for the Nusselt number, Eq. (3.4), and Sherwood number, Eq. (8.4), can be written as follows, respectively [193],

$$K_1 = CPr^{m_p}, \tag{8.11}$$

$$K_1 = C\,Sc^{m_p}, \tag{8.12}$$

where the constant C is the same for both equations, whereas the effect of the Prandtl and Schmidt numbers is taken into account by the respective factors in Eqs. (8.11) and (8.12).

In fact, the Levich's equation (8.9) applicable only for the asymptotic conditions Sc>>1 and Pr>>1 is a particular case of Eq. (8.12) at m_p=1/3 and C=0.62.

An application for the naphthalene sublimation technique means that Eqs. (8.11) and (8.12) will be used for the cases with the Prandtl and Schmidt numbers moderately less or larger than unity: Pr=0.7–0.74 for heat transfer in air, and Sc=2.28–2.5 for naphthalene sublimation in air. Hence, it is logical to assume that the coefficient C is equal to the constant K_1 at Sc=1 and Pr=1 for the boundary conditions T_w=*const* or C_w=*const* (see Table 8.1)

$$C = 0.3963. \tag{8.13}$$

In all the known applications of the naphthalene sublimation technique for the rotating-disk problems [24, 28, 29, 59, 64, 195], it was assumed that the exponent m_p is independent of the values of Pr and Sc, with the relation between the Nusselt and Sherwood numbers being written as

$$Nu/Sh = (Pr/Sc)^{m_p}. \tag{8.14}$$

Different authors recommended various values of the exponent m_p, which differ from each other by up to 45%. The values used were: m_p=1/3 [29]; m_p=0.4 [24, 29, 59, 96]; m_p=0.53 [64]; and m_p=0.58 [195].

It is obvious that an error in the choice of the value of m_p can lead to significant errors in the recalculations of the data obtained using the naphthalene sublimation technique to the case of heat transfer in air. It is therefore necessary to analyse the values of m_p suggested by different authors and develop recommendations regarding its most accurate value.

The values of the exponent m_p can be determined from the exact solution of the system of the Navier–Stokes and thermal boundary layer equations in a self-similar form (Tables 3.1 and 8.1). The data for m_p for a series of discrete values of the Prandtl (or Schmidt) numbers moderately different from unity are given in Table 8.4 [193]. One can conclude from these results that the exponent m_p decreases from the value m_p=0.5723 to m_p=0.5024 with the Prandtl or Schmidt numbers increasing from 0.7 to 2.5. Thus, the value m_p=0.53 offered in [64] is, in fact, the average over the range of its variation with the Prandtl or Schmidt numbers varying from Pr=0.7 to 2.5.

Table 8.4 Values of m_p in Eqs. (8.11), (8.12) and (8.14) based on the exact solution of Eqs. (2.37), (2.38), (2.39), (2.40) and (2.41) for laminar flow [193]

$Pr\,(Sc)$	0.5	0.6	0.7	0.71	0.72	0.8	0.9	0.95	0.99
m_p	0.5954	0.5827	0.5723	0.5714	0.5705	0.5638	0.5571	0.5551	0.5632
$Pr\,(Sc)$	1.05	1.1	1.5	2	2.28	2.4	2.5	3	4
m_p	0.5438	0.5424	0.5264	0.5123	0.5064	0.5041	0.5024	0.4949	0.4841
$Pr\,(Sc)$	5	10	20	50					
m_p	0.4765	0.4566	0.4411	0.4251					

Experimental data of different authors for the coefficient K_1 for naphthalene sublimation in air are given in Fig. 8.1 and explained in detail in its caption. In general, these data agree well with the results of the exact solution (see Table 8.3) with an exception for the too high value K_1=0.69 obtained for Sc=2.5 by the authors of [195]. Having recalculated the experimental data with the help of Eq. (8.14) at m_p=0.53 to the case of Pr=0.71, one can obtain [193]: K_1=0.318 [24]; K_1=0.325 [29]; K_1=0.328 [96, 97, 208]; K_1=0.331 [64]; K_1=0.321 [201]; K_1=0.322 [117]; K_1=0.354 [195]. Thus, data of [29, 64, 96, 97, 117, 201, 208] are in excellent agreement with the exact solution K_1=0.326 for Pr=0.71 and T_w=$const$ (see Table 3.1) and experimental data on heat transfer in these conditions (see Chap. 3). The value K_1=0.318 [24] is somewhat lower than the data of other authors, and the reason lies probably in the somewhat low value of K_1=0.59 obtained in [24] for naphthalene sublimation at Sc=2.28. The value K_1=0.354 [195] is too high, because the experimental value K_1=0.69 of these authors for naphthalene sublimation is also too high (see also Fig. 8.1).

With the value of m_p=0.4 recommended in [24, 29, 59, 96], one can obtain for the conditions at Pr=0.71 [193]: K_1=0.37 [24]; K_1=0.379 [29]; K_1=0.384 [96, 97, 208]; K_1=0.388 [64]; K_1=0.378 [201]; K_1=0.380 [117]; K_1=0.417 [195]. These values of the coefficient K_1 are too high in comparison with the exact solution K_1=0.326. It is especially important to accentuate this, because the experimental results [24, 29, 59] obtained relatively recently (as well as data of [117]) are accurate in the part of the mass transfer data.

Authors [29] recommended the value m_p=1/3. In this case, one can obtain for Pr=0.71 [193]: K_1=0.4 [24]; K_1=0.409 [29]; K_1=0.416 [96, 97, 208]; K_1=0.421 [64]; K_1=0.411 [201]; K_1=0.413 [117]; K_1=0.454 [195]. Thus, all the values of K_1 recalculated for the case of Pr=0.71 are even more overestimated in comparison with the exact solution.

Authors of the work [195] suggested the value m_p=0.58. In this case one can obtain for Pr=0.71 [193]: K_1=0.3 [24]; K_1=0.307 [29]; K_1=0.308 [96, 97, 208]; K_1=0.311 [64]; K_1=0.301 [201]; K_1=0.303 [117]; K_1=0.332 [195]. Hence, all the values of K_1 recalculated for the case of Pr=0.71 are too low in comparison with the exact solution except for the data [195]. Apparently, authors of the work [195] had to choose such a high value of m_p in order to obtain the recalculated coefficient K_1 agreeing well with the data for heat transfer at Pr=0.71. It is obvious, however, that the problem is that the value K_1=0.69 obtained in [195] for naphthalene sublimation is too high and disagree with the numerous data of other researchers.

A modification of Eq. (8.14) was offered in [201]

$$Nu/Sh_{\mathrm{Sc}=2.5} = f(Pr)Pr^{1/3} \tag{8.15}$$

This means that m_p=1/3, and all inaccuracies in the recalculation of the Nusselt number using the known value K_1=0.625 at Sc=2.5 for the conditions at other Prandtl or Schmidt numbers should be corrected by means of the function $f(Pr)$=0.576, 0.634, 0.737, 0.842 and 0.926 at Pr=0.1, 1, 2.5, 10 and 100, respectively. As a result, one can obtain that K_1=0.321, 0.396, 0.625, 1.134 and 2.686 for the same Prandtl numbers. This coincides with the exact solution at T_w=$const$ (see Table 8.1, n_*=0). With such an approach, the factor $Pr^{1/3}$ is in principle redundant, because the correction function $f(Pr)$ can be tabulated via a computation of the ratio $Nu/Sh_{\mathrm{Sc}=2.5}$ with the help of the exact solution over any range of Pr and Sc. An advantage of such a method is its higher accuracy and universality, while the disadvantage of Eq. (8.15) is its poorer obviousness in comparison with Eq. (8.14), because a tabulated function is always less convenient than an accurate approximating formula.

Thus, for the recalculation of the data for laminar mass transfer from a rotating disk for naphthalene sublimation in air to the case of laminar heat transfer in air, one can recommend to use Eq. (8.14) at the value of the power exponent m_p=0.53 offered in [64] or, alternatively, Eq. (8.15) (or its modification) with the appropriately tabulated function $f(Pr)$.

8.2 Transitional and Turbulent Flows for the Prandtl or Schmidt Numbers Moderately Different from Unity

In the current section, values of $Pr \leq 5$ and $Sc \leq 5$ are understood under the Prandtl or Schmidt numbers moderately different from unity. The main application of the results described in this section is development of recommendations applied to the experimental technique based on naphthalene sublimation in air at Sc=2.28–2.5 [193].

As mentioned above, only surface-averaged mass transfer coefficients for an *entire* disk have been determined with the help of naphthalene sublimation technique in the past [64, 96, 97, 201, 208]. Experimental data for the local coefficients of mass transfer for naphthalene sublimation from a surface of a rotating disk in transitional and turbulent flow regimes were apparently obtained for the first time in the recent investigations [24, 59].

Experimental data of different authors for local Sherwood numbers for naphthalene sublimation in air and their approximations by empirical equations for laminar, transitional and turbulent flows are shown in Fig. 8.2.

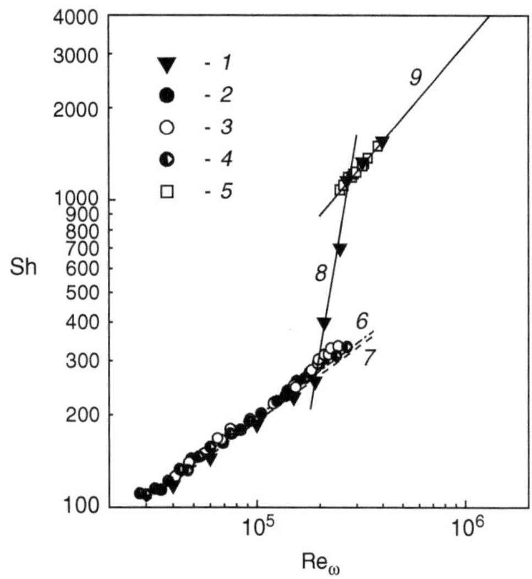

Fig. 8.2 Local Sherwood numbers for naphthalene sublimation in air [193]. Experiments: *1* – Sc=2.28 [24]; *2* – Sc=2.4 [97]; *3* – Sc=2.4 [208]; *4* – Sc=2.44 [64]; *5* – Sc not mentioned [59]. Empirical approximations of the experiments with the help of Eq. (8.4): *6* – laminar flow, n_R=1/2, K_1=0.625 [96, 97, 201, 208]; *7* – laminar flow, n_R=1/2, K_1=0.604 [29]; *8* – transitional flow, n_R=4, K_1=2×10^{-19}, Eq. (8.16) [24]; *9* – turbulent flow, n_R=0.8, K_1=0.0512, Eq. (8.17) [24]

It makes sense to rewrite Eqs. (3.13) and (3.29) approximating experimental data of the authors [24, 59] for transitional and turbulent flow regimes in a corrected (range of validity of Eq. (3.13)) or transformed (Eq. (3.29)) form more convenient for further analysis in the current section

$$Sh = 2.0 \times 10^{-19} \cdot Re_\omega^4 \quad \text{for } Re_\omega = (1.9 - 2.75) \times 10^5, \tag{8.16}$$

$$Sh = 0.0512 Re_\omega^{0.8} \quad \text{for } Re_\omega \geq 2.75 \times 10^5, \tag{8.17}$$

$$Sh = 0.0518 Re_\omega^{0.8} \quad \text{for } Re_\omega \geq 2.5 \times 10^5. \tag{8.18}$$

For a number of engineering applications, it is important to determine average Sherwood numbers Sh_{av} (or average Nusselt numbers Nu_{av}) of an *entire* disk, where regions of laminar and transitional flow or laminar, transitional and turbulent flow co-exist simultaneously.

Authors of works [64, 96, 97, 201, 208] obtained experimental data for the average Sherwood number of the *entire* disk where regions of laminar, transitional and, in some cases, turbulent flows existed simultaneously. In order to obtain a relation for Sh_{av} for the entire disk, authors of [97, 208] used an approach based on the Reynolds analogy between the processes of heat (and mass) transfer and fluid flow. Equations for the Nusselt and Sherwood numbers obtained as a result are inconvenient and contain parameters, whose determination in frames of the model used was quite complicated. The application of the Reynolds analogy seems to be forced by the lack of experimental data for the local Sherwood numbers that could have been used for constructing models for Sh_{av} for the *entire* disk in a more direct way.

Model [42] for Sh_{av} used in electrochemical applications looks much more perspective and justifiable. An extension of this model to the mass transfer for naphthalene sublimation allowed validations of the relatively recent data for the local values of the Sherwood numbers via comparisons with the huge data for Sh_{av} available for the *entire* disk [193].

The author [42] assumed that at simultaneous existence of the regions of laminar, transitional and turbulent flows on the disk, the span of the transitional zone is rather small, and the transition from laminar to turbulent flow occurs abruptly at a radial location r_{tr} for a Reynolds number $Re_{\omega,\mathrm{tr}}$. Then the integration in order to obtain the value Sh_{av} for the entire disk should be performed as

$$Sh_{\mathrm{av}} = \frac{2}{b}\left[\int\limits_0^{r_{\mathrm{tr}}} Sh_{\mathrm{lam}} dr + \int\limits_{r_{\mathrm{tr}}}^{b} Sh_{\mathrm{turb}} dr\right]. \tag{8.19}$$

The Sherwood numbers are defined by the first of Eqs. (8.4); the allowance should be made for the actual values $K_{1,\mathrm{lam}}$ and n_R=1/2 for laminar flow (subscript "lam"), and $K_{1,\mathrm{turb}}$ and n_R=0.8 for turbulent flow (subscript "turb").

As a result of the integration of Eq. (8.19), one can obtain

$$Sh_{\mathrm{av}} = K_{1,\mathrm{lam}} Re_{\omega,\mathrm{tr}}^{1/2}\left(\frac{Re_{\omega,\mathrm{tr}}}{Re_\varphi}\right)^{1/2} + \frac{2}{2n_R+1} K_{1,\mathrm{turb}} Re_\varphi^{n_R}\left[1 - \left(\frac{Re_{\omega,\mathrm{tr}}}{Re_\varphi}\right)^{n_R+1/2}\right]. \tag{8.20}$$

Equation (8.20) is valid for $Re_\varphi \geq Re_{\omega,\mathrm{tr}}$. If $Re_\varphi < Re_{\omega,\mathrm{tr}}$, the second term in Eq. (8.20) should be neglected. In the limiting case $Re_\varphi >> Re_{\omega,\mathrm{tr}}$, Eq. (8.20) reduces to the second of Eqs. (8.4) for developed turbulent flow existing over the entire disk surface, where

$$K_{2,\mathrm{turb}} = \frac{2}{2n_R+1} K_{1,\mathrm{turb}}. \tag{8.21}$$

Equation (8.20) coincides with Eq. (3.35), while Eq. (8.21) coincides with Eq. (3.45) under condition n_*=0 (which corresponds to the cases T_w=*const* and

$C_w = const$ being compared) with account for the relation $2n_R = 1 + m$ that follows from Eqs. (2.83) and (3.41).

The model of [42] can be generalized taking into account every region of the laminar, transitional and turbulent flows separately as suggested in [126]. Assuming that the transition starts at the radial coordinate r_{tr1} for the Reynolds number $Re_{\omega,\mathrm{tr1}}$, while the end of the transition takes place at the radial coordinate r_{tr2} for the Reynolds number $Re_{\omega,\mathrm{tr2}}$, then the integration to find Sh_{av} should be performed as follows:

$$Sh_{\mathrm{av}} = \frac{2}{b}\left[\int\limits_0^{r_{\mathrm{tr1}}} Sh_{\mathrm{lam}} dr + \int\limits_{r_{\mathrm{tr1}}}^{r_{\mathrm{tr2}}} Sh_{\mathrm{tran}} dr + \int\limits_{r_{\mathrm{tr2}}}^{b} Sh_{\mathrm{turb}} dr\right] \tag{8.22}$$

Newly introduced function Sh_{tran} is defined by the first of Eqs. (8.4) with allowance for experimentally determined values of $K_{1,\mathrm{tran}}$ and $n_{R,\mathrm{tran}}$ for transitional flow (subscript "tran").

As a result of the integration of Eq. (8.22), one can obtain

$$\begin{gathered} Sh_{\mathrm{av}} = K_{1,\mathrm{lam}} Re_{\omega,\mathrm{tr1}}^{1/2}\left(\frac{Re_{\omega,\mathrm{tr1}}}{Re_{\varphi}}\right)^{1/2} + \frac{2}{2n_{R,\mathrm{tran}}+1} K_{1,\mathrm{tran}} Re_{\omega,\mathrm{tr2}}^{n_{R,\mathrm{tran}}}\left(\frac{Re_{\omega,\mathrm{tr2}}}{Re_{\varphi}}\right)^{1/2} \\ \left[1-\left(\frac{Re_{\omega,\mathrm{tr1}}}{Re_{\omega,\mathrm{tr2}}}\right)^{n_{R,\mathrm{tran}}+1/2}\right] + \frac{2}{2n_R+1} K_{1,\mathrm{turb}} Re_{\varphi}^{n_R}\left[1-\left(\frac{Re_{\omega,\mathrm{tr2}}}{Re_{\varphi}}\right)^{n_R+1/2}\right]. \end{gathered} \tag{8.23}$$

Equation (8.23) is applicable for $Re_{\varphi} \geq Re_{\omega,\mathrm{tr2}}$. If $Re_{\varphi} < Re_{\omega,\mathrm{tr2}}$, the third term in Eq. (8.23) should be neglected, with the second term taking a simpler form:

$$\begin{gathered} Sh_{\mathrm{av}} = K_{1,\mathrm{lam}} Re_{\omega,\mathrm{tr1}}^{1/2}\left(\frac{Re_{\omega,\mathrm{tr1}}}{Re_{\varphi}}\right)^{1/2} + \frac{2}{2n_{R,\mathrm{tran}}+1} K_{1,\mathrm{tran}} Re_{\varphi}^{n_{R,\mathrm{tran}}} \\ \left[1-\left(\frac{Re_{\omega,\mathrm{tr1}}}{Re_{\varphi}}\right)^{n_{R,\mathrm{tran}}+1/2}\right] \end{gathered} \tag{8.24}$$

In the limiting case of $Re_{\varphi} >> Re_{\omega,\mathrm{tr2}}$, Eq. (8.23) reduces to the second of Eqs. (8.4) with $K_{2,\mathrm{turb}}$ being given by Eq. (8.21).

Authors [126], while determining the average Sherwood number for an entire disk, derived an expression that is a particular case of Eq. (8.23), because all its empirical constants are represented by their numerical values obtained in experiments [126] (electrochemical applications at high Schmidt numbers). Because of this, expression [126] cannot be directly applied for the case of naphthalene sublimation in air.

In order to make comparisons with the experimental data for Sh_{av}, let us substitute numerical values of the constants obtained in [24] (see Eqs. (8.16), (8.17) and caption in Fig. 8.1) into the general equations derived above. As a result, one can obtain for Eqs. (8.20), (8.21), (8.23) and (8.24), respectively:

(a) for Eq. (8.20)

$$Sh_{\text{av}} = 0.59Re_{\omega,\text{tr}}^{1/2}\left(\frac{Re_{\omega,\text{tr}}}{Re_{\varphi}}\right)^{1/2} + \frac{2}{2.6}0.512Re_{\varphi}^{0.8}\left[1-\left(\frac{Re_{\omega,\text{tr}}}{Re_{\varphi}}\right)^{1.3}\right] \quad (8.25)$$

(b) for Eq. (8.21)

$$K_{2,\text{turb}} = \frac{2}{2.6}K_{1,\text{turb}} = 0.0394 \quad (8.26)$$

(c) for Eq. (8.23)

$$Sh_{\text{av}} = 0.59 \times 1.9 \times 10^5 Re_{\varphi}^{-1/2} + \tfrac{4}{9}10^{-19}(2.75 \times 10^5)^{4.5} Re_{\varphi}^{-1/2}$$

$$\left[1-\left(\tfrac{1.9\times10^5}{2.75\times10^5}\right)^{4.5}\right] + 0.0394Re_{\varphi}^{0.8}\left[1-\left(\tfrac{2.75\times10^5}{Re_{\varphi}}\right)^{1.3}\right], \quad Re_{\varphi} \geq 2.75 \times 10^5, \quad (8.27)$$

(d) for Eq. (8.24)

$$Sh_{\text{av}} = 0.59 \times 1.9 \times 10^5 Re_{\varphi}^{-1/2} + \frac{4}{9}10^{-19}Re_{\varphi}^{4}\left[1-\left(\frac{1.9\times10^5}{Re_{\varphi}}\right)^{4.5}\right],$$

$$Re_{\varphi} = (1.9 - 2.75) \times 10^5. \quad (8.28)$$

The Reynolds number of the abrupt transition to turbulent flow $Re_{\omega,\text{tr}}$ in Eq. (8.25) was left as a unknown parameter that can be varied to reach a better agreement with the experiments.

A comparison of Eqs. (8.25), (8.26), (8.27) and (8.28) with experimental data of different authors is presented in Fig. 8.3. Experimental data *1*, *5* and curve *6* for Sh_{av} for developed turbulent flow were obtained in the present work by means of a recalculation of the experimental data [24, 59] and Eq. (8.17) with the help of Eq. (8.26). It should be remembered also that for laminar flow $K_{2,\text{lam}} = K_{1,\text{lam}}$ (curves *7* and *8*). Curve *9*, computed via combining Eqs. (8.27) and (8.28) and taking into account boundaries at the beginning and end of the transition to turbulent flow, agrees well with the experimental data [64, 97, 208] for Sh_{av} for the *entire* disk presented in Fig. 8.3.

Experimental points *1* for Sh_{av} for the *entire* disk presented in Fig. 8.3 were computed in the present work based on the data [24] for laminar, transitional and turbulent flow with the help of Eqs. (8.27) and (8.28). It is obvious that these points coincide with curve *9* for the corresponding values of the Reynolds number.

Substituting the Reynolds numbers on the lower and upper border of the transitional region (i.e. 1.9×10^5 and 2.75×10^5, respectively) as $Re_{\omega,\text{tr}}$ in Eq. (8.25) (abrupt transition to turbulent flow) resulted in curves *10* and *11*, which lie higher and lower than curve *9*, respectively. In order to agree with curve *9*, one should choose an average (or "effective") Reynolds number of the abrupt transition to tur-

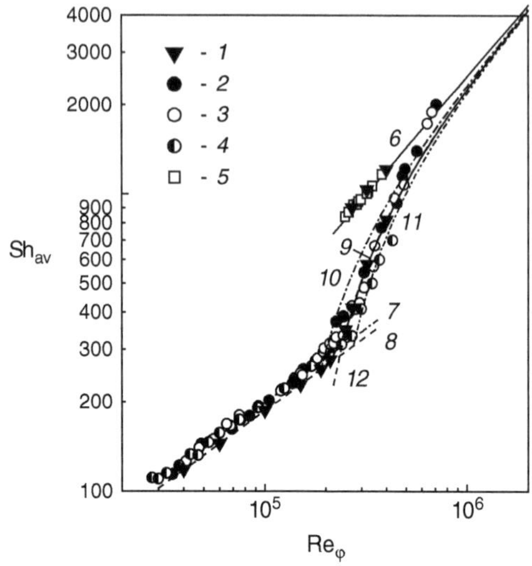

Fig. 8.3 Average Sherwood numbers for naphthalene sublimation in air [193]. Experiments: *1* – Sc=2.28 [24]; *2* – Sc=2.4 [97]; *3* – Sc=2.4 [208]; *4* – Sc=2.44 [64]; *5* – Sc not mentioned [59]. Calculation based on Eq. (8.4): *6* – turbulent flow region, n_R=0.8, K_2=0.0394, Eq. (8.26) [24]; *7* – laminar flow, n_R=1/2, K_1=0.625 [96, 97, 201, 208]; *8* – laminar flow, n_R=1/2, K_1=0.59 [24]. Calculation of Sh_{av} for the *entire* disk: *9* – Eqs. (8.27) and (8.28); *10* – Eq. (8.25) at $Re_{\omega,tr} = 1.9\times10^5$; *11* – Eq. (8.25) at $Re_{\omega,tr} = 2.75\times10^5$; *12* – Eq. (8.25) at $Re_{\omega,tr} = 2.35\times10^5$

bulent flow, which in this case is equal to $Re_{\omega,tr} = 2.35\times 10^5$ (curve *12*). This effective Reynolds number $Re_{\omega,tr}$ is, in fact, the arithmetic mean of the Reynolds numbers at the beginning $Re_{\omega,tr1}$ and end $Re_{\omega,tr2}$ of the transition to developed turbulent flow.

It should be pointed out that all curves *9–12* in the limiting case of $Re_\varphi \to \infty$ merge with curve *6* for developed turbulent flow over the entire disk.

Thus, Eqs. (8.23) and (8.24), based on the model taking into account simultaneous existence of the regions of laminar, transitional and turbulent flows, provide the highest accuracy in predicting the average Sherwood numbers for the *entire* disk. The simpler equation (8.20), which is based on the model [42], also provides the accuracy comparable with that of Eqs. (8.23) and (8.24) under the condition of correct determining of the "effective" Reynolds number $Re_{\omega,tr}$ of the abrupt transition to turbulent flow. One should also point out that the conditions at the beginning and end of the transition to turbulent flow are different in experiments of different authors; therefore, the "effective" value $Re_{\omega,tr}$ should be found out individually in every case.

Application to experimental determination of mass and heat transfer coefficients using naphthalene sublimation technique. The aforementioned analogy between heat transfer in air flow over a surface and mass transfer for naphthalene sublimation in air may be also applied to transitional and turbulent flow over a disk. However, one should correctly use correlations that follow from this analogy.

Like in the laminar regime of flow, a recalculation of the data for mass transfer to the respective relations for heat transfer is based on Eqs. (3.4) for the Nusselt number and Eqs. (8.4) for the Sherwood number, with coefficients K_1 in these equations being written as Eqs. (8.11) and (8.12), respectively [193]. In doing so, the factor C is considered to be equal to the value of K_1 at Sc=1, Pr=1 for the boundary conditions T_w=*const* or C_w=*const*.

It was pointed out above (in Sect. 8.1) that for laminar flow the exponent m_p, strictly speaking, is not constant, but varies depending on the Prandtl or Schmidt numbers (see Table 8.4). However, authors of the work [64] succeeded in determining a constant value m_p=0.53, which, being average over the range of the Prandtl and Schmidt numbers Pr=Sc=0.7–2.5, allows with good accuracy to recalculate mass transfer data into heat transfer data with the help of Eq. (8.14) that follows from Eqs. (8.11) and (8.12) under condition m_p=*const*.

In fact, authors [24, 59] assumed that, over the range of Pr=Sc=0.7–2.5, the exponent m_p is constant also for turbulent flow and is equal to m_p=0.4. This assumption was made based on the recommendations of [28]. As far as we know, no other author developed any recommendation regarding the value of m_p for transitional and turbulent flows. In Sect. 3.3.2, a recalculation was done by means of Eq. (8.14) of the values K_1 for Sc=2.28 given in explanations to Eq. (3.29) (or, which is the same, in Eqs. (8.17) and (8.18)) as applied to conditions of heat transfer of a rotating disk in air for Pr=0.72. It was obtained as a result that the value of K_1 at T_w=*const* should be equal to K_1=0.0323. However, in reality, at T_w=*const* and Pr=0.72, the coefficient K_1 is equal to 0.0188 according to experiments [11, 46, 146] (see Table 3.5) or to 0.0187 based on the theoretical model [173, 189].

So, it is again obvious also for turbulent flow that the aforementioned error in the choice of the value of m_p leads to the significant error in the recalculation of the data from the naphthalene sublimation technique for the case of heat transfer in air.

To remind, Eqs. (8.11) and (8.12) are supposed to be used for the Prandtl numbers moderately less than unity (for air Pr=0.7–0.74) and the Schmidt numbers moderately larger than unity (for naphthalene sublimation in air Sc=2.28–2.5). It is logical therefore to assume that the coefficient C in Eqs. (8.11) and (8.12) should be equal to the value of the constant K_1 for turbulent flow at Sc=1, Pr=1 for the boundary conditions T_w=*const* or C_w=*const* according to Table 3.7:

$$C = 0.0232 \tag{8.29}$$

Contrary to the laminar flow case, an exact self-similar solution for the turbulent flow regime does not exist at all. Hence, a determination of the exponent m_p should be based on experimental data. In fact, only experiments [24, 59] for naphthalene sublimation and their approximations in the form of Eqs. (8.17) and (8.18) are available for this purpose. In view of Eqs. (8.17) and (8.29), as well as the value K_1=0.0188 at T_w=*const* and Pr=0.72 [11, 46, 146], Eqs. (8.11) and (8.12) take the following form:

$$K_1 = 0.0232 Pr^{0.64} \quad \text{for } Pr \leq 1, \tag{8.30}$$

$$K_1 = 0.0232 Sc^{0.96} \quad \text{for } Sc \geq 1. \tag{8.31}$$

Thus, the exponent m_p for turbulent flow varies significantly depending on the Prandtl and Schmidt numbers. The relation between the Nusselt number for heat transfer in air and the Sherwood number for naphthalene sublimation in air looks therefore as follows:

$$Nu/Sh = Pr^{0.64}/Sc^{0.96} \tag{8.32}$$

As a development of the idea [201] expressed in Eq. (8.15), one can rewrite Eq. (8.32) as

$$Nu/Sh_{Sc=2.28} = f(Pr) \tag{8.33}$$

where the function $f(Pr)$ is equal to 0.367 at Pr=0.72.

Finally, in analogy to laminar flow, one can choose an effective value of m_p that allows using the relation between the Nu and Sh numbers in the form of Eq. (8.14)

$$Nu/Sh = (Pr/Sc)^{0.87} \tag{8.34}$$

The value of the exponent m_p in Eq. (8.34) is more than twice as high as the value 0.4 erroneously recommended in [24, 59]. One should remember, however, that the value m_p=0.87 must not be used in Eqs. (8.30) and (8.31), because this can cause significant errors in the calculations of the coefficient K_1.

Analysis for the transitional flow shows that the empirical equation (3.10) [146] for T_w=*const* and Pr=0.72 fits in the best way to compare with the case of C_w=*const* at hand. For transitional flow, Eq. (8.14) should be modified with allowance for Eqs. (3.10) and (8.16) in the following way:

$$Nu/Sh = (Pr/Sc)^{0.6} \tag{8.35}$$

The range of validity of Eq. (3.10), where Re_ω=(1.95–2.5)×10^5, is somewhat narrower in comparison with that of Eq. (8.16), where Re_ω=(1.9–2.75) ×10^5. However, this difference is rather small.

Thus, in order to recalculate the data for turbulent mass transfer for naphthalene sublimation from a disk into air to the case of heat transfer with air, one can recommend using any of Eqs. (8.32), (8.33) and (8.34), while for transitional flow Eq. (8.35) should be applied [193].

8.3 Transitional and Turbulent Flows at High Prandtl and Schmidt Numbers

For high values of the Prandtl and Schmidt numbers in the current section, the values peculiar for electrochemistry problems are understood. For example, in experiments [37, 39, 42, 48, 126], whose data are analysed below, Schmidt numbers were in the range of Sc=34–10,320. The main objective pursued in this section included comparisons, analysis and generalization of the experimental data and theoretical solutions of different authors and development of recommendations for their use [194].

Given below are empirical correlations for the Sherwood numbers in transitional and turbulent flow obtained by different authors experimentally as well as based on various semi-empirical models.

Author [42] measured average Sherwood numbers for the entire disk. Experimental data obtained for Re_φ=(0.278–1.8) $\times 10^6$, Sc=930–10,320 agree well with the approximating equation [42]

$$Sh_{\rm av} = Sc^{1/3} Re_\varphi^{-1/2} [0.62 Re_{\omega,\rm tr} + 1.08 \times 10^{-2} (Re_\varphi^{1.37} - Re_{\omega,\rm tr}^{1.37})]. \tag{8.36}$$

Equation (8.36) was obtained based on Eq. (8.20) at $K_{1,\rm lam}=0.62 \cdot \mathrm{Sc}^{1/3}$, $K_{1,\rm turb}=0.0148 \cdot \mathrm{Sc}^{1/3}$, n_R=0.87 and agrees well with the experiments at $Re_{\omega,\rm tr}=2.78\times 10^5$. According to [42], transitional region existed at Re_ω=(2.3–2.9) $\times 10^5$. Relation (8.36) lies somewhat lower than the experimental data [42] in the region of transitional flow (similar to curve *12* in Fig. 8.3) that is caused by the simplified assumptions accepted in model (8.20) and discussed in Sect. 8.2.

An asymptotic form of Eq. (8.36) for developed turbulent flow looks as follows [42]:

$$Sh_{\rm av} = 1.08 \times 10^{-2} Re_\varphi^{0.87} Sc^{1/3}. \tag{8.37}$$

Based on Eq. (8.37) with allowance for Eq. (8.21), one can obtain a relation for the local Sherwood numbers

$$Sh = 1.48 \times 10^{-2} Re_\varphi^{0.87} Sc^{1/3}. \tag{8.38}$$

Average Sherwood numbers for the entire disk were measured also in work [37] over the range $Re_\varphi=5\times 10^4$–1.8×10^6, Sc=345–6450. Transitional flow was observed at Re_φ=(2.3–2.9) $\times 10^5$. For developed turbulent flow, i.e. for $Re_\varphi=3\times 10^5$–1.8×10^6, authors [37] obtained the following asymptotic correlation:

$$Sh_{\rm av} = 0.0725 Re_\varphi^{0.9} Sc^{0.33}. \tag{8.39}$$

Experimental data for $Sh_{\rm av}$ for the entire disk as well as the data for the local Sherwood numbers at laminar, transitional and turbulent flow over the range $Re_\omega=4\times 10^4$–2.2×10^6, Sc=680–7200 were obtained by the authors [39]. In experiments [39], transitional flow existed for Re_ω=(2.2–3.0) $\times 10^5$. For developed turbulent flow at $Re_\omega=3.0\times 10^5$–2.2×10^6, authors [39] obtained the following relations:

$$Sh = 1.09 \times 10^{-2} Re_\omega^{0.91} Sc^{1/3}, \tag{8.40}$$

$$Sh_{\rm av} = 7.67 \times 10^{-3} Re_\varphi^{0.91} Sc^{1/3}. \tag{8.41}$$

In order to approximate experimental data [39] for the average Sherwood numbers of an entire disk, the following relation was developed by the author of this monograph in analogy to Eq. (8.36)

$$Sh_{av} = Sc^{1/3} Re_{\varphi}^{-1/2} [0.62 Re_{\omega,tr} + 7.67 \times 10^{-3} (Re_{\varphi}^{1.41} - Re_{\omega,tr}^{1.41})], \tag{8.42}$$

where the Reynolds number of abrupt transition was assumed to be equal to the value $Re_{\omega,tr}=2.78\times10^5$ as suggested in the work [42].

Experimental data [48] for Sh_{av} for an entire disk were obtained for $Re_{\varphi}=10^4$–1.18×10^7, $Sc=34$–1400. For the developed turbulent flow region at $Re_{\varphi}=8.9\times10^5$–$1.18\times10^7$, authors [48] obtained the following empirical correlation:

$$Sh_{av} = 1.17 \times 10^{-2} Re_{\varphi}^{0.896} Sc^{0.249}. \tag{8.43}$$

One should point out significant deviations of the experimental data [48] from the approximating dependence (8.43).

Original experimental data for the local Sherwood numbers at transitional flow over the range $Re_{\omega}=(2.0$–$3.0)\ \times10^5$ and $Sc=1192$–2465 were obtained in [126] and approximated by empirical equation (3.14), which for the sake of convenience is rewritten in the current section

$$Sh = 3.4 \times 10^{-14} Re_{\omega}^{3} Sc^{1/3}. \tag{8.44}$$

Authors [126] presented also their own approximation of the experimental data [37] for developed turbulent flow

$$Sh_{av} = 7.8 \times 10^{-3} Re_{\varphi}^{0.9} Sc^{1/3}. \tag{8.45}$$

Authors [126] also obtained an equation for the average Sherwood numbers at simultaneous existence of laminar and transitional flow that is in fact a particular case of Eq. (8.24) based on Eq. (8.9) for laminar flow and original equation (8.44):

$$Sh_{av} = Sc^{1/3} Re_{\varphi}^{-1/2} [0.89 \times 10^5 + 9.7 \times 10^{-15} Re_{\varphi}^{3.5}]. \tag{8.46}$$

For the average Sherwood numbers of the entire disk, where regions of laminar, transitional and turbulent flow existed simultaneously, authors [126] obtained an equation

$$Sh_{av} = Sc^{1/3} Re_{\varphi}^{-1/2} [7.8 \times 10^{-3} Re_{\varphi}^{1.4} - 1.3 \times 10^5], \tag{8.47}$$

which is in fact a particular case of Eq. (8.23) with Eq. (8.9) being used for the laminar flow, Eq. (8.44) for transitional regime and Eq. (8.45) for turbulent flow.

Theoretical correlations for local Sherwood numbers for developed turbulent flow and high Schmidt numbers obtained with the help of different model assumptions in [81, 216] look as follows, respectively,

$$Sh_{av} = 7.07 \times 10^{-3} Re_{\varphi}^{0.9} Sc^{1/3}, \tag{8.48}$$

$$Sh_{av} = 5.93 \times 10^{-3} Re_{\varphi}^{0.91} Sc^{0.34}. \tag{8.49}$$

A solution for the average Sherwood numbers that coincides with experimental equation (8.45) was obtained in the theoretical investigation [102]. A similar solution was obtained using a theoretical model in [141] with the coefficient at the Reynolds number equal to 6.43×10^{-3}.

Authors [125] proposed a relation intended to take into account the effect of the Schmidt number on the $Sh_{\rm av}$ over the wide range of variation of Sc=0.72 to ∞

$$Sh_{\rm av} = \frac{0.0267 Sc Re_{\varphi}^{0.8}}{1.3 + 3.19(Sc - 1)Sc^{-1/3} Re_{\varphi}^{-0.1}}. \tag{8.50}$$

Several empirical and theoretical equations agree well with each other while calculating the Sherwood numbers for the developed turbulent flow as well as for the average Sherwood numbers for the entire disk at simultaneous co-existence of laminar, transitional and turbulent flows over the disk. As seen from Fig. 8.4, empirical equations (8.37) and (8.39), as well as the theoretical equation (8.48), practically coincide. Theoretical solution (8.49) also agrees well with Eqs. (8.37), (8.39) and (8.48). Approximation (8.45) exceeds noticeably the original experimental data [37]. Theoretical solution in the form of Eq. (8.45) with the coefficient 6.43×10^{-3} [141] provides the data that are 9% lower than those obtained by Eq. (8.48). Very good agreement of empirical equations (8.37) and (8.39) is thus an evidence in favour of reliability of these experimental data.

As the data in Fig. 8.5 show, Eq. (8.41) for turbulent flow and Eq. (8.42) for the entire disk exceed significantly Eqs. (8.37) and (8.36), respectively.

Empirical equation (8.43) obtained in the work [48] is in fact the only one where the power exponent 0.249 at the Schmidt number differs noticeably from 1/3. Such a significant disagreement with the results of all the other experimental and theoretical investigations known in the literature, together with the aforementioned deviation of the experimental data from the approximation curve [48], is most probably the evidence of the erroneous value of the exponent 0.249 suggested in [48]. As the data in Fig. 8.6 show, differences among the curves that one can obtain based on Eq. (8.43)

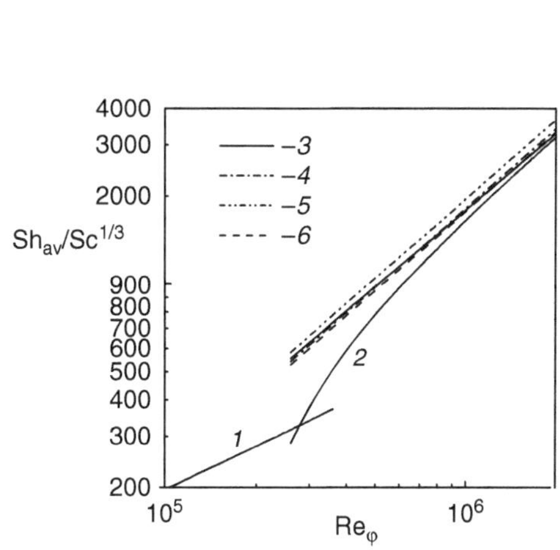

Fig. 8.4 Average Sherwood numbers at high Schmidt numbers. Approximation of experiments: *1* – laminar flow, Levich's equation (8.9); *2* – Eq. (8.36) for an *entire* disk [42]; *3* – Eq. (8.37) [42]; *4* – Eq. (8.39) [37]; *5* – Eq. (8.45) [102, 126]. Theoretical solution: *6* – Eq. (8.48) [81]

Fig. 8.5 Average Sherwood numbers at high Schmidt numbers, approximation of experiments: *1* – laminar flow, Levich's equation (8.9); *2* – Eq. (8.36) for an *entire* disk [42]; *3* – Eq. (8.42), *entire* disk; *4* – Eq. (8.51), *entire* disk; *5* – Eq. (8.37) [42]; *6* – Eq. (8.41) [39]

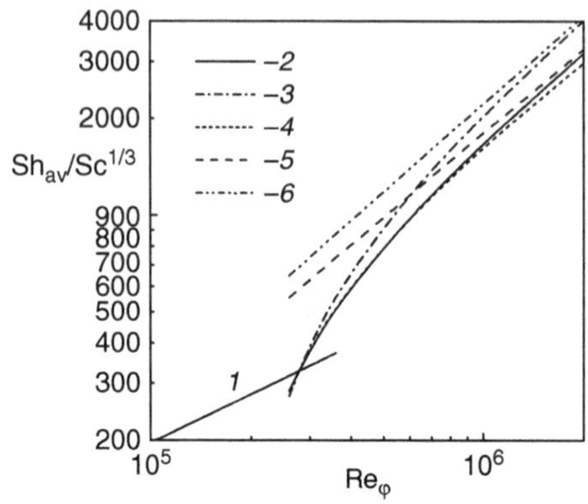

Fig. 8.6 Average Sherwood numbers at high Schmidt numbers, approximation of experiments: *1* – laminar flow, Levich's equation (8.9); *2* – Eq. (8.36) for an *entire* disk [42]; *3* – Eq. (8.37) [42]; *4* – Eq. (8.43), Sc=34 [48]; *5* – (8.43), Sc=109 [48]; *6* – Eq. (8.43), Sc=1300 [48]

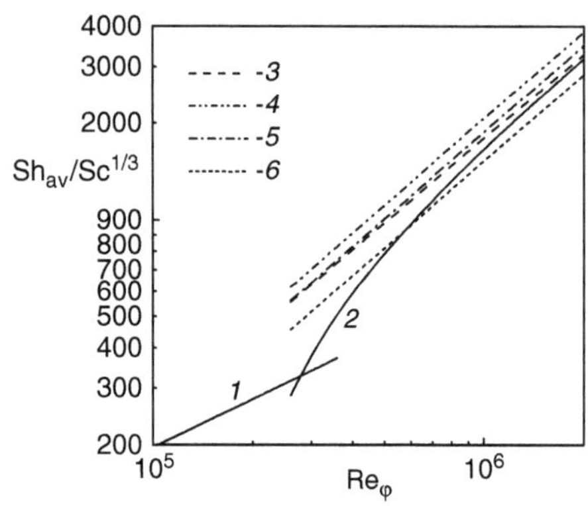

in the coordinates $Sh_{\mathrm{av}}/Sc^{1/3}$ versus Re_{φ} for different Schmidt numbers (over the range of variation of Sc observed in experiments [48]) are rather significant. Therefore, Eq. (8.43) should be considered unreliable.

Based on the analysis done with the help of Figs. 8.4, 8.5 and 8.6, one can conclude that the most reliable empirical relations obtained via approximations of the experimental data for developed turbulent flow and an *entire* disk are Eqs. (8.36), (8.37), (8.38) and (8.39).

The range of the Schmidt numbers investigated in electrochemical experiments [37, 42], the data which we consider the most reliable, is within Sc=345–10,320. Reliable experimental data [24, 59] for the average and local Sherwood numbers for naphthalene sublimation in air at Sc=2.28 were published significantly later. It is therefore interesting to find out whether Eqs. (8.36), (8.37), (8.38) and (8.39) obtained for the asymptotic case of very high Schmidt numbers can be applied to the case of the moderate values Sc=2.28–2.5.

Exponent at the Reynolds number in Eq. (8.38) for developed turbulent flow is different from the respective exponents in Eqs. (8.17) and (8.18) for the case Sc=2.28. For the convenience of comparisons, let us transform Eq. (8.38) in such a way that the power exponent at the Reynolds number is equal to 0.8, like in Eq. (8.17).

Given in Fig. 8.5 is curve *2* obtained by Eq. (8.36), as well as curve *4* computed using the following correlation obtained by the author of the present work:

$$Sh_{\mathrm{av}}Sc^{-1/3} = 0.62Re_{\omega,\mathrm{tr}}^{1/2}\left(\frac{Re_{\omega,\mathrm{tr}}}{Re_{\varphi}}\right)^{1/2} + \frac{2}{2.6}0.0365Re_{\varphi}^{0.8}\left[1-\left(\frac{Re_{\omega,\mathrm{tr}}}{Re_{\varphi}}\right)^{1.3}\right] \tag{8.51}$$

Here again, like in Eq. (8.36), $Re_{\omega,\mathrm{tr}}=2.78\times10^5$. Curves *2* and *4* practically coincide for $Re_{\varphi}\leq 9.0\times10^5$. Differences between curves *2* and *4* become noticeable for the Reynolds numbers increasing over the range $Re_{\varphi}>9.0\times10^5$.

For the developed turbulent flow, asymptotical forms of Eq. (8.51) for local and average Sherwood numbers are

$$Sh = 3.65\times10^{-2}Re_{\omega}^{0.8}Sc^{1/3}, \tag{8.52}$$

$$Sh_{\mathrm{av}} = 3.65\times10^{-2}\frac{2}{2.6}Re_{\varphi}^{0.8}Sc^{1/3} = 2.81\times10^{-2}Re_{\varphi}^{0.8}Sc^{1/3}. \tag{8.53}$$

Analysis of the data for naphthalene sublimation in air fulfilled in Fig. 8.7 shows that empirical equation (8.38) [42] used at $Sc=2.28$ allows computing Sherwood numbers close enough to the experimental data [24, 59] and their approximation equation (8.17). Equations (8.38) (curve *10*) and (8.17) (curve *9*) agree well with each other at sufficiently high Reynolds numbers $Re_{\omega}=(0.6\text{–}2.0)\times10^6$. As could be expected, curve *11* plotted using Eq. (8.52) is close to empirical curve *10* at lower Reynolds numbers $Re_{\omega}\leq 7.0\times10^5$. Using Eq. (8.52), one can obtain the value of $K_1=0.048$ at $Sc=2.28$, that is only 6.7% lower than the value of $K_1=0.0512$ in Eq. (8.17).

Curve *12* plotted based on Eq. (8.40) exceeds significantly experimental data in Fig. 8.7 and their approximation curve *9* (Eq. (8.17)), as well as curves *10, 11* plotted using Eqs. (8.38) and (8.52) (similar to Fig. 8.5).

Empirical equation (8.44) [126] for transitional flow region at $Sc=2.28$ (curve *13* in Fig. 8.7) agrees very well with Eq. (8.16) and experimental data [24].

Theoretical equation (8.50) is very inaccurate. According to Eq. (8.50), for the power exponent at Re_{φ} equal to 0.8, one can obtain the value $K_2\approx0.029$ at $Sc=2.28$ noticeably different from the experimental value $K_2=0.0394$ [24], Eq. (8.26). For high Schmidt numbers, Eq. (8.50) yields $K_2/Sc^{1/3}\approx 0.035$, which is much higher than the respective value 0.0281 in Eq. (8.53). For air at $Pr=0.72$, Eq. (8.50) gives $K_2\approx0.018\text{–}0.019$ that is significantly different from the reliable experimental values $K_2=0.0145\text{–}0.015$ [41, 138, 173] (see Table 3.5).

Thus, Eqs. (8.36), (8.37), (8.38) and (8.39) should be recognized as the most reliable empirical equations for the computation of the Sherwood numbers for the cases of high Schmidt numbers at the developed turbulent flow and for an entire disk. As the analysis has shown, these equations with the maximal inexactitude less than 7% can allow estimating mass transfer intensity also at moderate Schmidt numbers

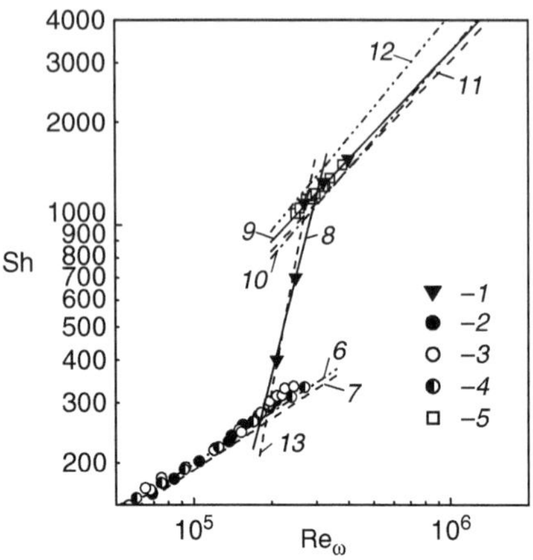

Fig. 8.7 Local Sherwood numbers for naphthalene sublimation in air. Data *1–9*: see caption in Fig. 8.2. Developed turbulent flow, $Sc=2.28$: *10* – Eq. (8.38) [42]; *11* – Eq. (8.52); *12* – Eq. (8.40) [39]. Transitional flow, $Sc=2.28$: *13* – Eq. (8.44) [126]

of $Sc=2.28$. For transitional flow, empirical equation (8.44) originally obtained by the authors of the work [126] at high Schmidt numbers $Sc=1192$–2465 remains valid also for $Sc=2.28$.

8.4 An Integral Method for Modelling Heat and Mass Transfer in Turbulent Flow for the Prandtl and Schmidt Numbers Larger than Unity

8.4.1 Prandtl and Schmidt Numbers Moderately Different from Unity

In analogy to what was said in Sect. 8.2, understood under the Prandtl and Schmidt numbers moderately different from unity are values of $Pr \leq 5$ or $Sc \leq 5$. However, real validation of the solution described below is possible only via comparisons with the data on mass transfer in naphthalene sublimation in air at $Sc=2.28$–2.5.

Analysis performed below is based on the present integral method for the problem of heat transfer over a free rotating disk using power-law approximations of the velocity and temperature profiles, as well as power laws for surface friction and heat transfer described in Sect. 2.4. As mentioned in Sect. 2.4, it is possible to obtain an analytical solution of the heat transfer problem with boundary conditions (2.34), (2.35) and (2.36) under assumptions that (a) relative thermal boundary layer thickness $\Delta=\delta_T/\delta$ is constant and independent of the radial coordinate r, (b) exponents in power-law approximations of the profiles of velocity components (2.42), (2.43), (2.44), (2.45), (2.46) and (2.47) and temperature (2.58) are equal to each other $n=n_T$ and (c) under condition that $Pr=const.$

Finally, Nusselt numbers obtained as a result of the solution can be calculated by Eqs. (3.4). Under condition that $\Delta \geq 1$ (thermal boundary layer thickness δ_T exceeds momentum boundary layer thickness δ), the constants K_1 for the local Nusselt numbers and K_2 for the average Nusselt numbers in Eqs. (3.4) as applied to the case of a free rotating disk can be calculated by Eqs. (3.38)–(3.49) presented in Sect. 3.3.3. In order to compute Δ for the case where $\Delta \leq 1$, one should numerically solve transcendental equation (3.50) instead of using Eqs. (3.43), (3.44) and (3.48). Recommendations for calculation of the exponent n_p at the Prandtl number for the case of $\Delta \leq 1$ were not developed in Sect. 3.3.3, because in air cooling systems, i.e. at Pr=0.7–0.74, condition $\Delta \geq 1$ always holds.

As mentioned in Sect. 8.2, for the case of mass transfer in naphthalene sublimation in air, local Sherwood numbers measured in experiments [24, 59] were approximated by Eqs. (8.17) or (8.18). In these formulas, the exponent at the Reynolds number is equal to 0.8, which corresponds to that in the solutions (3.38)–(3.49) at $n=n_T=1/7$ (see Table 3.6). Technique of naphthalene sublimation in air is used for experimental determination of heat transfer rate of the disk in air at $T_w=const$ (see Eqs. (8.29), (8.30), (8.31), (8.32), (8.33), (8.34) and (8.35)). This is an evidence of the fact that the approximate analytical solution (3.38)–(3.49) for the problem of heat transfer of a rotating disk at $n=n_T$ and $\Delta=const$, which has proved its efficiency for the Prandtl numbers less than unity, can be also applied for modelling heat and mass transfer problems for the Prandtl and Schmidt numbers moderately different from unity.

It should be found out first of all, which of Eqs. (3.43) or (3.50) for the relative thickness Δ of the thermal boundary layer can be applied for the case at hand. Numerical values of Δ for the Prandtl numbers Pr=0.72, 1.0 and 2.28 at different values of the exponent n_* in the boundary condition (2.35) calculated by Eqs. (3.43) and (3.50) are given in Table 8.5. The value of the exponent n_p at the Prandtl number for Pr=2.28 was chosen in such a way that the constant K_1 in Eq. (3.4) for the local Nusselt number calculated by Eq. (3.44) for $\Delta \geq 1$ for an isothermal disk ($n_*=0$) is equal to $K_1=0.0512$. This corresponds to the value of the constant $K_1=0.0512$ in Eq. (8.17) for the Sherwood number obtained at Sc=2.28 in experiments [24]. It was obtained as a result at Pr=2.28 that

$$n_p = 0.042/(1 - K_V). \tag{8.54}$$

Analysis of data in Table 8.5 reveals that at Pr=0.72–1, the value of the relative thickness Δ of the thermal boundary layer exceeds unity for $n_*<3$. However, Eq. (3.43), which is formally valid only for the case $\Delta \geq 1$, allows also over the range of n_*=3–4 to calculate values of $\Delta \leq 1$ that coincide with the data predicted by Eq. (3.50). This is another confirmation of the fact that Eq. (3.44) for the coefficient K_1 at Pr=0.72–1 and its analogues for more complicated cases of fluid flow considered in Chaps. 5 and 6 can be used over the entire range of the values n_*=-1.5 to 4 studied in this monograph.

For the value of the Prandtl number Pr=2.28 one can find out that $\Delta \geq 1$ at $n_*<2$. In analogy to what was said above, Eq. (3.43) also for n_*=2–4 allows predicting values of $\Delta \leq 1$ that in fact coincide with the results obtained with the help of Eq. (3.50).

Table 8.5 Values of Δ given by Eqs. (3.43) and (3.50) at n=n_T=1/7

Pr	Equations	n_*=–1.5	n_*=–1	n_*=–0.5	n_*=0	n_*=1	n_*=2	n_*=3	n_*=4
0.72	(3.43)	60.664	15.924	7.0271	4.0322	1.984	1.2841	0.9563	0.7728
0.72	(3.50)	3.3117	2.9253	2.6237	2.3549	1.81476	1.2819	0.9563	0.7720
1.0	(3.43)	32.677	9.5524	4.5472	2.7600	1.4679	1.0	0.7723	0.6413
1.0	(3.50)	3.1437	2.7462	2.4200	2.1093	1.4530	1.0	0.7715	0.6364
2.28	(3.43)	27.983	8.4144	4.0845	2.5149	1.3640	0.9412	0.7335	0.6131
2.28	(3.50)	3.0991	2.6972	2.3621	2.0361	1.3582	0.9412	0.7320	0.6064

As applied to the range of Δ≥1, Eq. (3.50) also predicts values of Δ exceeding unity; however, their numerical values significantly deviate from those predicted by Eq. (3.43). These deviations increase very strongly with the decreasing values of the exponent n_*, in particular, when n_* becomes negative. Therefore, Eq. (3.50) becomes inapplicable for prediction of the values of Δ over the range Δ≥1, which means that in this case Eqs. (3.43) and (3.50) cannot substitute each other.

The fact that thermal boundary layer thickness can still exceed momentum boundary layer thickness also for the Prandtl numbers larger than unity from the first look seems to be paradoxical. Apparently, the reason for this phenomenon is that for the Prandtl or Schmidt numbers moderately larger than unity the effect of Pr and Sc values is insufficiently strong to overcome the effect of the radial temperature gradient, which results in the values of Δ≥1 at n_*<2. However, the effect of the radial temperature gradient on the disk surface weakens with the increased Prandtl numbers, which results in that the values of n_* that make possible conditions with Δ≥1 at Pr>1 gradually decrease (see Table 8.5).

Thus, the present integral method developed in Chap. 2 and validated in detail for the case of a free rotating disk in air in Chap. 3 remains still applicable for the cases where Prandtl and Schmidt numbers moderately exceed unity. One should use in this case the corrected value of the exponent n_p at the Prandtl or Schmidt number, which for naphthalene sublimation in air can be calculated by Eq. (8.54). It is also important to point out that based on the value of n_p found for the conditions with n_*=0 the integral method allows modelling heat transfer of a rotating disk at all the other thermal boundary conditions on its surface and, in particular, for arbitrary values of n_* in the boundary condition (2.35). Values of the coefficient K_1 given by Eq. (3.44) with account for Eq. (8.54) at different values of n_* and Pr=2.28 are presented in Table 8.6.

For higher values of Pr and Sc (for instance, for heat transfer in water or technical oil, etc.), the numerical coefficient in Eq. (8.54) apparently accepts other values, which should be determined via comparisons with experimental data. Such experimental data can, in particular, allow pinpointing Prandtl and Schmidt numbers, at which thermal/diffusion boundary layer thickness under conditions T_w=$const$ or C_w=$const$ becomes less than momentum boundary layer thickness.

Ratio Δ of the thicknesses of the thermal/diffusion and momentum boundary layers causes significant effect on peculiarities of the model used in frames of the

Table 8.6 Values of the coefficient K_1 given by Eqs. (3.44) and (8.54) at $n=n_T=1/7$ and $Pr=2.28$

n_*	$n_*=-1.5$	$n_*=-1$	$n_*=-0.5$	$n_*=0$	$n_*=1$	$n_*=2$	$n_*=3$	$n_*=4$
K_1	0.0363	0.0431	0.0478	0.0512	0.0559	0.0590	0.0611	0.0627

present integral method. As will be shown below, parameter Δ at very high Schmidt numbers is very small and becomes itself a function of the local rotational Reynolds number Re_ω. Temperature/concentration profiles in wall coordinates also become functions of the local Reynolds number Re_ω. It is of significant interest to determine limiting Prandtl or Schmidt numbers, for which fundamental assumptions $\Delta=const$ and $T^+ \equiv T^+(y^+)$ (see Eq. (2.66)) of the present integral method become invalid. However, such investigation was not included in objectives pursued in frames of this monograph.

8.4.2 High Prandtl and Schmidt Numbers

Model with a constant value of $\boldsymbol{\Delta<<1}$**.** At very high Prandtl (or Schmidt) numbers, the ratio Δ of thicknesses of the thermal (or diffusion) and momentum boundary layers becomes very small. Owing to this, transcendental equation (3.50) obtained under assumptions $\Delta=const$, $n=n_T$ and $T^+ \equiv T^+(y^+)$ can be simplified, because all the terms in the parentheses in the left-hand side except for a_* become negligibly small:

$$\Delta^{2n+1} a_* = \frac{4+m}{4+m+n_*}\left(a_* - 2b_* + c_*\right) Pr^{-n_p}. \tag{8.55}$$

From Eq. (8.55), one can obtain an analytical solution for Δ:

$$\Delta = \left[\frac{4+m}{4+m+n_*}\left(1 - \frac{2D_3}{C_2}\right)\right]^{\frac{1}{2n+1}} Pr^{-\frac{n_p}{2n+1}}, \tag{8.56}$$

where constants C_2 are D_3 are defined in explanations to Eqs. (2.73) and (2.74).

Based on Eqs. (3.42) and (8.56), the constant K_1 can be expressed as

$$K_1 = K_3\left[\frac{4+m}{4+m+n_*}\left(1 - \frac{2D_3}{C_2}\right)\right]^{\frac{-n}{2n+1}} Pr^{1+n_p\left(\frac{n}{2n+1}-1\right)}. \tag{8.57}$$

As shown in Sect. 8.3, the exponent at the Prandtl number in Eq. (8.57) for very high values of Pr should be equal to 1/3. This leads to the following expression for n_p:

$$n_p = \frac{2}{3}\cdot\frac{2n+1}{n+1}. \tag{8.58}$$

Finally, the solutions for the constants K_1 and K_2 with account for Eq. (3.45) are

$$K_1 = K_3 \left[\frac{4+m}{4+m+n_*} \left(1 - \frac{2D_3}{C_2} \right) \right]^{\frac{-n}{2n+1}} Pr^{1/3}, \tag{8.59}$$

$$K_2 = K_3 \left[\frac{4+m}{4+m+n_*} \left(1 - \frac{2D_3}{C_2} \right) \right]^{\frac{-n}{2n+1}} \frac{n_*+2}{2+n_*+m} Pr^{1/3}. \tag{8.60}$$

To make comparisons with electrochemical experiments, we will consider Sherwood numbers instead of the Nusselt numbers, thus substituting also Pr with Sc.

Comparisons of the solution (8.60) for Sh_{av} (at n_*=0) with empirical equation (8.37) [42] and theoretical formula (8.48) [81] are performed in Fig. 8.8. As may be concluded, curves *5* and *6* computed by Eq. (8.60) at n=1/7 and 1/9 lie 20–30% lower than curves *3* and *4* plotted based on Eqs. (8.37) and (8.48), respectively. Such deviations of the theory from experiments are unacceptable. Besides, the slope angle of curves *5* and *6* (the exponent at Re_φ is equal to 0.8 and 0.833, with the constant K_2 being equal to 0.0207 and 0.0126, respectively) is visibly different from the slope angle of curves *3* and *4* (the exponent at Re_φ is 0.87 and 0.9, whereas the constant K_2 is equal to 0.0207 and 0.126, respectively). This is the evidence of the partial inadequacy of the model assumptions of the present integral method as applied to the conditions of convective heat and mass transfer at high Pr and Sc values. Therefore, it is necessary to revise these assumptions to improve the agreement with the empirical equations and solutions obtained using other theoretical models. Analysis of Eq. (3.42) for the constant K_1 shows that increase in the exponent at the Reynolds number in the relation for the Sherwood number can be reached, if the parameter Δ becomes variable decreasing with increase in the local Reynolds number Re_ω.

Model with a variable value of Δ. The present integral method is based on the two-layer model of velocity and temperature profiles in the boundary layer. In the viscous sub-layer of the momentum boundary layer, the first of Eqs. (2.67) is in force, while in the heat conduction sub-layer the temperature profile is used that is described by the second of Eqs. (2.67). Outside of the viscous sub-layer, tangential

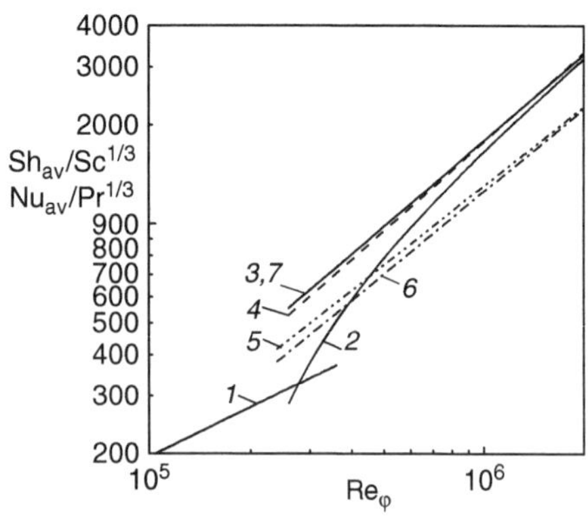

Fig. 8.8 Average Sherwood numbers for high Schmidt numbers. Approximation of experiments: *1* – laminar flow, Levich's equation (8.9) [105]; *2* – Eq. (8.36) for an *entire* disk [42]; *3* – Eq. (8.37) [42] Theoretical solutions: *4* – Eq. (8.48) [81]; *5* – Eq. (8.60) for n=1/7; *6* – Eq. (8.60) for n=1/9; *7* – Eq. (8.79)

and radial velocity components of the fluid are approximated by the power-law profiles analysed in detail in Sects. 2.3.1, 2.3.2 and 2.4.2, as well as in Sect. 3.3.3. In the cases of convective heat and mass transfer at the Prandtl and Schmidt numbers moderately different from unity (less or larger than unity), the thickness of the thermal/diffusion boundary layer is larger or moderately less than that of the momentum boundary layer (see Table 8.5). Hence, to obtain a solution of the integral equation (2.28) for the thermal boundary layer, the integral in its right-hand side or, in other words, the enthalpy thickness δ_T^{**} has been found using profiles (2.42), (3.37) for the radial velocity component v_r across the *entire* momentum boundary layer that was only somewhat thicker or somewhat thinner than the thermal boundary layer. Viscous and heat conduction sub-layers were ignored at the integration, since their thickness is negligibly small and does not change the value of the definite integral. Integration of the velocity profiles in the boundary layer equations (2.22), (2.23) and (2.24) is usually done in the very same way [41, 80, 138, 173, 189].

The situation is principally different for very high Prandtl and Schmidt numbers. Here thermal/diffusion boundary layers are very thin and develop within the viscous sub-layer of the momentum boundary layer, where the profile of v_r is a linear function of the coordinate z orthogonal to the disk. This fact was noticed and used in the development of theoretical models of convective heat and mass transfer of a rotating disk at large *Pr* and *Sc* numbers in the works [39, 81, 102, 105, 141, 199].

Linear profile of v_r in the vicinity of the wall can be written as

$$v_r = \frac{\tau_{wr}}{\mu} z = \frac{\tau_w \alpha}{\mu(1+\alpha^2)^{1/2}} z = \frac{\rho V_*^2 \alpha}{\mu(1+\alpha^2)^{1/2}} \frac{c_f}{2} z = \alpha(1+\alpha^2)^{1/2} \omega A_c Re_\omega^{n_R} z. \tag{8.61}$$

where the exponent n_R is given in Eq. (3.41).

Based on the definition of the constant z_1^+, one can derive the following formula:

$$\frac{z_1}{\delta} = \frac{z_1^+}{\gamma(1+\alpha^2)^{1/2} A_c^{1/2} Re_\omega^{1/(1+3n)}}. \tag{8.62}$$

According to Eq. (8.62), linear model (8.61) for the radial velocity profile holds up to the values of z/Δ from $z/\delta \le 0.01$ to 0.02. As analysis in Fig. 8.9 confirms, linear model (8.61) approximately up to $z/\delta_\varphi^{**} \le 0.2$ or $z/\delta \le 0.02$ agrees well with the experiments and power-law profiles of v_r. Thus, the value of $\Delta = \delta_T/\delta = 0.02$ outlines the limit up to which the linear equation (8.61) can be used to model the radial velocity profile.

In frames of the power-law model used, the Stanton number characterizing heat transfer rate on the disk surface is described by Eq. (2.69). In Sect. 2.4.3, the model assumption $(z_{1T}^+/z_1^+)^{n_T-1} Pr^{-n_T} = Pr^{-n_p}$ is used for the last two factors in Eq. (2.69) (where exponent n_p remains unknown to be found via comparisons with experimental data for the Nusselt number). Such assumption is justifiable for the cases considered above, where thermal/diffusion boundary layer thickness exceeded or was somewhat less than momentum boundary layer thickness. At high

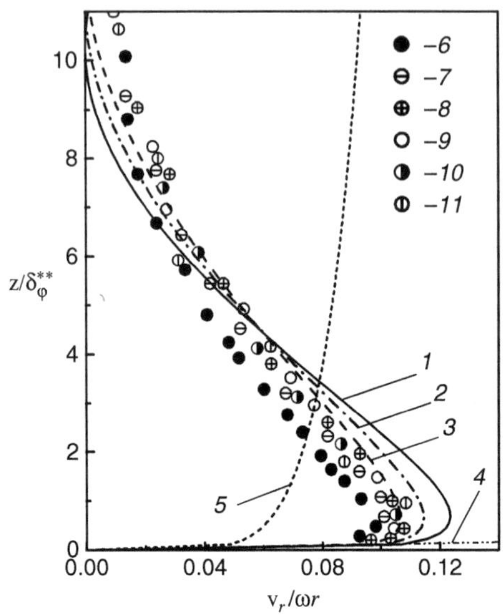

Fig. 8.9 Radial velocity profiles in the turbulent boundary layer over a free rotating disk. For notations *1–3*, *6–11* see caption in Fig. 2.2; *4* – Eq. (8.61) at $Re_\omega = 1.0 \times 10^6$; *5* – Eq. (3.37) at $\sigma = 0$, $n = 1/7$ (see Fig. 3.8)

Prandtl or Schmidt numbers, another boundary layer structure takes place: entire temperature/concentration profile in very thin thermal/diffusion boundary layers develops within the viscous sub-layer of the momentum boundary layer. It is therefore deemed to be logical to suppose that the interrelation between coordinates of the boundary of the viscous sub-layer (point z_1^+) and heat conduction sub-layer (point z_{1T}^+) has a more complicated form

$$(z_{1T}^+/z_1^+)^{n_T-1} Pr^{-n_T} = K_\alpha Pr^{-n_p}. \tag{8.63}$$

Constants K_α and n_p in Eq. (8.63) should be found via comparisons with experimental data for the *Nu* or *Sh* numbers. Therefore, Eqs. (2.71) and (2.72) for the Stanton and Nusselt numbers at $n = n_T$ can be written as

$$St = (c_f/2)\Delta^{-n} Pr^{-n_p} K_\alpha = A_c Re_\omega^{-2n/(3n+1)} \Delta^{-n} Pr^{-n_p} K_\alpha, \tag{8.64}$$

$$Nu = St Re_\omega Pr (1+\alpha^2)^{1/2} = A_c (1+\alpha^2)^{1/2} Re_\omega^{n_R} \Delta^{-n} Pr^{1-n_p} K_\alpha. \tag{8.65}$$

Finding the definite integral in the left-hand side of Eq. (2.25) with allowance for the linear model (8.61) for v$_r$ and power-law function (2.58) for Θ, and substituting Eqs. (8.64) and (8.65) into the right-hand side of Eq. (2.25), one can reduce the integral equation of the thermal boundary layer (2.25) to the following form:

$$\frac{n}{2(n+2)} \alpha\omega \frac{d}{dr}\left[r\delta^2 \Delta^2 Re_\omega^{n_R} \Delta T\right] = K_\alpha \Delta^{-n} Pr^{-n_p} Re_\omega^{n_R} \nu \Delta T \tag{8.66}$$

where, to remind, $\Delta T = T_w - T_\infty$, $n = n_T$, and the parameter Δ is given by Eq. (2.82).

It was assumed in the present integral method as applied to the problems of convective heat and mass transfer at the Prandtl and Schmidt numbers moderately larger than unity that $\Delta = const$. This condition is inapplicable to Eq. (8.66), because, given $\Delta = const$, exponents at the variable r in both sides of Eq. (8.66) do not coincide.

One can assume by analogy to Eq. (2.82) that parameter Δ is a power-law function of the radial coordinate r:

$$\Delta = C_\Delta r^k. \tag{8.67}$$

Substituting Eq. (8.67) into Eq. (8.66), one can derive with account for Eqs. (2.35), (2.82), (2.83) and (3.41) that

$$\Delta = C_{\Delta *} Re_\omega^{k/2}, \tag{8.68}$$

$$C_{\Delta *} = C_{\Delta **}\ Pr^{-n_p/(2+n)}, \tag{8.69}$$

$$C_{\Delta **} = \left[\frac{K_\alpha 2(n+2)/n}{\alpha\gamma^2(1 - nk + n_* + 2n_R)}\right]^{1/(n+2)}.$$

$$k = -2m/(2+n). \tag{8.71}$$

As a result, Eqs. (8.65) for the local Nusselt number and the respective relation for the average Nusselt number Nu_{av} take the following form:

$$Nu = K_1 Re_\omega^{n_{R*}}, \tag{8.72}$$

$$Nu_{\text{av}} = K_2 Re_\varphi^{n_{R*}}, \tag{8.73}$$

$$n_{R*} = n_R + mn/(2+n), \tag{8.74}$$

$$K_1 = K_\alpha K_3 C_{\Delta **}^{-n} Pr^{1/3}, \tag{8.75}$$

$$K_2 = 2K_1/(2n_{R*} + 1), \tag{8.76}$$

$$n_p = (2+n)/3. \tag{8.77}$$

Relation (8.77) for the exponent n_p was obtained in view of the requirement that the cumulative exponent at the Prandtl number in Eq. (8.75) should be equal to 1/3.

Thus, in Eqs. (8.72) and (8.73), the cumulative exponent n_{R*} at the Reynolds number given in Eq. (8.74) increases in comparison with Eq. (3.41) by the additional term $mn/(2+n)$, owing to that *parameter* Δ *becomes variable* decreasing with the increasing local Reynolds number (i.e. local coordinate r) in accordance with Eq. (8.68).

At n=1/7 one has n_{R*}=0.84, and at n=1/9 one can obtain n_{R*}=0.868, which practically coincides with the exponent 0.87 at the Re_φ number in empirical equation

(8.38) generalizing experimental data [42]. In order for Eqs. (8.72) and (8.73) at n=1/9 and n_{R*}=0.868, as applied to Sherwood and Schmidt numbers, to agree with Eqs. (8.38) and (8.37), respectively, one should choose the value K_α=1.254, that finally entails for n_*=0

$$Nu = 1.52 \times 10^{-2} Re_\omega^{0.868} Pr^{1/3}, \tag{8.78}$$

$$Nu_{\rm av} = 1.11 \times 10^{-2} Re_\varphi^{0.868} Pr^{1/3}, \tag{8.79}$$

$$\Delta = 18.31 Re_\omega^{-0.3158} Pr^{-1/3}. \tag{8.80}$$

Curve *7* plotted in Fig. 8.8 based on Eq. (8.79) coincides with curve *3* depicted based on empirical equation (8.37) [42]. Equations (8.78) and (8.38) also practically coincide.

Experiments [42] were performed at Sc=930–10,320, Re_φ=(0.278–1.8) $\times 10^6$. According to Eq. (8.80), one can obtain the value Δ=0.036 at minimal values Sc=930 and Re_ω=0.278$\times 10^6$ [42]. Parameter Δ decreases with an increase in the Schmidt and Reynolds numbers Re_ω. At Sc=10,320 and Re_ω=0.278$\times 10^6$, one has Δ=0.015; at Sc=930 and Re_ω=1.8$\times 10^6$ one has Δ=0.02. These Δ values agree well with the aforementioned limit $\Delta \leq 0.02$, at which the linearity of the radial velocity profile remains in force.

Model with a variable value Δ and profile T^+ depending on Re_ω. It was obtained in works [81, 102, 105, 141] based on the analysis of models for eddy-viscosity in the near-wall region at high Prandtl numbers that the Nusselt and Stanton numbers are determined by the following equations:

$$Nu = K_N (1+\alpha^2)^{1/2} (c_f/2)^{1/2} Re_\omega Pr^{1/3}, \tag{8.81}$$

$$St = K_N (c_f/2)^{1/2} Pr^{-2/3}, \tag{8.82}$$

where the constant K_N assumes different values depending on the model assumptions used by the authors.

In the above-mentioned works, power-law dependence, Eq. (2.87), was used for the friction coefficient at n=1/7. Then the relations for the local and average Nusselt numbers with allowance for Eq. (8.61) reduce to the form of Eqs. (8.72) and (8.73), respectively, where

$$K_1 = K_N (1+\alpha^2)^{1/2} A_c^{1/2} Pr^{1/3}, \tag{8.83}$$

$$n_{R*} = (2n+1)/(3n+1), \tag{8.84}$$

and the interrelation between the constants K_1 and K_2 is given in Eq. (8.76). For n=1/7, Eq. (8.84) results in n_{R*}=0.9.

As shown in Sect. 8.3, Eq. (8.48) is the most accurate among the correlations obtained in [81, 102, 105, 141] based on model (8.81) (presented in a transformed form of an equation for average Sherwood numbers). Then, agreeing Eqs. (8.81) and (8.48) with allowance for Eq. (8.76), one can obtain the value K_N=0.05986.

Solving the thermal boundary layer equation using assumption (8.67), like it was done for *the model with a variable value of* Δ, one can come again to the solution (8.68) for Δ, where

$$C_{\Delta *} = C_{\Delta **} Pr^{-1/3}, \tag{8.85}$$

$$C_{\Delta **} = \left[\frac{K_N \; 2(n+2)/n}{\alpha \gamma^2 (1 + 2m + 2k + n_* + 2n_R) A_c^{1/2}} \right]^{1/(n+2)} \tag{8.86}$$

$$k = (2n-1)/(3n+1). \tag{8.87}$$

Substituting values n=1/7 and n_*=0 into Eqs. (8.68), (8.85), (8.86) and (8.87) results in

$$\Delta = 12.54 Re_{\omega}^{-1/4} Pr^{-1/3} \tag{8.88}$$

One can estimate values Δ, which can be obtained by Eq. (8.88) with allowance for Eqs. (8.85), (8.86) and (8.87) for the conditions of experiments [42] with Sc=930–0320, Re_{φ}=(0.278–1.8) $\times 10^6$. For the minimal values of Sc=930 and $Re_{\omega}=0.278\times10^6$ [42], the relative thermal boundary layer thickness Δ in accordance with Eq. (8.88) is equal to Δ=0.037 and decreases with both Sc and Re_{φ} increasing. At Sc=10,320 and $Re_{\omega}=0.278\times10^6$, one can obtain Δ=0.016, while at Sc=930 and $Re_{\omega}=1.8\times10^6$ one has Δ=0.023. These Δ values exceed insignificantly the data obtained by Eq. (8.80) in frames of *the model with a variable value of* Δ, and they also agree rather well with the limiting value $\Delta \le 0.02$, at which the linear model of the radial velocity profile holds.

Let us consider the temperature profile in wall coordinates T^+. In frames of *the models with a constant and variable value of* Δ, profile T^+ in the region where the power law (2.28) at $n=n_T$ is valid remains independent of the Reynolds number:

$$T^+ = (z^+)^n (1+\alpha^2)^{-n/2} \gamma^{-n} A_c^{-(n+1)/2} Pr^{n_p} K_{\alpha}^{-1} \tag{8.89}$$

This agrees with suggestions of the authors of works [71, 203].

Model based on Eqs. (8.81) and (8.82) leads to the profile of T^+ dependent on Re_{ω}:

$$T^+ = (z^+)^n (1+\alpha^2)^{-n/2} \gamma^{-n} A_c^{-n/2} C_{\Delta *}^{-n} K_N^{-1} Pr^{2/3} Re_{\omega}^{-0.5(2n^2+n)/(3n+1)}. \tag{8.90}$$

Thus, the investigations performed above resulted in the following accomplishments [194].

Empirical/theoretical relations for Sherwood numbers available in the literature were thoroughly validated, and the most reliable were distinguished.

A novel approach to modelling temperature/concentration profiles for the values of Pr and Sc significantly larger than unity was proposed. An original integral method developed based on this allows estimating relative thickness Δ of the thermal/diffusion boundary layers that has not been accomplished by theoretical models of the other authors. It was shown that namely decrease in the values of Δ with the increasing local radii entails the additional increase in the exponent at the Reynolds number in the expression for the *Nu* or *Sh* numbers in comparison with air flows. As a result, solutions obtained for the surface heat and mass transfer agree well with the selected empirical formulas.

References

1. Abrahamson S, Lonnes S (1993) An integral method for turbulent boundary layers on rotating disks. Trans ASME J Fluids Eng 115 (4): 614–619.
2. Abrahamson S, Lonnes S (1994) An improved model for radial injection between corotating disks. Trans ASME J Fluids Eng 116 (2): 255–257.
3. Alden J (1994) Computational Electrochemistry. D. Phil. thesis. Chapter 7. Oxford University, Oxford, UK.
4. Ariel PD (1996) The flow near a rotating disk: an approximate solution. Trans ASME J Appl Mech 63 (2): 436–438.
5. Astarita T, Cardone G (2008) Convective heat transfer on a rotating disk with a centred impinging round jet. Int J Heat Mass Transfer 51 (7–8) 1562–1572.
6. Awad MM (2008) Heat transfer from a rotating disk to fluids for a wide range of Prandtl numbers using the asymptotic model. Trans ASME J Heat Transfer 130 (1): 014505-1–014505-4.
7. Baibikov AS (1998) Method of calculation of a turbulent flow in an axial gap with a variable radius between a rotating disk and an axisymmetric casing. J Eng Phys Thermophys 71 (6): 1072–1081.
8. Barcia OE, Mangiavacchi N, Mattos OR, Pontes J, Tribollet B (2008) Rotating disk flow in electrochemical cells: a coupled solution for hydrodynamic and mass equations. J Electrochem Soc 155 (5): D424–D427.
9. Baughn JW, Ireland PT, Jones TV, Saniei N (1989) A comparison of the transient and heated-coating methods for measurement of local heat transfer coefficients on pin fin. Trans ASME J Heat Transfer 111 (4): 877–881.
10. Bejan A (1993) Heat transfer. John Wiley and Sons, Inc., New York.
11. Bogdan Z (1982) Cooling of a rotating disk by means of an impinging jet. Proc. 7th International Heat Transfer Conference "Heat Transfer 1982", Munich (Germany) 3: 333–336.
12. Bons J (2009) Transient method for convective heat transfer measurement with lateral conduction—Part I: application to a deposit-roughened gas turbine surface. Trans ASME J Heat Transfer 131 (1): 011301-1–011301-7.
13. Bons J, Fletcher D, Borchert B (2009) Transient method for convective heat transfer measurement with lateral conduction—Part II: application to an isolated spherical roughness element. Trans ASME J Heat Transfer 131 (1): 011302-1–011302-7.
14. Brdlik PM, Savin VK (1966) Heat transfer in the vicinity of the stagnation point in an axisymmetric jet flowing over flat surfaces normal to the flow. J Eng Phys Thermophys 10 (4): 241–245.
15. Brodersen S, Metzger DE, Fernando HJS (1996) Flows generated by the impingement of a jet on a rotating surface. P. I. Basic flow patterns. Trans ASME J Fluids Eng 118 (1): 61–67.
16. Buschmann MH (2002) A solution for the flow between a cone and a plate at low Reynolds number. J Thermal Sci 11 (4): 289–295.

I.V. Shevchuk, *Convective Heat and Mass Transfer in Rotating Disk Systems*, Lecture Notes in Applied and Computational Mechanics 45, DOI 10.1007/978-3-642-00718-7_BM2,

© Springer-Verlag Berlin Heidelberg 2009

17. Buschmann MH, Dieterich P, Adams NA, Schnittler H-J (2005) Analysis of flow in a cone-and-plate apparatus with respect to spatial and temporal effects on endothelial cells. Biotechnol Bioeng 89 (5): 493–502.
18. Buznik VM, Artemov GA, Bandura VN, Kardashev YuD, Fedorovskiy AM (1966) Heat transfer of a flat disk rotating in quiescent environment. Izvestiya vuzov. Energetika 1: 84–86 (in Russian).
19. Cardone G, Astarita T, Carlomagno GM (1996) Infrared heat transfer measurements on a rotating disk. Opt Diagn Eng 1 (2): 1–7.
20. Cardone G, Astarita T, Carlomagno GM (1997) Heat transfer measurements on a rotating disk. Int J Rotating Machinery 3 (1): 1–9.
21. Case P (1966) Measurements of entrainment by a free rotating disk. J Roy Aero Soc 71: 124–129.
22. Cebeci T, Bradshaw P (1984) Physical and computational aspects of convective heat transfer. Springer-Verlag, Berlin-Heidelberg-New York-Tokyo.
23. Cham T-S, Head TR (1969) Turbulent boundary-layer flow on a rotating disk. J Fluid Mech 37 (1): 129–147.
24. Chen Y-M, Lee W-T, Wu S-J (1998) Heat (mass) transfer between an impinging jet and a rotating disk. Heat Mass Transfer 34 (2–3) 101–108.
25. Chen JX, Gan X, Owen JM (2001) Heat transfer from air-cooled contra-rotating disks. Trans ASME J Turbomach 119 (1): 61–67.
26. Cheng W-T, Lin H-T (1994) Unsteady and steady mass transfer by laminar forced flow against a rotating disk. Wärme und Stoffübertragung 30 (2): 101–108.
27. Chin D, Litt M (1972) An electrochemical study of flow instability on rotating disk. J Fluid Mech 54: 613–625.
28. Cho HH, Rhee DH (2001) Local heat/mass transfer measurement on the effusion plate in impingement/effusion cooling systems. Trans ASME J Turbomach 123 (3): 601–608.
29. Cho HH, Won CH, Ryu GY, Rhee DH (2002) Local heat transfer characteristics in a single rotating disk and co-rotating disks. Microsyst Technol 9 (6–7): 399–408.
30. Cho HH, Wu SJ, Kwon HJ (2000) Local heat/mass transfer measurements in a rectangular duct with discrete ribs. Trans ASME J Turbomach 122 (3): 579–586.
31. Clarkson MH, Chin SC, Shacter P (1980) Flow visualization of inflexional instabilities on a rotating disk. AIAA Paper No 80–0279: 1–14.
32. Cobb EC, Saunders OA (1956) Heat transfer from a rotating disk. Proc Roy Soc A 236: 343–349.
33. Cochran WG (1934) The flow due to a rotating disk. Proc Cambridge Phil Soc 30: 365–375.
34. Cooper AJ, Carpenter PW (1997) The stability of rotating-disc boundary-layer flow over a compliant wall. Part 1. Type I and II instabilities. J Fluid Mech 350: 231–259.
35. Corke TC, Knasiak KF (1998) Stationary travelling cross-flow mode interactions on a rotating disk. J Fluid Mech 355: 285–315.
36. Dennis RW, Newstead C, Ede AJ (1970) The heat transfer from a rotating disc in an air crossflow. Proc. IV International Heat Transfer Conf. Paris-Versailles (France) 3: Paper No FC 7.1.
37. Daguenet M (1968) Etude du transport de matière en solution, a l'aide des électrodes a disque et a anneau tournants. Int J Heat Mass Transfer 11 (11) 1581–1596.
38. Derevich IV (2005) Modeling of the motion of particles in a rotary crusher. Theor Foundations Chem Eng 39 (2): 213–219.
39. Deslouis C, Tribollet B, Viet L (1980) Local and overall mass transfer rates to a rotating disk in turbulent and transition flows. Electrochimica Acta 25 (8): 1027–1032.
40. Dong Q, Santhanagopalan S, White RE (2008) A comparison of numerical solutions for the fluid motion generated by a rotating disk electrode. J Electrochem Soc 155 (9): B963–B968.
41. Dorfman LA (1963) Hydrodynamic resistance and the heat loss of rotating solids. Oliver and Boyd, Edinburgh, UK.

42. Dossenbach O (1976) Simultaneous laminar and turbulent mass transfer at a rotating disk electrode. Berichte der Bunsen-Gesellschaft-Physical Chemistry Chemical Physics 80 (4): 341–343.
43. Dyban YP, Kabkov VIa (1979) Experimental study of flow of air in a gap formed by two rotating disks. Fluid Mech-Soviet Res 8 (4): 99–104.
44. Dyban YP, Mazur AI (1982) Convective heat transfer in jet flows past bodies. Naukova Dumka, Kiev, USSR (in Russian).
45. Eckert ERG, Goldstein RJ (1976) Measurements in heat transfer, 2nd edn. Hemisphere Publishing Corporation, Inc., Washington, New York, London.
46. Elkins CJ, Eaton JK (1997) Heat transfer in the rotating disk boundary layer. Stanford University, Department of Mechanical Engineering, Thermosciences Division Report TSD-103. Stanford University (USA).
47. Elkins CJ, Eaton JK (2000) Turbulent heat and momentum transfer on a rotating disk. J Fluid Mech 402: 225–253.
48. Ellison BT, Cornet I (1971) Mass transfer to a rotating disk. J Electrochem Soc 118 (1): 68–72.
49. Fedorov BI, Plavnik GZ, Prokhorov IV, Zhukhovitskii LG (1976) Transitional flow conditions on a rotating disk. J Eng Phys 31: 1448–1453.
50. Fewell ME, Hellums JD (1977) The secondary flow of Newtonian fluids in cone and plate viscometers with small gap angles. Trans Soc Rheol 21 (4): 535–565.
51. Filippov IF (1986) Heat transfer in electrical machines. Energoatomizdat, Leningrad, USSR (in Russian).
52. Goldstein S (1935) On the resistance to the rotation of a disc immersed in a fluid. Proc Cambridge Phil Soc 31: 232–241.
53. Gregory N, Stuart JT, Walker WS (1955) On the stability of three-dimensional boundary layers with application to the flow due to a rotating disk. Phil Trans Roy Soc Ser A 248: 155–199.
54. Gregory N, Walker WS (1960) Experiments on the effect of suction on the flow due to a rotating disk. J Fluid Mech 9: 225–234.
55. Ginzburg IP (1970) Theory of resistance and heat transfer. Izd. Leningrad. Univ., Leningrad, USSR (in Russian).
56. Hall P (1986) An asymptotic investigation of the stationary modes of instability of the boundary layer on a rotating disk. Proc Roy Soc Lond A 406 1830: 93–106.
57. Han B, Goldstein RE (2001) Jet-impingement heat transfer in gas turbine systems. Heat transfer in gas turbine systems. Ann N Y Acad Sci 234: 147–161.
58. Hartnett JP, Deland EC (1961) The influence of Prandtl number on the heat transfer from rotating nonisothermal disks and cones. Trans ASME J Heat Transfer 83 (1): 95–96.
59. He Y, Ma LX, Huang S (2005) Convection heat and mass transfer from a disk. Heat Mass Transfer 41 (8): 766–772.
60. Holman JP (1981) Heat transfer. McGraw-Hill, Inc., New York.
61. Indinger T, Shevchuk IV (2004) Transient laminar conjugate heat transfer of a rotating disk: theory and numerical simulations. Int J Heat Mass Transfer 47 (14–16): 3577–3581.
62. Isachenko VP, Osipova VA, Sukomel AS (1977) Heat transfer. Mir Publishers, Moscow, USSR.
63. Itoh M, Hasegawa I (1994) Turbulent boundary layer on a rotating disk in infinite quiescent fluid. JSME Int J Ser B 37 (3): 449–456.
64. Janotková E, Pavelek M (1986) A naphthalene sublimation method for predicting heat transfer from a rotating surface. Strojnícky časopis 37(3): 381–393 (in Czech).
65. Janotková E, Pavelek M (1993) Application of interferometry to temperature field measurements in neighbourhood of rotating discs. Proc. 11th International Congress Chemical and Process Engineering CHISA 93, Prague (Czech Republic). Paper No 0015.
66. Jarre S, Le Gal P, Chauve MP (1996) Experimental study of rotating disk instability. I. Natural flow. Phys Fluids 8 (2): 496–508.

67. Jarre S, Le Gal P, Chauve MP (1996) Experimental study of rotating disk instability. II. Forced flow. Phys Fluids 8 (11): 2985–2994.
68. Johnson MC (1980) Turbulent heat transfer to a rotating disk: a review and extension of Dorfman. Trans ASME J Heat Transfer 102 (4): 780–781.
69. Kabkov VIa (1974) Characteristics of turbulent boundary-layer on a smooth disk rotating in a large volume. In: Teplofizika i Teplotekhnika, Naukova Dumka, Kiev 28: 119–124 (in Russian).
70. Kabkov VIa (1978) Heat transfer at air flow in a gap between a disk and a deflector rotating together with it. In: Teploobmen v Energeticheskikh Ustanovkakh, Naukova Dumka, Kiev: 121–124 (in Russian).
71. Kader BA (1981) Temperature and concentration profiles in fully turbulent boundary layers. Int J Heat Mass Transfer 24 (9): 1541–1544.
72. Kapinos VM (1964) Hydraulic resistance and heat transfer of a free disc with a cob. Izvestiya vuzov. Energetika 11: 85–92 (in Russian).
73. Kapinos VM (1964) Heat transfer of a turbine rotor with a radial coolant flow. Inzhenerno-Fizicheskiy Zhurnal 7(1): 3–11 (in Russian).
74. Kapinos VM (1965) Heat transfer from a disc rotating in a housing with a radial flow of coolant. J Eng Phys Thermophys 8 (1): 35–38.
75. Kapinos VM (1965) Heat transfer of a disk rotating in a housing. Izvestiya vuzov. Aviatsyonnaya Tekhnika 2: 76–86 (in Russian).
76. Kapinos VM (1965) Effect of a radial gradient of a relative circumferential velocity component on heat transfer in forced flow between two rotating disks. Izvestiya vuzov. Energetika 3: 111–120 (in Russian).
77. Kapinos VM, Pustovalov VN, Rud'ko AP (1971) Heat transfer in a coolant flow from a center to a periphery between two rotating disks. Izvestiya vuzov. Energetika 6: 116–124 (in Russian).
78. Karabay H, Chen J-X, Pilbrow R, Wilson M, Owen JM (1999) Flow in a "cover-plate" preswirl rotor-stator system. Trans ASME J Turbomach 121 (1): 161–166.
79. Karabay H, Wilson M, Owen JM (2001) Predictions of effect of swirl on flow and heat transfer in a rotating cavity. Int J Heat Fluid Flow 22: 143–155.
80. Karman Th von (1921) Über laminare und turbulente Reibung. Z Angew Math Mech 1 (4): 233–252.
81. Kawase Y, De A (1982) Turbulent mass transfer from a rotating disk. Electrochimica Acta 27 (10): 1469–1473.
82. Kays WM, Crawford ME (1980) Convective heat and mass transfer. McGraw-Hill, Inc., New York.
83. Kelson N, Desseaux A (2000) Note on porous rotating disk flow. ANZIAM J 42 (E): C837–C855.
84. Kempf G (1924) Über Reibungswiderstand rotierender Scheiben. Vorträge auf dem Gebiet der Hydro- und Aerodynamik, Innsbrucker Kongress, 1922. Berlin: 22.
85. Khalatov AA, Avramenko AA, Shevchuk IV (1996) Heat transfer and fluid flow in the fields of centrifugal forces. Volume 1: Curvilinear flows. Publ. Institute of Engineering Thermophysics, National Academy of Sciences of Ukraine, Kiev, Ukraine (in Russian).
86. Khalatov AA, Avramenko AA, Shevchuk IV (1996) Heat transfer and fluid flow in the fields of centrifugal forces. Volume 2: Rotating systems. Publ. Institute of Engineering Thermophysics, National Academy of Sciences of Ukraine, Kiev, Ukraine (in Russian).
87. Khalatov AA, Avramenko AA, Shevchuk IV (2000) Heat transfer and fluid flow in the fields of centrifugal forces. Volume 3: Swirl flows. Publ. Institute of Engineering Thermophysics, National Academy of Sciences of Ukraine, Kiev, Ukraine (in Russian).
88. Khalatov AA, Avramenko AA, Shevchuk IV (2000) Heat transfer and fluid flow in the fields of centrifugal forces. Volume 4: Engineering and technological equipment. Publ. Institute of Engineering Thermophysics, National Academy of Sciences of Ukraine, Kiev, Ukraine (in Russian).
89. Khalatov AA, Shevchuk IV (2002) Law of heat transfer near a disk rotating in a quiescent ambience. Dopividi NAN Ukrainy 8: 84–88 (in Russian).

90. Kilic M, Owen JM (2003) Computation of flow between two disks rotating at different speeds. Trans ASME J Turbomach 125 (2): 394–400.
91. Kingsley-Rowe JR, Lock GD, Owen JM (2005) Transient heat transfer measurements using thermochromic liquid crystal: lateral-conduction error. Int J Heat Fluid Flow 26 (2): 256–263.
92. Kobayashi R, Kohama Y, Takamadate Ch (1980) Spiral vortices in boundary layer transition regime on a rotating disk. Acta Mechanica 35 (1–2): 71–82.
93. Kochin NE, Kibel IA, Roze NV (1964) Theoretical hydrodynamics. Interscience, New York.
94. Koh JCY, Price JF (1967) Non-similar boundary layer heat transfer of a rotating cone in forced flow. Trans ASME J Heat Transfer 89: 139–145.
95. Kohama Y (1983) Study on boundary layer transition of a rotating disk. Acta Mechanica 50 (3–4): 193–199.
96. Koong S-S, Blackshear PL Jr (1965) Experimental measurement of mass transfer from a rotating disk in a uniform stream. Trans ASME J Heat Transfer 85: 422–423.
97. Kreith F, Taylor JH, Chong JP (1969) Heat and mass transfer from a rotating disk. Trans ASME J Heat Transfer 81: 95–105.
98. Kuznetsov AL (1962) Experimental investigation of heat transfer from a disk rotating in quiescent environment. In: Trudy Leningrad. Korablestroit. Inst. 38: 183–186 (in Russian).
99. Lallave JC, Rahman MM, Kumar A (2007) Numerical analysis of heat transfer on a rotating disk surface under confined liquid jet impingement. Int J Heat Fluid Flow 28 (4): 720–734.
100. Lam CKG, Bremhorst KA (1981) Modified form of the k–ε model for predicting wall turbulence. Trans ASME J Fluid Eng 103 (3): 456–460.
101. Launder BE, Sharma BI (1974) Application of the energy dissipation model of turbulence to the calculation of the flow near a spinning disc. Lett Heat Mass Transfer 1: 131–138.
102. Law CG Jr, Pierini P, Newman J (1981) Mass transfer to rotating disks and rotating rings in laminar, transition and fully-developed turbulent flow. Int J Heat Mass Transfer 24 (5): 909–918.
103. Le Palec G (1989) Numerical study of convective heat transfer over a rotating rough disk with uniform wall temperature. Int Commun Heat Mass Transfer 16 (1): 107–113.
104. Le Palec G, Nardin P, Rondot P (1990) Study of laminar heat transfer over a sinusoidal-shaped rotating disk. Int J Heat Mass Transfer 33 (6): 1183–1192.
105. Levich VG (1962) Physicochemical hydrodynamics. Prentice-Hall, Inc., Englewood Cliffs, NJ.
106. Lin H-T, Lin L-K (1987) Heat transfer from a rotating cone or disk to fluids at any Prandtl number. Int Commun Heat Mass Transfer 14 (3) 323–332.
107. Lin DTW, Li HY, Yan WM (2008) Estimating the wall heat flux of unsteady conjugated forced convection between two corotating disks using an inverse solution scheme. Trans ASME J Heat Transfer 130 (12): 121702-1–121702-8.
108. Lingwood RJ (1995) Absolute instability of the boundary layer on a rotating disk. J Fluid Mech 299: 17–33.
109. Lingwood RJ (1996) An experimental study of absolute instability of the rotating-disk boundary-layer flow. J Fluid Mech 314: 373–405.
110. Lingwood RJ (1997) Absolute instability of the Ekman layer and related rotating flows. J Fluid Mech 331: 405–428.
111. Littel HS, Eaton JK (1994) Turbulence characteristics of the boundary layer on a rotating disk. J Fluid Mech 266: 175–207.
112. Lock GD, Daguze F, Syson BJ, Owen JM (1996) The application of thermal imaging techniques to determine heat transfer coefficients for slow thermal transients. Proc. 13th Symp. on Measuring Techniques for Transonic and Supersonic Flows in Cascades and Turbomachines, Zurich (Switzerland): 1–7.
113. Loitsyanskii LG (1966) Mechanics of liquids and gases. Pergamon, Oxford, UK.
114. Lokai VI, Bodunov MN, Zhuikov VV, Shchukin AV (1985) Heat transfer in cooled parts of aircraft gas-turbine engines. Izdatel′stvo Mashinostroenie, Moscow, USSR (in Russian).

115. Luikov AV (1968) Analytical heat diffusion theory. Academic Press, New York.
116. Mabuchi I, Tanaka T, Sakakibara Y (1971) Studies of convective heat transfer from a rotating disk (5th Report, Experiment on the laminar heat transfer from a rotating isothermal disk in a uniformed forced stream). Bull JSME 14 (72): 581–589.
117. Mabuchi I, Kotake Y, Tanaka T (1972) Studies of convective heat transfer from a rotating disk (6th Report, Experiment on the laminar mass transfer from a stepwise discontinuous naphthalene disk rotating in a uniformed forced stream). Bull JSME 15 (84): 766–773.
118. Mager A (1952) Generalization of boundary-layer momentum-integral equations to three-dimensional flows including those of rotating system. NACA Rep 1067: 1–37.
119. Malik MR (1986) The neutral curve for stationary disturbances in rotating-disk flow. J Fluid Mech 164: 275–287.
120. Malik MR, Wilkinson SP, Orzag SA (1981) Instability and transition in a rotating disc. AIAA J 19 (9): 1131–1138.
121. Martinez-Botaz RF, Lock GD, Jones TV (1994) Heat transfer measurements in an annular cascade of transonic gas turbine blades using the transient liquid crystal technique. ASME Paper 94–GT-172: 1–8.
122. McComas ST, Hartnett JP (1970) Temperature profiles and heat transfer associated with a single disc rotating in still air. Proc. IV International Heat Transfer Conference. Paris-Versailles (France) 3: Paper FC 7.7.
123. Metzger DE, Bunker RS, Bosch G (1991) Transient liquid crystal measurements of local heat transfer on a rotating disk with jet impingement. Trans ASME J Turbomach 113 (1): 52–59.
124. Minagawa Y, Obi S (2004) Development of turbulent impinging jet on a rotating disk. Int J Heat Fluid Flow 25 (5): 759–766.
125. Mishra P, Singh PC (1978) Mass transfer from spinning disks. Chem Eng Sci 33 (11): 1449–1461.
126. Mohr CM, Newman J (1976) Mass transfer to a rotating disk in transitional flow. J Electrochem Soc 123 (11): 1687–1691.
127. Mooney M, Ewart RH (1934) The conicylindrical viscometer. Physics 5: 350–354.
128. Morse AP (1988) Numerical prediction of turbulent flow in rotating cavities. Trans ASME J Turbomach 110 (2): 202–215.
129. Morse AP (1991) Assessment of laminar-turbulent transition in closed disc geometries. Trans ASME J Turbomach 113 (2): 131–138.
130. Newman BG (1983) Flow and heat transfer on a disk rotating beneath a forced vortex. AIAA J 21 (8): 1066–1070.
131. Newman JS (1991) Electrochemical systems, 2nd edn. Prentice-Hall, Inc., Englewood Cliffs, New York.
132. Nikitenko NI (1963) Experimental investigation of heat exchange of a disk and a screen. J Eng Phys 6 (6): 1–11.
133. Northrop A, Owen JM (1988) Heat transfer measurements in rotating-disc systems. Part 1: The free disc. Int J Heat Fluid Flow 9 (1): 19–26.
134. Northrop A, Owen JM (1988) Heat transfer measurements in rotating-disc systems. Part 2: The rotating cavity with a radial outflow of cooling air. Int J Heat Fluid Flow 9 (1): 27–36.
135. Northrop A (1984) Heat transfer in a cylindrical rotating cavity. D. Phil. thesis. University of Sussex, Brighton, UK.
136. Ong CL, Owen JM (1991) Computation of the flow and heat transfer due to a rotating disc. Int J Heat Fluid Flow 12 (2): 106–115.
137. Oehlbeck DL, Erian FF (1979) Heat transfer from axisymmetric sources at the surface of a rotating disk. Int J Heat Mass Transfer 22 (6): 601–610.
138. Owen JM, Rogers RH (1989) Flow and heat transfer in rotating-disc systems. Volume 1: Rotor-stator systems. Research Studies Press Ltd., Taunton (Somerset, England).
139. Owen JM, Rogers RH (1995) Flow and heat transfer in rotating-disc systems. Volume 2: Rotating cavities. Research Studies Press Ltd., Taunton (Somerset, England).

140. Parthasarathy RN (2002) Analysis of the turbulent boundary layer on a rotating disk. Microsyst Technol 8 (4–5): 278–281.
141. Paterson JA, Greif R (1973) Transport to a rotating disk in turbulent flow at high Prandtl or Schmidt number. Trans ASME J Heat Transfer 95 (4): 566–568.
142. Pavelek M, Janotková E (1989) Surface temperature calculation for applying naphthalene sublimation method on rotating disks. Proc. 6th Conference on Thermogrammetry and Thermal Engineering. Budapest (Hungary): 111–115.
143. Pavelek M, Janotková E (1996) An experimental study of heat transfer from co-rotating disks. Proc. 12th International Congress Chemical and Process Engineering CHISA 96. Prague (Czech Republic). Paper G 8.48.
144. Pier B (2003) Finite-amplitude crossflow vortices, secondary instability and transition in the rotating-disk boundary layer. J Fluid Mech 487: 315–343.
145. Pilbrow R, Karabay H, Wilson M, Owen JM (1999) Heat transfer in a 'cover-plate' pre-swirl rotating-disc system. Trans ASME J Turbomach 121 (2): 249–256.
146. Popiel CzO, Boguslawski L (1975) Local heat-transfer coefficients on the rotating disk in still air. Int J Heat Mass Transfer 18 (1): 167–170.
147. Popiel CO, Boguslawski L (1986) Local heat transfer from a rotating disk in an impinging round jet. Trans ASME J Heat Transfer 108 (2): 357–364.
148. Prata AT, Pilichi CDM, Ferreira RTS (1995) Local heat transfer in axially feeding radial flow between parallel disks. Trans ASME J Heat Transfer 117 (1): 47–53.
149. Rahman MM, Lallave JC (2007) A comprehensive study of conjugate heat transfer during free liquid jet impingement on a rotating disk. Nume Heat Transfer A Appl 51 (11): 1041–1064.
150. Rahman MM, Lallave JC (2008) Thermal transport during liquid jet impingement from a confined spinning nozzle. AIAA J Thermophys Heat Transfer 22 (2): 210–218.
151. Roy RP, Xu G, Feng J (2001) A study of convective heat transfer in a model rotor-stator disk cavity. Trans ASME J Turbomach 123 (3): 621–632.
152. Roy KER, Prasad BVSSS, Murthy SS, Gupta NK (2005) Conjugate heat transfer in an arbitrary shaped cavity with a rotating disk. Heat Transfer Eng 25 (8): 69–79.
153. Sahin I (1988) Applications of various coordinate transformations for rotating disk flow stability. AIAA J 26 (3): 368–370.
154. Saniei N, Yan X (1998) Effect of jet location on impingement cooling of a rotating disk. Proc. ASME Heat Transfer Division, HTD 361–1: 203–210.
155. Saniei N, Yan XT, Schooley WW (1998) Local heat transfer characteristics of a rotating disk under jet impingement cooling. Proc. 11th International Heat Transfer Conference "Heat Transfer 1998", Kyongju (Korea) 5: 445–450.
156. Saniei N, Yan XT (2000) An experimental study of heat transfer from a disk rotating in an infinite environment including heat transfer enhancement by jet impingement cooling. J Enhanced Heat Transfer 7: 231–245.
157. Şara ON, Erkmen J, Yapici S, Çopur M (2008) Electrochemical mass transfer between an impinging jet and a rotating disk in a confined system. Int Commun Heat Mass Transfer 35 (3): 289–298.
158. Schlichting G (1968) Boundary-layer theory. McGraw-Hill Book Company, New York.
159. Schmidt W (1921) Ein einfaches Messverfahren für Drehmomente. Z VDI 65: 411–444.
160. Schultz DL, Jones TV (1973) Heat transfer measurements in short duration hypersonic facilities. Aeronautical RD AGARDOGRAPH (NATO Advisory Group) 165.
161. Sdougos HP, Bussolari SR, Dewey CF (1984) Secondary flow and turbulence in a cone-and-plate device. J Fluid Mech 138: 379–404.
162. Shchukin VK (1980) Heat transfer and hydrodynamics of pipe flows in mass-force fields. 2nd revised and enlarged edition. Izdatel'stvo Mashinostroenie, Moscow, USSR (in Russian).
163. Shevchuk IV, Khalatov AA (1997) Integral method for calculating the characteristics of a turbulent boundary layer on a rotating disk: quadratic approximation of the tangent of the flow swirl angle. Heat Transfer Res 28 (4–6): 402–413.

164. Shevchuk IV (1998) Heat transfer in underswirled radial turbulent outflow in the gap between co-rotating discs of gas turbine rotors. Promyshlennaya Teplotekhnika 20 (1): 54–58 (in Russian).
165. Shevchuk IV (1998) Simulation of heat transfer in a rotating disk: the effect of approximation of the tangent of the angle of flow swirling. High Temperature 36 (3) 522–524.
166. Shevchuk IV, Khalatov AA, Karabay H, Owen JM (1998) Heat transfer in turbulent centrifugal flow between rotating discs with flow swirling at the inlet. Heat Transfer Res 29 (6–8): 383–390.
167. Shevchuk IV (1998) The effect of distribution of radial velocity in the flow core on heat transfer under conditions of centrifugal flow in a gap between parallel rotating disks. High Temperature 36 (6): 972–974.
168. Shevchuk IV, Khalatov AA (1999) The integral method of calculation of heat transfer in the turbulent boundary layer on a rotating disk: quadratic approximation of the tangent of the flow-swirl angle. Ind Heat Eng 1 (1): 139–144.
169. Shevchuk IV (1999) Integral method of calculation of a turbulent centrifugal underswirl flow in a gap between parallel rotating disks. Heat Transfer Res 30 (4–6): 238–248.
170. Shevchuk IV (1999) Effect of wall-temperature distribution on heat transfer in centrifugal flow in the gap between parallel rotating disks. J Eng Phys Thermophys 72 (5): 896–899.
171. Shevchuk IV (1999) Prediction of heat transfer and fluid flow over a rotating disk: effect of the radial velocity profile. Proc. 4th International Symp. on Experimental and Computational Aerothermodynamics of Internal Flows. Dresden (Germany): 2: 184–189.
172. Shevchuk IV (1999) Simulation of heat transfer and hydrodynamics over a free rotating disk using an improved radial velocity profile. J Thermal Sci 8 (4): 243–249.
173. Shevchuk IV (2000) Turbulent heat transfer of rotating disk at constant temperature or density of heat flux to the wall. High Temperature 38 (3): 499–501.
174. Shevchuk IV (2001) Effect of the wall temperature on laminar heat transfer in a rotating disk: an approximate analytical solution. High Temperature 39 (4): 637–640.
175. Shevchuk IV (2001) An improved enthalpy-thickness model for predicting heat transfer of a free rotating disk using an integral method. Proc. 35th National Heat Transfer Conference ASME NHTC'01. Anaheim (California, USA). Paper NHTC2001–20195: 1–8.
176. Shevchuk IV, Saniei N, Yan XT (2001) Effect of the wall temperature distribution in direct and inverse problems of laminar heat transfer over a free rotating disk. Proc. 35th National Heat Transfer Conference ASME NHTC'01. Anaheim (California, USA). Paper NHTC2001–20196: 1–8.
177. Shevchuk IV (2001) Solution of direct and inverse problems of heat transfer over a free rotating disk using an improved enthalpy-thickness model. Proc. 5th International Symposium on Experimental and Computational Aerothermodynamics of Internal Flows, Gdansk (Poland) 2: 591–598.
178. Shevchuk IV (2001) Turbulent heat transfer over a free rotating disk: analytical and numerical predictions using solutions of direct and inverse problems. Proc. International Mechanical Engineering Congress and Exposition IMECE'01, New York (USA): 1–8.
179. Shevchuk IV (2002) Numerical modeling of turbulent heat transfer in a rotating disk at arbitrary distribution of the wall temperature. J Eng Phys Thermophys 75 (4): 885–888.
180. Shevchuk IV (2002) Laminar heat transfer in a rotating disk under conditions of forced air impingement cooling: approximate analytical solution. High Temperature 40 (5): 684–692.
181. Shevchuk IV, Saniei N, Yan XT (2002) Impinging jet heat transfer over a rotating disk: exact solution and experiments. Proc. 8th AIAA/ASME Joint Thermophysics and Heat Transfer Conference. St. Louis (Missouri, USA). Paper AIAA 2002–3015: 1–11.
182. Shevchuk IV, Buschmann MH (2002) Laminar swirling flow between cone and plate. Proc. Colloquium Fluid Dynamics 2002, Prague (Czech Republic): 155–158.
183. Shevchuk IV (2003) Exact solution of the heat transfer problem for a rotating disk under uniform jet impingement. Fluid Dynamics 38 (1): 18–27.

184. Shevchuk IV, Saniei N, Yan XT (2003) Impingement heat transfer over a rotating disk: integral method. AIAA J Thermophys Heat Transfer 17 (2): 291–293.
185. Shevchuk IV (2004) A self-similar solution of Navier-Stokes and energy equations for rotating flows between a cone and a disk. High Temperature 42 (1): 95–100.
186. Shevchuk IV (2004) Laminar heat transfer of a swirled flow in a conical diffuser: self-similar solution. Fluid Dynamics 39 (1): 42–46.
187. Shevchuk IV (2004) Unsteady-state laminar heat transfer in a rotating disk: self-similar solution. High Temperature 42 (4): 592–595.
188. Shevchuk IV, Delas NI (2005) Aerodynamics and turbulent flow heat exchange in the rotary disk air cleaner. Heat Transfer Res 36 (1–2): 104–113.
189. Shevchuk IV (2005) A new type of the boundary condition allowing analytical solution of the thermal boundary layer equation. Int J Thermal Sci 44 (4): 374–381.
190. Shevchuk IV, Buschmann MH (2005) Rotating disk heat transfer in a fluid swirling as a forced vortex. Heat Mass Transfer 41 (12): 1112–1121.
191. Shevchuk IV, Indinger T (2005) Transient conjugate heat transfer of a disk rotating in still air. Proc. 4th International Conference on Computational Heat and Mass Transfer. Paris (France). Paper ICCHMT'05–150: 1–6.
192. Shevchuk IV (2006) Unsteady conjugate laminar heat transfer of a rotating non-uniformly heated disk: application to the transient experimental technique. Int J Heat Mass Transfer 49 (19–20): 3530–3537.
193. Shevchuk IV (2008) A new evaluation method for Nusselt numbers in naphthalene sublimation experiments in rotating-disk systems. Heat Mass Transfer 44 (11): 1409–1415.
194. Shevchuk IV (2009) An integral method for turbulent heat and mass transfer over a rotating disk for the Prandtl and Schmidt numbers much larger than unity. Heat Mass Transfer 45(10): 1313–1321.
195. Shimada R, Naito S, Kumagai S, Takeyama T (1987) Enhancement of heat transfer from a rotating disk using a turbulence promoter. JSME Int J Ser B 30 (267): 1423–1429.
196. Shvets IT, Dyban EP (1974) Air cooling of gas turbine parts. Naukova Dumka, Kiev, Ukraine (in Russian).
197. Smith NH (1946) Exploratory investigation of laminar boundary layer oscillations on a rotating disc. NACA Tech. Note 1227: 1–21.
198. Soong CY, Chyuan CH (1998) Similarity solutions of mixed convection heat and mass transfer in combined stagnation and rotation-induced flows over a rotating disk. Heat Mass Transfer 34 (2–3): 171–180.
199. Sparrow EM, Gregg JL (1959) Heat transfer from a rotating disc to fluids at any Prandtl number. Trans ASME J Heat Transfer 81: 249–251.
200. Sparrow EM, Geiger GT (1985) Local and average heat transfer characteristics for a disk situated perpendicular to a uniform flow. Trans ASME J Heat Transfer 107 (2): 321–326.
201. Sparrow EM, Chaboki A (1982) Heat transfer coefficients for a cup-like cavity rotating about its own axis. Int J Heat Mass Transfer 9 (25): 1334–1341.
202. Stepanov GY, Zitser IM (1986) Inertial air purifiers. Mashinostroenie, Moscow (in Russian).
203. Suga K (2007) Computation of high Prandtl number turbulent thermal fields by the analytical wall-function. Int J Heat Mass Transfer 50 (25–26): 4967–4974.
204. Szeri AZ, Giron A (1984) Stability of flow over a rotating disk. Int J Numer Methods Fluids 4 (10): 989–996.
205. Tadros SE, Erian FF (1983) Heat and momentum transfer in the turbulent boundary layer near a rotating disk. Paper ASME 83–WA/HT–6: 1–8.
206. Tadros SE, Erian FF (1982) Generalized laminar heat transfer from the surface of a rotating disk. Int J Heat Mass Transfer 25 (11): 1651–1660.
207. Theodorsen Th, Regier A (1944) Experiments on drag of revolving discs, cylinders and streamline rods at high speeds. NASA Report No 793.
208. Tien CL, Campbell DT (1963) Heat and mass transfer from rotating cones. J Fluid Mech 17: 105–112.

209. Tien CL, Tsuji IJ (1964) Heat transfer by laminar forced flow against a non-isothermal rotating disk. Int J Heat Mass Transfer 7 (2): 247–252.
210. Tien CL, Tsuji IJ (1965) A theoretical analysis of laminar forced flow and heat transfer about a rotating cone. Trans ASME J Heat Transfer 87 (2): 184–190.
211. Tifford AN, Chu ST (1954) On the flow and temperature fields in forced flow against a rotating disc. Proc. 2nd US National Congress of Applied Mechanics, Ann Arbor (Michigan, USA): 793–800.
212. Truckenbrodt E (1954) Die turbulente Strömung an einer angeblasenen rotierenden Scheibe. Z Angew Math Mech 34: 150–162.
213. Viskanta R (1993) Heat transfer to impinging isothermal gas and flame jets. Exp Thermal Fluid Sci 6 (2): 111–134.
214. Vogel G, Weigand B (2001) A new evaluation method for transient liquid crystal experiments. Proc. 35th National Heat Transfer Conference ASME NHTC'01. Anaheim (California, USA). Paper NHTC2001–20250: 1–6.
215. Wagner G, Kotulla M, Ott P, Weigand B, von Wolfersdorf J (2005) The transient liquid crystal technique: influence of surface curvature and finite wall thickness. Trans ASME J Turbomach 127 (1): 175–182.
216. Wasan DT, Tien CL, Wilke CR (1963) Theoretical correlation of velocity and eddy viscosity for flow close to a pipe wall. AIChE J 9 (4): 567–569.
217. Watanabe T (1985) Stability of boundary layers along a rotating disk. Trans JSME Ser B 51 (470): 3344–3347.
218. Watanabe T (1989) Effect of surface roughness on boundary layer transition on a rotating disk. Trans JSME Ser B 55 (515): 1842–1846.
219. Weisstein EW. Hypergeometric function. MathWorld. A Wolfram Web Resource.
220. aus der Wiesche S (2004) LES study of heat transfer augmentation and wake instabilities of a rotating disk in a planar stream of air. Heat Mass Transfer 40 (3–4): 271–284.
221. aus der Wiesche S (2007) Heat transfer from a rotating disk in a parallel air crossflow. Int J Thermal Sci 46 (8): 745–754.
222. Wilkinson SP, Malik MR (1985) Stability experiments in the flow over a rotating disk. AIAA J 23 (4): 588–595.
223. von Wolfersdorf J (2007) Influence of lateral conduction due to flow temperature variations in transient heat transfer measurements. Int J Heat Mass Transfer 50 (5–6): 1122–1127.
224. Wu X, Squires KD (2000) Prediction and investigation of the turbulent flow over a rotating disk. J Fluid Mech 418: 231–264.
225. Yan Y, Gord MF, Lock GD, Owen JM (2003) Fluid dynamics of a pre-swirl rotor-stator system. Trans ASME J Turbomach 125 (4): 641–647.
226. Zoueshtiagh F, Ali R, Cooley AJ, Thomas PJ, Carpenter PW (2003) Laminar-turbulent boundary-layer transition over a rough rotating disk. Phys Fluids 15 (8): 2441–2444.

Index

A
Abscissa, 65, 187
Accessory equation, 75
Acoustic measurements, 38
Aerodynamics
  flow, 172
  model, 169
Air cleaner, 1, 168–177
  scaling of, 176
Air cooling systems, 1
Air ingress, region of, 153
Air jets, 119
Algebraic model of turbulent viscosity, 17
Aluminium disk, 99
Analytical modelling, 72
Angular velocity, 170
Approximate solution, inaccuracies of, 116–117
Asymptotic solution, 36, 59, 197–198
Asymptotic theoretical solution, 198
Automatic mesher, 85
Axial component profile, 187
Axial velocity component, 172, 188

B
Biot number, 84, 88, 92–93
Boundary layer equation, 54
Bypass transition, 38

C
Cartesian coordinate system, 5, 6, 9
Cebeci–Smith model, 54
Centrifugal forces, 3–4, 169
  projection of, 12
CFD code, 16, 18, 85, 90, 167, 169
Co-axial impingement, 133
Computational curves, 160
Computational mesh, 18, 171
Cone–disk system, 181
Cone rotation, 187, 189, 192
Conical diffuser, 1, 179, 192
Conical gaps, 185
Convection diffusion equation, 9, 194
Convective heat transfer, 24
Coriolis force, 3–5, 12, 39
Co-rotating
  cover plate, 151
  disk, 21, 31, 91, 152
  fluid, 117, 187
Curvilinear coordinate system, 5
Curvilinear flow, *see* swirl flow
Cylindrical polar coordinate system, 6, 7, 9

D
Differential equations, 4–9, 11–15, 54, 86, 106, 182–183
  approximate analytical method, 17
  continuity, momentum, and heat transfer, 4–8
  convective diffusion, 9
  differential boundary layer equation, 13–14
  integral boundary layer equation, 14–15
  Navier–Stokes and energy equation, 11–12
  numerical method, 17–18
  self-similar solution, 15–16
  swirl parameter, 150
Diffuser, 164, 181, 189–190
Diffusion coefficient, 9, 193, 198
Dimensionless temperature profiles, 127
Disk rotation, 1, 142, 144, 146, 175, 192
  in a fluid, 101–118
    laminar flow, 106–118
    turbulent flow, 101–106
Disk rotational velocity, 175
Dorfman, L. A. 22, 36, 44–46, 73–74, 104
Dorfman method, 44, 104

Dorfman's equation, 51–53, 54, 58, 60, 68, 70, 72, 195
drawbacks of, 59
Downstream disk, 150, 164–165
Dust collector, 169

E
Eigenvalues, 79, 93
Ekman layers, 147–148, 155, 166
Ekman-type layers, 29, 150–156, 165, 168
zone of the, 156
Electrochemical application, 203–204
Electrochemistry problem, 198
Electromagnetic field, 2
Elliptic solver, 164
Endothelium cells, 179
Enthalpy thickness, 18, 23–24, 58, 103, 114, 160
Excessive mass force, 3
Experimental log-book, 154
Experimental measurements, systematic error of, 57

F
Faraday constant, 198
Finite-difference method, 17–18, 171
Flow direction, effect of, 188
Flow overswirl, 164–168
Fluid mechanics, 163, 168–169
Fourier number, 78, 84, 94, 96–98
Fourier series, 91
Fourth-order polynomials, 112
Free rotating disk, 17, 19, 21, 25, 33–34, 48, 51, 66, 76, 103, 114, 121, 126, 136, 143–144, 162, 172, 187–188, 194
Free-vortex law, 150, 152, 165
Frozen rotor domain interface, 85

G
Gas turbine, 1, 164, 168, 172
rotors, 118, 148
Generalized analytical solution, 58–61
Geometric parameters of the rotating air cleaner, 171
Gravitational forces, 2, 4, 11

H
Heat and mass transfer methods, 37
Heat conduction equation, 116
Heat flux, 13, 18, 24–25, 54, 77, 163–164, 192
meters, 54
Heat transfer, 34
Heat transfer coefficient, 77–81, 84, 91–92, 96–98, 100
transient values of, 97–98
Heat transfer for laminar flow model, 103
Heat transfer investigation, 77, 185
Heat transfer regime theory, 92, 99–100
High-performance technique for cooling or heating, 118
High Prandtl and Schmidt numbers
model with constant value of $\Delta << 1$, 217–218
model with variable value of Δ, 218–222
model with variable value Δ and profile T^+ depending on Re_ω, 222–224
Hydrodynamic parameters, 139
Hydrodynamics, 85, 119

I
Impingement domination regime, 133
Impingement-to-rotation velocity, ratios of, 142
Impinging jet, 1, 118, 120, 123, 131, 150, 160, 164, 167
Industrial application, 1
Inertial forces, 2, 4
Inertial mass forces, 2
Integral equations, 4, 11–15
continuity, momentum, and heat transfer, 4–8
differential boundary layer equation, 13–14
diffusion, 9–10
integral boundary layer equation, 14–15
Navier–Stokes and energy equation, 11–12
Integral methods of modelling, 23–29
integral equations, 27–29
structure, 23–24
surface friction model, 25–27
turbulent flow, 24–25
Integral methods of solution, 18–23
momentum and boundary layer, 18–22
thermal boundary layer, 22–23
Inter-disk cavity, 176
Inter-disk spacing, 175, 176
Isothermal disk, 35–36, 47, 58, 81, 84, 94–95, 97, 99–100
Isothermal rotating disk
numerical solution, 85–91
computational domain and grid, 85–86
steady-state fluid flow, 86–88
unsteady fluid flow, 88–91
surface heat transfer solutions, 82–85
Isothermal slab, 84

J
Jet cooling, 1
Jet-to-disk distance, 120

K
k–∈ turbulence model, 18, 171
Kaolin layer, 37
Karman, Th von, 19, 25, 42, 44, 48, 154
Karman's linear approximation, 163
Karman's method, 41–42

L
Lamé coefficients, 5
Laminar boundary layers, 62, 139
Laminar flow, 5, 6, 12, 15, 22–24, 33–36, 40, 47, 58–60, 62, 66, 70, 72–73, 75, 78, 81, 87, 90, 102, 109, 110, 114, 120, 130, 136, 140, 144–145, 147, 181–182, 193–200, 203, 205, 207–208
  parameters, 108, 109
Laminar fluid flow, 11, 17, 86, 121
  model, 17
Laminar heat transfer, 88, 117, 201
Laminar mass transfer, 201
Laminar–turbulent transition, 37
Large eddy simulation (LES), 18, 145
Launder, B. E., 164–165
Launder-Sharma model, 164
Laws of momentum, 4
Levich, V. G., 198–199
Levich's equation, 198–199
L'Hospital rule, 64
Linear transformation, 81, 182–183
Local mass transfer coefficients, 199
Local modelling theory, 22, 46, 72–76, 163
Logarithm, 98
Logarithmic approximations, 45
Longitudinal field, 4

M
Marching, 182
Mass forces, 1, 2, 3, 4, 5, 6
Mass transfer coefficients, 83, 193, 202
MathCAD software, 16, 33, 81, 83, 92, 106, 126, 182, 184, 194
Models for the velocity profiles, 18, 109
Moment coefficient, 34
Momentum boundary layer, tangential equation of, 58
Momentum transfer, equation of, 5
Morse, A. P., 164–165

N
Naphthalene sublimation, 46, 51, 121, 193, 199–203, 206–208
  mass transfer in, 145
Naphthalene sublimation technique, 193, 199, 202, 206–207, 215
Navier–Stokes equations, 18, 58, 80–81, 168–184, 192–193
Newton's law, 4
Non-swirling flow, 160, 164, 190, 192
Non-swirling radial flow, 189
Non-uniformly rotating heating disk, 91–100
  heat conduction problem, 93–94
  self-similar solution, 92–93
  unsteady heat conduction, 94–100
Nusselt number, 17, 21, 24, 35, 39, 46–48, 52–54, 57–59, 60–61, 63, 66–68, 70, 72, 74–75, 78, 82–83, 88–90, 104, 106, 115, 117, 120–121, 127, 133–134, 142, 145–146, 151–152, 157–161, 162–164, 166–167, 185–190, 192, 194, 199, 202, 206–207, 214–215
  calculation of, 70
  curves of, 164, 190

O
Occasional perturbations, 4
One-dimensional conduction, 87
One dimensional heat conduction theory, 77
Overswirled flow, 164, 166–168, 175

P
Parallel co-rotating disk, 148–164
  flow structure, 148–152
  Nusselt numbers, 157–161
  radial variation computation, 152–157
  surface temperature, 161–164
Phoenics code, 169–170
Plexiglas, 78, 82, 86, 88, 94–95, 98–100
Polynomial approximations, 69
Power-law
  approximation, 19, 22, 24, 41–42, 57, 147, 163
  distribution, 53, 61, 92, 95
  function, 22
  profiles model, 148
Prandtl number, 23, 26, 35–36, 45–46, 51–52, 58, 60, 66, 72, 73, 84, 104, 114, 127, 130, 140–142, 185, 193–195, 201, 207, 208–224
  deviation of, 116
  See also ***High Prandtl and Schmidt numbers***

Pressure drop, 175–176
Pressure gradient, 3, 74, 124
Pulsations, 4
Pumping effect, 175, 198

Q

Quasi-conjugate heat transfer solution, 91
Quasi-isothermal zones, 91

R

Radial component, 185
  effects, 98
Radial distributions, arbitrary, 163
Radial flow acceleration, 118–143
  flow impingement, 118–122
  laminar flow, 125–143
  turbulent flow, 123–125
Radial flow velocity, 170
Radial heat conduction, 16, 89, 94, 96, 99
Radial pressure gradient, 16
Radial surface temperature gradients, 98
Radial temperature gradient, 56
Radial variation law, 28, 161, 164
Radial velocity, 19, 42, 44, 49, 120, 123, 127, 168, 174, 176, 185, 190, 192
  component, 49, 123, 168, 172, 185, 192
  profile, 48, 103, 189
Recirculation flow, 30, 164, 167–168, 189–190
  emergence of, 168
  modelling, 168–169
Regular heat transfer regime theory, 98
Regular regime of heating, 84, 91
Reynolds analogy, 23, 58, 63, 74, 104, 114, 124, 130, 203
  parameter, 23, 58, 63, 74, 104, 114, 124, 130
Reynolds number, 18–19, 37–44, 47–48, 56–57, 66, 67, 74, 92, 131, 133–135, 142, 145–146, 154, 186, 187, 203–206, 210–213, 217, 218, 221–224
  arithmetic mean of, 206
Ring-like jet, 164, 167
Rotating air cleaners, 169
Rotating cavities, 164
Rotating disk, 1, 9, 11, 15, 17–18, 31, 35–36, 39–40, 46, 54, 57, 68, 72, 78–79, 82, 88, 91, 109, 118–119, 121, 125, 127–128, 130–133, 144–145, 148, 161, 181–182, 184–189, 187, 193, 202
  cavities, 147–148, 150, 168, 172
  cone rotation, 185–187
  co-rotating disk and cone, 188
  counter-rotating disk and cone, 188–189
  electrode, 193, 198
  numerical values, 184–185
  stationary cone, 187–188
  systems, 22–24, 54, 104, 118, 184–187
Rotating-disk air cleaner, 18, 168–169, 176
  aerodynamics and heat transfer, 168–177
    computational scheme parameters, 171
    geometrical and regime parameters, 169–171
    problem characteristics, 168–169
    results, 171–177
Rotational velocity of the disks, 175
Round jets, 118
Runge-Kutta method, 28, 153, 158

S

Schmidt number, 36, 40, 46, 121, 193–194, 196–199, 201, 204, 207, 208–224
Self-similar Navier–Stokes equations, 79–82
Self-similar solution, 128
Semi-infinite slab, 77, 94, 97–98
Seventh-order polynomial, 55, 159, 165
Sharma, B. I., 164–165
Shchukin, V. K, 3
Shear stresses, 13, 22, 24–26, 108, 111, 127, 137, 166
Sherwood number, 46, 83, 121, 145, 194, 199, 202–204, 206–207, 209–213
Shooting method, 16, 184
Sinusoidal waves of perturbations, 37
Slezkin-Targ method, 17
Smoothing curves, 161
Solid-body rotation, 29, 103, 117, 148
Source region, 150–153, 155–165
Spacing, 176–177
Spinning effect, 150
Spiral vortices, 37, 40
  emergence of, 37–38
Stagnation pressure, 119
Stanton number, 23, 26, 48, 73–74, 104
Static pressure, 13, 170, 172, 190
  drop, 191
Stationary conical diffuser, 189–192
Stationary diffuser, 189
Stationary thermal boundary layer, equation of, 13
Streamline curvature, 2, 3
Stress tensor, 5
Superimposed radial flow, effect of, 152
Surface forces, 1
Surface friction, 17, 26, 40, 176
  coefficient, 17, 26

Swirl coefficient, 149
Swirl flow, 1–2, 181–182
Swirling effect, 168, 175
Swirling flows, 101
Swirling jet impinging, 115
Swirl parameter, 108–109, 115, 152–153, 156, 173–175
effect of, 113
model, 151

T
Tangential velocities, 25, 41, 185
component, 41, 127, 153, 163, 172–173, 176, 184, 190
profiles, 191
Taylor's series, 51
Temperature profile, power-law model, 48
Thermal and hydrodynamic conditions, 55
Thermal boundary conditions, 57, 65, 67, 72, 79–80, 85, 185
Thermal boundary layer equation, 16, 23, 27, 61, 72, 80–81, 123, 181–182, 200
Thermal diffusivity coefficient, 9
Thermoanemometry, 37
Thin-slab approximation, 94
Transcendental equation, 51, 103, 123
Transformed boundary conditions, 184
Transient experimental technique, 77–79
Transient technique, 62, 78–79, 98, 100
Transitional and turbulent flow, 44, 47, 201–208
at high Prandtl and Schmidt numbers, 208–214
Transverse force, 4
Turbomachinery, 77
Turbulent boundary layer, 19, 62, 139, 163
Turbulent flow, 4, 5, 6, 7, 9, 12, 18–19, 22–24, 37–57, 60, 65, 68, 70, 72, 103, 109–110, 114, 123–125, 130, 136, 140, 145, 147–148, 161, 176, 193–194, 199, 203, 205–207
heat and mass transfer, Prandtl and Schmidt numbers, 214–224
numerical computation, 54–57
parameters of, 41–45
radial velocity profile, 48–53
surface heat transfer, 45–48
See also Transitional and turbulent flow
Turbulization effect of protrusions, 158
Turbulization flow, 145
Two-dimensional heat conduction, 91–92, 100

U
Underswirl of fluid flow, 149
Unsteady surface temperature distributions, 94–95

V
Velocity components, dimensionless profiles of, 126
Velocity profiles, logarithmic model, 44
Viscous dissipation, 4, 182
Visualization, 37
Volt–ampere characteristics, analysis of, 198
Von Karman's method, 25, 41–42, 44, 48, 154, 163

W
Wall-jets, 126
Wall temperature distribution restoration, 61–72

MIX
Papier aus verantwortungsvollen Quellen
Paper from responsible sources
FSC® C105338

If you have any concerns about our products,
you can contact us on
ProductSafety@springernature.com

In case Publisher is established outside the EU,
the EU authorized representative is:
Springer Nature Customer Service Center GmbH
Europaplatz 3, 69115 Heidelberg, Germany

Printed by Libri Plureos GmbH
in Hamburg, Germany